Advanced Nondestructive Det
in Food

Quansheng Chen • Hao Lin • Jiewen Zhao

Advanced Nondestructive Detection Technologies in Food

 Springer

Quansheng Chen 🆔
College of Food and Biological
Engineering
Jimei University
Xiamen, Fujian, China

Hao Lin 🆔
School of Food and Biological
Engineering
Jiangsu University
Zhenjiang, Jiangsu, China

Jiewen Zhao
School of Food and Biological
Engineering
Jiangsu University
Zhenjiang, Jiangsu, China

ISBN 978-981-16-3362-1   ISBN 978-981-16-3360-7   (eBook)
https://doi.org/10.1007/978-981-16-3360-7

This Springer imprint is published by the registered company Springer Nature Singapore Pte Ltd.
The registered company address is: 152 Beach Road, #21-01/04 Gateway East, Singapore 189721,
Singapore

Preface

Nondestructive detection technology, developed in the late twentieth century, is an emerging technology for fast testing of external and internal food quality. With the rapid development of food industry, sensing technology, optical technology, nanotechnology, information science, and other advanced technologies are more and more applied to the field of food detection by using nondestructive methods.

Nondestructive testing technology is a frontier research in the field of food testing, involving interdisciplinary knowledge. In recent years, with the rapid development of nanomaterial, intelligent sensing, network communication, and big data processing technology, a variety of new nondestructive detection technologies are emerging. Nondestructive detection technology is becoming an important technical support to promote the food industry toward information and intelligence due to its fast and nondestructive advantages. Therefore, it is necessary to introduce the food nondestructive technology systematically. The authors of this book have been engaged in food quality and safety rapid nondestructive testing researches for more than 30 years. Therefore, this book is a systematic summary of the authors' research for many years and also a prospect of the future development trend of nondestructive detection technology.

This book focuses on in situ detection of food quality and safety using nondestructive method, including principles, signal processing and analysis of data, nondestructive detection system, and application in food industry for each technology. First, this book introduces principles and characteristics for various nondestructive methods in food nondestructive detection. As for nondestructive measurement, big data is always obtained for each testing. Therefore, this book also describes in detail signal and big data processing for each nondestructive method. Meanwhile, this book also introduces the rapid portable detection equipment for food and agricultural products developed in recent years, as well as the intelligent monitoring equipment in the process of food processing. Furthermore, this book also introduces the relevant application cases. In this way, the reader may better understand how to apply nondestructive technology for food quality detection.

In noninvasive measurement of food quality, this book has a systematic introduction of detection principle, data processing, rapid detection system, and in-field detection case studies. This book can be used as a professional textbook for undergraduates majoring in food science and engineering. It can also be used as a reference book for scientific research and technical personnel engaged in the field of food quality and safety detection.

We are especially grateful to Professor Lu of the University of Mississippi for his suggestions and amendments to this book. We express our special thanks to Dr. Wencui Kang for her serious revising and edits of this book. We also thank Dr. Yi Xu, Peihuang He, Hanyu Zheng, Afan Zhu, Xiaodan Ding, Fuyun Wang, and Yue Liang for revising this book. Thanks to Dr. Hassan, Dr. Zareef, Dr. Waqas, and Dr. Selorm for their revising and polishing works in this book. We also acknowledge the financial support by the National Natural Science Foundation of China [31972154] and the Project of Faculty of Agricultural Equipment of Jiangsu University.

Xiamen, Fujian, China
Zhenjiang, Jiangsu, China

Quansheng Chen
Hao Lin
Jiewen Zhao

Contents

About the Authors

Quansheng Chen is a professor in the School of Food and Biological Engineering of Jiangsu University. He has been engaged in the study of nondestructive detection of food quality and safety for nearly 20 years. His main research interest includes using NIR, computer vision, spectral imaging, and multi-sensor fusion technologies for food quality and safety, nondestructive detection, and real-time monitoring of food quality using nondestructive methods.

Hao Lin is an associate professor in the School of Food and Biological Engineering of Jiangsu University. He has been engaged in the study of nondestructive detection of food quality and safety for 15 years. His main research interest includes using NIR, bionic sensors, colorimetric sensors, acoustic sensor technologies for food quality and safety, nondestructive detection, and the development of intelligent equipment.

 **Jiewen Zhao** is a professor in the School of Food and Biological Engineering of Jiangsu University. He has been engaged in the study of nondestructive detection of food quality and safety for more than 30 years. His main research interest includes using NIR, computer vision, bionic sensors, spectral imaging and multi-sensor fusion technologies for food quality and safety, nondestructive detection, and online grading for food quality using nondestructive methods.

Chapter 1
Introduction

Currently, driven by rigid demand and consumption upgrading, the food industry continuously adjusted and optimized the industrial structure and accelerated the pace of transformation. The food manufacturing industry continues to play an important supporting role in ensuring people's livelihood, stimulating consumption, and promoting economic and social development. The rapid growth of the total output value of the food industry continuously increases with updating the technology of the food processing and manufacturing industry, while the traditional food enterprises add challenges due to high energy and low efficiency as well as extensive processing methods. A relatively lagging detection process restricts the modernization and upgradation of the food industry in food storage and processing. Recently, the processing and manufacturing industry is pushing gradually for high-end products by utilizing intelligent sensors, modern networks, and big data. Nondestructive detection (NDD) with the merits of high speed, low cost, and easy to set up in-line system has become a growing technique in the field of food quality monitoring and assurance to modernize the food industry. To date, with the rapid development of the food industry, NDD introduces a strong innovation impetus for their upgradation and bottleneck overcoming. Simultaneously, NDD with environmental and efficient characteristics provides crucial scientific and technological support for the sustainable development of the food industry, leading the food industry to the direction of intelligence and information.

Abbreviation	Full name
AIS-1	First airborne imaging spectrometer
CCD	Charge-coupled device
CT	Computerized tomography
ECR	Electron cyclotron resonance
HSI	Hyper-spectral imaging
NDD	Nondestructive detection
NIR	Near-infrared

(continued)

Abbreviation	Full name
NMR	Nuclear magnetic resonance
MRI	Magnetic resonance imaging
UV	Ultraviolet
UV-vis	Ultraviolet and visible

1.1   Summary

1.1.1   Technical Connotation

NDD is to detect the structure, quality, state, and defects of the tested objects with modern technology and equipment without damaging or affecting their performance or the internal organization, based on the structure change causing the response discrepancy of heat, sound, light, electricity, and magnetism. In allusion to various testing objects, the technology inputs several forms of physical energy, such as light, force, sound, electricity, magnetism, and the corresponding response signals varying with the counterparts. Specifically, food materials reflect the distinctive difference appearance, internal composition, structure, and other characteristics by the several signals between input and output.

For example, food energy somewhat varies after effects from external light; it is usually manifested as absorbing light of a certain frequency, transitioning from low-energy to high-energy or from an excited state to a ground state before radiating the light of a certain frequency. Moreover, the frequency of absorbing and radiating light is related to the properties of the molecules and atoms comprising the substance, with high selectivity. Hence, the appearance or internal quality characteristics of food materials can be determined by the optical properties. Generally speaking, the light in the wavelength range from ultraviolet to visible light can excite the electronic energy level of food, and the light in the infrared region stimulates the vibrational energy and rotational energy between the molecules. The mechanical method, based on mechanical characteristics of vibration, is to detect the quality by applying vibration or torsional excitation to the sample, causing vibration amplitude and phase motion attenuation. Likewise, acoustic analysis utilizing external excitation, the internal and external structure, and other quality characteristics of food can be judged according to the transformation of acoustic signals emitted by the excited food. Moreover, the electromagnetic characteristics of food can be divided into active and passive electromagnetic methods. The former utilizes electromagnetic properties of the tested object, such as bionic sensing or bioelectricity. The latter reflects the distribution characteristics of electric field and current in the space occupied by food, and influences the behavior of food and agricultural products in the electric field, which can be shown as the electrical conductivity and dielectric constant of food.

NDD is an applied science-related technology developed in the late twentieth century, based primarily on the knowledge of physics, information science, sensing technology, optoelectronic technology, data processing, and computer application. Since the input and output of NDD for food is physical energy, there is not any nuisance to the testing object. Additionally, it is applied as the fast and high-throughput detection of quality and safety of food materials in the modern food industry. That is to say, the quality testing of the tested objects can be completed in a short time. Consequently, with the rapid growth of the Chinese food industry, NDD has become an influential technology as a primary quality monitoring system.

1.1.2   History

Typical applications of NDD can also be found in foreign developed countries. For example, the United States applied electronic sorting machines to separate soybeans and peanuts in the 1930s [1]. Additionally, the Russell Research Center of the United States Department of Agriculture made a practical infrared detector for melon ripening according to the various degrees of infrared light absorption by the sugar content of melons and fruits. In 1953, the United States began to use commercial electronic color sorters. In the 1960s, a high-density spectrophotometer connected to the computer was invented to measure the appearance quality of hosts of agricultural products by Norris [2]. Since then, Ben-Gera used near-infrared spectroscopy to detect fat and moisture in meat, achieving excellent result [3]. Afterwards, an in-depth study on photoelectric NDD technology, the principle, and application of scattering characteristics in food and agricultural by-products has been conducted. Prior studies suggested that photoelectric technology has great potential in the quality inspection and classification of food and agricultural products.

The detection and research on optical characteristics have been carried out in Japan since the 1970s. A sorting method had been developed by Zhong for tomatoes, bananas, oranges, and raw tea, utilizing the delayed luminescence characteristics of agricultural products. Additionally, a rice quality tester has been invented by the Japanese Institute of Industry and Technology; the light penetration rate and the spectral ratio of rice combined with a mathematical model were used to detect the maturity, disease, and color grade of rice. Besides, the Japanese stagnant Chemical Industry Company has invented a liquid sugar content detector based on optical fiber technology.

Furthermore, Israel, the Netherlands, France, Germany, Italy, Spain, etc. have also developed NDD-based detection systems for food and agriculture products. For example, Israel has developed a fruit classifier, utilizing reflectance spectrum characteristics to analyze fruit surface defects and a weighing system to comprehensively evaluate the information collected by the computer to grade the fruit.

The research on NDD technology of food and agricultural products in China started relatively late. In the 1970s, several institutions have begun to explore the feasibility of NDD to detect and analyze the quality of food. For instance, the Electronic Technology Application Group of Beijing Institute of Grain Science and Technology (1977) has applied the near-infrared reflectance signal to quickly detect the composition of grains and oilseeds. Since the 1980s, some universities and scientific research institutes have adopted NIR spectroscopy to investigate and detect the composition, freshness, and maturity of cereals, eggs, and fruits. Besides, bioelectric characteristics has been adopted to identify spermatozoa and azoospermia in eggs, and fluorescence technology has been utilized to detect and analyze the toxins of agricultural products such as peanuts.

In the twenty-first century, food NDD gets further development, hosts of new NDDs emerging as new bionic sensing, high (multi) spectral imaging, odor imaging, and multi-information sensor fusion. While, traditional NDD, such as machine vision and NIR spectroscopy, was constantly updated and expanded their application. NDD plays a crucial role in the quality and safety monitoring of food purchase, transportation and storage, as well as the quality inspection and monitoring of food and agricultural product processing.

1.1.3 Trends

To establish a more comprehensive understanding of the basic research status and development trend of NDD, the author makes a full-scale search on the evolution trend of international basic research of NDD. Sprouted in the 1950s, few relative papers were published until the end of the 1970s. Figure 1.1 shows the evolution trend of basic research on international food NDD after the 1980s, and several basic research papers have been published from the 1980s to the 1990s. After the

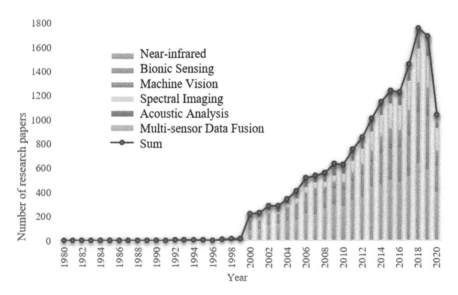

Fig. 1.1 The evolution trend of basic research of international food nondestructive detection technologies after 1980s

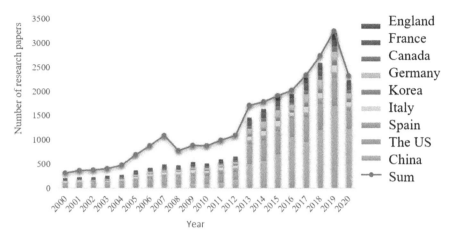

Fig. 1.2 The distribution trend of the main countries (regions) in the basic research of food nondestructive testing technology in the world after the twenty-first century

twenty-first century, there has been a breakthrough growth in the quantities of papers published in this field. Specifically, more than 200 papers can be retrieved in 2000, exceeding 1000 papers was published in 2013, and over 1600 papers every year after 2017. Among the published papers, NIR spectroscopy has become the top technique in the field of nondestructive testing for food quality and safety. Furthermore, there are various researches on bionic sensing, machine vision, and spectral imaging.

Figure 1.2 presents the distribution trend in the main countries (regions) on the basic research of NDD after the twenty-first century. Figure 1.2 shows that the United States dominated the leading position of the research at the beginning of NDD research. France, Germany, and other developed countries possessed somewhat foundation. In the preliminary stages, few relative articles were published by scholars in China until 2000, which then increased rapidly year by year. By 2013, the number of Chinese scholars engaged in the field of NDD has exceeded that of the United States, becoming the country with the largest number of scientific and technological papers published in the field. Generally speaking, the number of papers published by Chinese and American scholars occupies an obvious advantage in the world, and scholars from Germany, France, Spain, and other EU countries also published numerous basic research papers. Additionally, scholars from South Korea, India, Canada, and Brazil have also started fundamental scientific research and published correlative scientific papers in the field of NDD.

Figure 1.3 shows the distribution of basic research on food NDD technology in various continents; European scholars account for the highest proportion in the field of food NDD, reaching 38%, Asians approximating it with 36%. The papers published are mainly by the scholars from China, Japan, South Korea, India, and other countries. Additionally, the United States, Canada, and other North American countries have more relevant research in the field.

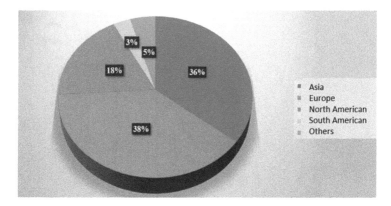

Fig. 1.3 Distribution of basic research on food nondestructive detection technology in different continents

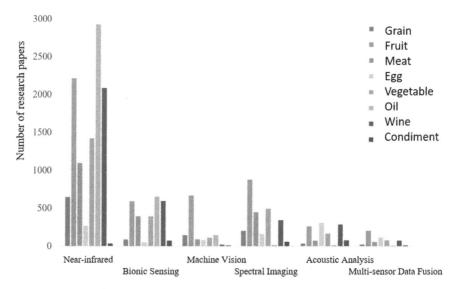

Fig. 1.4 Distribution of research papers on food nondestructive detection technology for different testing objects

Figure 1.4 shows the distribution trend of research papers on the application of different food NDD to various testing objects. From this perspective, the research distribution of NDD technology varies with testing objects, as well as the applicable testing objects of different technologies are also different. It is shown that the number of research papers on the detection of vegetable oil by near-infrared spectroscopy is the highest, mainly attributing to the distinct varieties of the chemical composition of vegetable oil in the region of NIR spectroscopy. Besides, NIR spectroscopy detection is widely utilized in the quality detection of fruits, grains, and poultry. Moreover, numerous NDD techniques are involved in bulk agricultural products, fruits, grains, meat, and other quality and safety testing.

1.2   Characteristics

The advancement of science and technology has greatly expanded the range of NDD. Particularly, the advanced technologies have been constantly updated, such as sensing technology, information science, and nanotechnology. Hence, various NDD instruments and equipment appropriate to the food industry with high automation and precision have been continuously developed. Not only does it shorten the analysis time and reduces human errors, but it also greatly improves the speed, sensitivity, and accuracy of food analysis and testing. The following is a brief introduction of several novel technologies for rapid NDD of food quality and safety.

1.2.1   Optical Characteristic Detection Technology

Optics is a physical discipline that studies the behavior and properties of light (electromagnetic waves) and the interaction between light and matter. When light illuminates on the material, some of the parts reflected by the surface of the material, and the rest is refracted into the material tissue. Besides, the light entering the material is absorbed and converted into heat, some is scattered in all directions, and the rest passes through the material. Furthermore, the optical properties of food materials contain light reflection, absorption, transmission, and photoluminescence of food materials, the properties varying with the materials, applied to particle size measurement, quality evaluation, chemical analysis, grade differentiation, maturity, safety, and freshness discrimination. Recently, with the development of optical technology, NDD underlying optical properties has become an exceedingly active and extensive technology in the field of rapid NDD for food quality and safety.

1. **Spectral Detection Technique**

   (a) Ultraviolet/visible spectrum detection technology

      Both ultraviolet (UV) and visible (Vis) absorption spectra pertain to molecular spectra, caused by the transition of valence electrons. The composition, content, and structure of the substance could be analyzed, determined, and inferred via the absorption spectrum of UV/Vis. M (atom or molecule) absorbing UV/Vis is excited to the excited state, returning to the ground state via a radiative or nonradiative relaxation process, and the relaxation could be accomplished by M* decomposition into new components, namely photochemical reaction. The UV/Vis absorption spectra of substances are determined by the transition of valence electrons in molecules. That is to say, the composition of molecules, especially the properties of valence electrons, varies with the absorption spectra. Therefore, a correlation could be established between the wavelength of the absorption peak and the studied substances to identify the functional groups in the molecule. Essentially, the compounds containing absorption functional groups could be quantitatively determined by UV/Vis absorption spectra. While utilizing the UV/Vis spectrum, food with various components has a unique absorption spectrum in the region. Hence, the evaluation could be accomplished by the detection of corresponding components via the change of absorption spectrum. Consequently, UV/Vis spectrum analysis has the preponderance of simplicity, rapidity, low detection limit, high sensitivity, and so on.

(b) Near-infrared spectroscopy detection technology

Near-infrared spectrum (NIR) is an electromagnetic spectrum between Vis and medium infrared light (MIR), with the wavenumber between 780 and 2500 nm. The absorption of molecules in the NIR region is mainly attributed to the change of the vibrational state. From the point of view of quantum mechanics, NIR spectrum is caused by the transition of molecular vibration from ground state to high energy on account of the nonresonance of molecular vibration. At room temperature, numerous molecules are in the vibrational ground state ($V = 0$). Fundamental frequency transition being from the vibrational ground state to the vibrational first excited state ($V = 1$), the radiation absorption produced by this transition is fundamental frequency absorption. NIR records the frequency doubling and combining information of chemical bond fundamental frequency vibration in molecules, dominated by the overlap of hydrogen-containing groups C, N, and O [4]. Therefore, in the range of NIR, the frequency doubling and combining absorption of XH vibrations of hydrogen-containing groups was measured. In the detection of food by NIR, appropriate multivariate correction methods could be selected to make a correlation between the corrected samples from NIR with their component concentration or property data and establish the relationship between the absorption spectrum of the corrected sample and the concentration or properties of its components. Hence, unknown samples could be classified by correction model and the absorption spectra. Additionally, the feature of NIR could be extracted by selecting the appropriate pattern recognition method for the establishment of the corresponding classification model. Likewise, the application of the established classification model and the absorption spectra of unknown samples could be utilized in the classification. Specifically, the analysis technology of NIR spectroscopy is distinctively different from other conventional analysis techniques. Modern NIR, an indirect analysis technique, accomplished the qualitative or quantitative analysis of unknown samples via the establishment of a correction model. Compared with the traditional physical and chemical analyses, NIR has the merits of high speed and efficiency, low cost, reproducibility, no pretreatment of the samples, convenient, and online monitoring of the production process. Besides, multiple components or properties of the sample could be detected simultaneously via the spectral measurement and the corresponding correction [5–7].

(c) Raman spectroscopy detection technology

Raman spectrum is a kind of scattering spectrum based on the Raman dispersion effect discovered by Indian scientist C. V. Raman, with analysis of the scattering spectrum which is different from the incident light frequency to obtain the information of molecular vibration and rotation, and applied to the **study** of molecular structure. Generally, to enhance feeble Raman spectrum of food molecules, surface coarsening of the electrode could be utilized to obtain 10^4–10^7 times intensity for the surface-enhanced Raman scattering/spectroscopy (SERS). SERS is a kind of molecular vibration spectrum. Under the irradiation of incident light, a series of physical or chemical changes perform on the rough surface of the substrate, which enhanced the Raman signal. There are a variety of Raman spectrum extension technologies, among which SERS is the earliest discovered and the most widely used one.

In the rapid detection of food and agricultural products, trace amounts of substances could be determined by utilizing the fingerprint of SERS with high sensitivity. For instance, it could be realized by using the characteristics that the concentration of the substance to be tested and the intensity of the Raman spectrum follow Beer's law. However, generally, there is noise and incineration background in the original spectrum obtained by the SERS spectrometer in addition to the measured sample. Assuming that the original spectrum can directly be classified and identified or quantitatively modeled, the accuracy of the results would be seriously reduced due to the feeble Raman signal composed of noise and background. Hence, appropriate mathematical models and analytical methods underlying chemometrics can be established by slightly modifying the SERS signal through the pretreatment method to obtain satisfactory quantitative results [8, 9].

2. **Modern Imaging Detection Technology**

The essence of modern imaging technology resembles human vision via the charge-coupled device (CCD) to capture electromagnetic waves in a certain wavelength range to convert optical signals into electrical signals with the records of quantification into computer processing. With the development of science and technology, the elements of modern imaging technology are continuously updated. Furthermore, the advancement of computers and modern sensors expands the scale of the application of modern imaging. Since the 1960s, modern imaging has been gradually applied to the automatic detection of food and agricultural products, increasingly vital to the promotion of agricultural modernization and automation. Prospectively, the automatic detection of food and agricultural products could be provided with multifaceted solutions and ideas by the application of imaging.

(a) Color Imaging Technology

The wavelengths of electromagnetic waves directly perceived by human eyes range from 400 to 780 nm, with various colors formed in the brain. The colors red (R), green (G), and blue (B) are called tricolor, stimulating the corresponding photoreceptor cells, and the brain analyzes and recombines to form various colors. The tricolor imaging in the visible band is the technology closest to the principle of human visual. Specifically, the images are only plotted in gray level by a single CCD lens. With the three CCD lens, tricolor could be separated by a light splitter, such as a filter, respective intensities being captured. It could be restored to the color of the real object after the transformation of the color model. Conventional color imaging refers to acquisition of images within the wavelength range of electromagnetic waves, for extraction of information of an object captured by human vision. Particularly, converting the captured object into an image signal via image capture devices (CCD) and transmitting it to a special image processing system to identify the detected object, conventional color imaging is the exceedingly mature modern imaging technology to date. With the preponderance of convenience and low cost, it has occupied the majority of the market of food imaging technology detection. Presently, VIS has been successfully utilized in the detection, grading, and evaluation of fruits, vegetables, meat, fish, and so on. For instance, VIS imaging is utilized to accomplish the rapid detection of appearance quality indexes such as size, shape, color, texture, and defects, improving the accuracy and speed of online inspection. Thus, several techniques, such as mechanical sorting and automatic classification, are available by VIS. Additionally, real-time monitoring of color changes in food processing could be implemented by extracting and analyzing color to achieve online quality control in food processing. Recently, computer vision technology has shifted from rapid online detection of food appearance quality to rapid NDD of internal properties and components. Moreover, with the three-dimensional visualization, the variation of food appearance quality, nutritional composition, and processing quality could be analyzed intuitively and objectively. Especially in the evaluation of the distribution density of nutrients, the analysis of internal organization, and the monitoring of organizational structure transformation in the processing, the results that traditional methods could not achieve would be obtained [10, 11].

(b) High (Multi) Spectral Imaging Technique

Spectral imaging is a novel technology developed in the 1980s, integrating advanced technologies in the fields of optics, optoelectronics, electronics, information processing, and computer science, with organic combination of the traditional two-dimensional imaging with spectrum. Hyper-spectral imaging (HSI) has the characteristics of multi-band, high spectral resolution, and spectral integration. In 1983, the American Jet Propulsion Laboratory developed the first airborne imaging spectrometer (AIS-1), showing its potential in image acquisition and analysis. Subsequently, Canada, Australia, France, Germany, and other countries invested billions of dollars in the development and application of HSI. Initially, the technology was mainly applied in space remote sensing. Nowadays, it has been extended to the fields of medical diagnosis, drug and food analysis, progress in spectral image data acquisition, spectral calibration, three-dimensional data reconstruction, processing analysis, and pattern recognition.

Spectral imaging is to display, process, analyze, and interpret the image by multiple spectral channels and various sensitivity of spectral reflection (absorption) rate of the target in various bands. With a specific light source or filtering equipment, the wavelength range of the light source could be selected to enhance the image features of various parts of the target object, beneficial to the quality detection. Spectral imaging integrates the analysis of image and spectrum, with unique advantages in detection. Additionally, the spectral image is a three-dimensional data block composed of hosts of two-dimensional images under continuous wavelengths obtained by the spectroscopic system in a specific range. At each specific wavelength, it could be provided that two-dimensional image information via spectral data from the grayscale of the same pixel at various wavelengths. Furthermore, the image represents the appearance features such as size, shape, and color, and the spectrum reflects the internal structure, composition content, and other characteristics. Thus, it could be indicated that spectral imaging technology is available to visually analyze the internal and external quality characteristics of food [12, 13].

The development of science and technology expanded the scale of the accuracy of spectral resolution for imaging spectrum. Depending on the disparity of spectral resolution, spectral images could be divided into three categories, such as multispectral, hyper-spectral, and ultra-spectral images. Generally, the image with spectral resolution in the order of $10^{-1}\lambda$ is called multispectral image, hyper-spectra is the image with spectral resolution in the range of $10^{-2}\lambda$, and ultra-spectral within the scale of $10^{-3}\lambda$. Spectral imaging with various resolutions could be selected underlying accuracy and requirements of detection. In general, HSI is suitable for the detection of food and agricultural products [14–17].

Recently, spectral imaging has been increasingly crucial for application in the detection of food quality and safety, especially in the detection of internal organization, structure, and surface fine features.

(c) X-ray Imaging Detection Technology

X-ray belongs to electromagnetic radiation, resembling visible light, but its wavelength is much shorter than visible light, between ultraviolet and gamma rays. The frequency of X-ray is about 10^3 times that of visible light, and its photon energy is higher than that of visible light. For its short wavelength and high photon energy, X-ray with strong penetration produces reflection, refraction, scattering, interference, diffraction, polarization, and absorption, similar to other electromagnetic waves. Penetrating through the material, it is partially absorbed, and its intensity attenuation becomes weaker. The degree of absorption is related to the composition, density, and thickness of the material. The X-ray absorptivity or transmittance of the sample is at rest with the composition and ratio of the material. Hence, numerous samples vary with opacity coefficients to X-ray. Therefore, the grayscale image shows the disparity of the detected object in density or material thickness.

It could be accomplished that the internal information of objects is difficult to detect by visible light, via X-ray Line Scan. Generally, the wavelength range of X-ray is within 0.1–100 nm, in which the electromagnetic wave with a wavelength less than 1 nm is called hard X-ray, and the wavelength of more than 1 nm is called soft X-ray. The hard X-ray with strong energy has the relatively intense penetration ability, while the soft X-ray energy is relatively small, with weak counterpart ability, vulnerable to being absorbed by the object. Accordingly, the soft X-ray shows thorough internal details, with less hidden danger to the target. Therefore, soft X-ray linear scanning imaging technology is more widely applied, such as detecting the internal quality defects of fruits (water core, browning, abrasion, decay, insect pests), judging whether there is pest infection in wheat grains, identifying the internal foreign bodies of poultry and livestock products, and so on. However, there are some limitations in X-ray scanning imaging, like the overlap, because it projects the internal shape of the object on the two-dimensional plane, increasing the difficulty of judging. Moreover, the safety of X-rays on the human body is still controversial, and there are restrictions in the application of X-rays for food [18].

(d) Others

Computerized tomography scanner (CT) is a novel imaging technology introduced to address the limitation of projection overlap in X-ray linear scanning imaging. Underlying the principle of interaction between rays and matter, initially, hosts of ray absorption data are obtained by scanning around the object. Next, the cross-sectional digital image is acquired by projection reconstruction. Compared with the conventional X-ray linear scanning imaging, CT shows the morphological characteristics of a certain section of the object with higher sensitivity and resolution. Generally, the density resolution of CT is 20 times higher than the conventional X-ray linear scanning imaging. The cross-sectional image obtained by CT has accurate layer thickness and clear image. Additionally, the cross-sectional images required for diagnosis, such as coronal plane and sagittal plane, could be obtained by processing and reconstruction of computer software, with the three-dimensional reconstruction of the measured object by digital processing. Hence, in the analysis of the internal subtle characteristics of food, with the predominance that other technologies are difficult to replace, CT image analysis has been applied in the detection of aquatic products, livestock and poultry products, fruits and vegetables, and others with the predominance that other technologies are difficult to replace.

Magnetic resonance imaging (MRI) produces cross-sectional images with high spatial resolution using strong nuclear magnetic resonances, gradient fields, and hydrogen atoms inside the human body. Via the spatial coding technology, the nuclear magnetic resonance signal emitted in the electromagnetic form is detected and received by the detector, and input to the computer. Subsequently, the image is finally formed after data processing and conversion. For example, on nuclear magnetic resonance imaging, brightness is correlated with water of tissue structure. Further, MRI occupies accurate results compared with CT. It could be detected that the internal information of agricultural products without external damage. Recently, MRI has been widely utilized for internal quality detection of fruits as well as the identification of grain germplasm. Consequently, MRI, cutting the layer directly in any direction, has no radiation harm to the human body, with the prospect in the application.

Ultrasonic imaging is an image diagnosis technology that applies ultrasonic detection technology to the tested object and detects the tissue of the object. It could be applied in the detection of the fat content of animal products, with high discrimination of soft tissue, for the merits of high frequency, short wavelength, concentrated energy, excellent directivity, strong penetration ability, safe and noninvasive, etc.

To sum up, the above imaging technologies obtain hosts of information about the detection object with excellent results, while the limitations, such as large equipment, inconvenient movement, high cost constrain the application.

1.2.2 Force/Acoustic Detection Technology

The term "wave" can be defined as the propagation phenomenon of the mechanical vibration of an object in a medium with particles and elasticity. Specifically, sonic waves could cause sound sensation in the auditory organ of the individuals' ear. The mechanical properties of food can be revealed by vibrating sound waves. The vibration dynamic refers to the signal produced by food under external excitation [19]. Specifically, the vibration signal is used to analyze the reflection, scattering, transmission, and absorption of agricultural product via the basic law of the interaction between vibration signals and agricultural products. According to the difference of spectral resolution, spectral images can be divided into three categories: multispectral, hyperspectral, and ultra-spectral images. The evaluation could be accomplished by utilizing the action of percussion vibration response signal characteristics under the action of sonic waves, such as reflection characteristics, scattering characteristics, transmission characteristics, absorption characteristics, attenuation coefficient, and propagation velocity, as well as their own acoustic impedance and natural frequency. Generally, agricultural products with diverse structural strengths produce corresponding sound signals under external excitation; a clear sound illustrates as large structural strength of objects or vice versa. Accordingly, it can be detected by analyzing the difference of sound signal caused by external shock. The acoustic characteristics may vary with the internal organization of food, by which the internal quality be able to detect with classification. Presently, by and large, the nondestructive testing device for the acoustic characteristics is composed of an acoustic generator, acoustic sensor, charge amplifier, dynamic signal analyzer, computer, and printout equipment. Specifically, during the detection, the sound waves emitted by the acoustic generator are continuously thrown at the tested material, and the acoustic signals reflected, scattered, or transmitted from the material are received by the acoustic sensor and transmitted to the dynamic signal analyzer and computer after amplification. Consequently, the relevant acoustic characteristics display on the output equipment. For example, in NDD, acoustic characteristic detection can be applied to evaluate the maturity of watermelon, the crack of poultry eggs, and the hardness of pears.

1.2.3 Bionic Sensing Detection Technology

The sensory quality mainly includes various indexes, such as color, aroma, taste, shape, quality, and so on. Presently, the sensory quality evaluation is completed via artificial cognition by individuals' organs, that is, vision, smell, taste, touch, and other sensory organs before the evaluation of the brain according to experience. However, it is often blamed for obvious defects: firstly, experiential behavior and related personnel need long-term training. Second, artificial sensory evaluation is vulnerable to external factors, subjectivity, and repeatability. The individual sensitivity differences, such as sensory organs, geographical and gender disparity, personal preferences, emotional and physical conditions, and other external factors, are affecting the test results. Therefore, it is necessary to develop an intelligent sensory evaluation system for food to avoid the above problems, acting as a crucial role to guide the production process, ensure the quality, and increase food added value. Hence, food bionic sensors become an alternative.

In artificial sensory inspection, the color and shape are obtained by visual correlated organs, synthesized and balanced by the brain, and compared with that based on the knowledge to evaluate sensory indexes, such as color, aroma, taste, and shape comprehensively; accordingly, vision, smell, and taste act as a significant role in food sensory testing. The development of computers, microelectronics, and material science opens a new way for bionic sensor technologies, such as vision, smell, and taste, laying the foundation for intelligent sensory testing. Electronic vision technology being a part of computer vision, electronic smell, and taste is also known as "electronic nose" and "electronic tongue." It is composed of sensor array, signal processing, and pattern recognition modules. And it simulates the functions of the human nose and tongue respectively to evaluate flavor and taste quality by instrument of smell and taste [20–22]. Since the 1980s, bionic sensing evaluation has been the focus of scholars at home and abroad, with the increasing appreciation in the evaluation of food flavor quality. Unlike ordinary chemical analysis, a bionic sensor does not obtain quantitative or qualitative results, but the overall response of multiple components is obtained by numerous sensors simultaneously, and the overall information of the object is known as characteristic information. Without specific physical meaning, it is relied on the results of the expert sensory evaluation or conventional physical and chemical analysis, harnessing chemometrics to establish a model. Via comparative analysis, intelligent evaluation, such as quality classification and authenticity identification, can be achieved by the functions resembling human perceptual organs.

Recently, with the continuous advancement of bionic sensing, the form of the sensor is constantly changing, and its connotation and extension are constantly expanding. The expression form of the sensor is one-dimensional electrical signal and image. Olfactory visualization technology is a novel type of bionic sensing technology, emerged in 2000, first proposed by Professor Kenneth S. Suslick of the University of Illinois at Urbana-Champaign in the United States. Utilizing metal porphyrin as a sensor for the detection of volatile substances for the qualitative and quantitative detection of volatile chemicals lays the foundation for the application of odor imaging detection technology. The technology is a new visualization method that imitates the olfactory system of individuals and mammals, based on the color variation of chemical chromogenic agents after the reaction between chemical chromogenic agents and detected gases. Compared with the traditional electronic nose and electronic tongue, the bionic sensing technology strongly depends on strong forces, such as covalent bond, ionic bond, and oxygen bond. Additionally, it can minimize the defects of strong anti-interference ability to interference factors such as water vapor in the environment that exist in biological, chemical, or physical sensor technology. The odor imaging sensor is composed of dyes with specific recognition ability. After the interaction between dyes and the detected substances, the color changes significantly with the specific RGB digital signal formed after computer processing. Consequently, the RGB data are regressed with the characteristic chemical index substances related to odor by the analysis of the corresponding pattern recognition. The detected substance can be analyzed qualitatively and quantitatively. The advancement of material processing technology and the continuous improvement of computer data processing and odor imaging detection technology expanded the scale of application in environmental monitoring, food and beverage quality monitoring, disease diagnosis, and other fields. To date, odor imaging detection technology has been gradually applied to detect volatile foods such as vinegar and liquor.

1.2.4 Electromagnetic Detection Technology

The elements of food are composed of electrons, protons, and neutrons, with somewhat electromagnetic properties, which are divided into two categories, active and passive. Active electrical characteristics caused by the existence of energy sources in food materials mainly expressed as bioelectric potential are rarely used in detection because the active electrical characteristic is exceedingly weak requiring precise instruments. Previous studies have demonstrated that bioelectric potential has a crucial impact on the preservation of food. The passive one mainly refers to the behavior of food under the external electromagnetic field. After the sample being affected by the electromagnetic field, it reacts to the external environment, such as nuclear magnetic resonance (NMR) and electron cyclotron resonance (ECR). To date, electromagnetic properties detection has been used widely, such as the evaluation of the quality of fresh fruits and eggs by constructing correlation between the dielectric ratio and the quality characteristics, the tests of water content in food and agricultural products via the correlation between impedance value and moisture, and the determination of states of free atoms and transition metal ions by electron cyclotron resonance. Moreover, it is suitable for the analysis of reactive intermediates containing peroxides formed in the early stage of automatic oxidation of oil. Besides, the magnetic resonance characteristics are utilized to detect and analyze the water distribution and water transfer of grains and fruits. Nonetheless, the rapid detection being hard to accomplish in large quantities, or requires high cost, the electromagnetic characteristics detection are not widely applied in food quality and safety testing.

1.2.5   Multi-sensor Data Fusion Detection Technology

Multi-sensor data fusion (MSDF) was first developed in the military field. In the 1970s, the US department of defense attached great significance to research of acoustic signal recognition and attempted to fuse multiple independent and continuous signals to detect enemy submarines in a certain sea area. The technology began to appear accordingly; it is originally defined as a multi-level and multifaceted process of processing, detecting, interconnecting, correlating, estimating, and combining multi-source information and data. The definition mainly emphasizes three aspects of multi-sensor information fusion: first, the content of MSDF mainly includes processing, detection, interconnection, correlation, estimation, and combination of information. Second, MSDF processes multi-source information at several levels, each of which represents different levels of information abstraction. Third, the results of MSDF include low-level state, identity estimation, and high-level global situation assessment at the whole tactical level [23].

The indicators that reflect food are various, including quality and safety, appearance, and internal quality, while a single detection technology has limited information, which, inevitably, affects the accuracy and stability of the test results. For example, computer vision technology represents the external features such as size, color, shape, and texture of food, but it is usually powerless to analyze the internal tissue characteristics and composition content, while NIR is the opposite of that. Hence, utilizing the strengths of various testing methods, combining each other to improve the comprehensiveness, reliability, and sensitivity of testing is a new development trend of rapid NDD for food quality. To date, scholars at home and abroad have put forward novel issues underlying MSDF, such as the fusion of visible light imaging technology, NIR spectroscopy, electronic nose, electronic tongue, and other sensing information technology to evaluate the quality of wine, fruit, beverages, meat, etc. Compared with single-sensor, MSDF has the merits of numerous information, medium tolerance, resembling human cognitive process, which distinguishes food comprehensively [24–27].

As mentioned above, the advancement of various information, materials, computers, sensing, and other technologies expanded the scale of food rapid NDD as a cross-discipline. The application in food evaluation varies with the diverse technical characteristics. However, the objective of the book was to focus on the technologies that are active in the field of food detection in recent decades, guiding the intelligence of the food industry in the future, such as computer vision, spectral analysis, spectral sensing technology, bionic sensing technology, CSA, acoustic characteristic detection, multi-information fusion, and so on.

1.3   The Instrumentation and Its Trends

NDD has attracted much attention in the field of food quality and safety due to its advantages of no damage, quick speed, and easy to store and transfer analysis data. Notably, it has been used as an effective measure to guarantee food quality and safety when it was applied in the in situ safety inspection of food collection–storage engineering and food processing. Compared with conventional methods, it greatly shortened analysis time, reduced the risk of human errors, and improved the speed, sensitivity, and accuracy of food analysis and detection. With the rapid development of NDD, the corresponding rapid detection equipment has also been explored. Nowadays, the relatively commonly used equipment for food quality and safety include portable NIR spectrometer, portable electronic nose, portable industrial devices, portable multispectral device, near-infrared-based imaging device, as well as portable acoustic vibrations analyzer. And most of them are under commercial production, which has the ability to realize the rapid detection for food quality and safety. In addition, some other nondestructive testing equipment, such as hyperspectral testing equipment, odor imaging testing equipment, multi-information fusion testing equipment, also begin to step out of the laboratory and step into the process of practical application.

1.3.1   Instrumentation for Monitoring of Food During Storage

The nutrients of agricultural products can occur in some changes caused by both changes from the external environment and the function of their own enzymes in the process of acquisition, storage, and transportation, resulting in a significant decrease in their quality. Besides, if there were no effective prevention and control measures, it will further lead to a significant reduction in the quality of food and into a serious food safety problem, especially with the assistant of some harmful material that exists in the outside environment. Recently, NDD has showed great advantages in real-time monitoring of the quality and safety information in the process of food storage due to its fast and nondestructive testing characteristics. Some companies like Thermo, Brimrose, and Viavi have developed various small and cost-effective NIR equipment, which can satisfy the demand of secondary development from users. They have been applied to the rapid, accurate, and nondestructive detection of fruit soluble solids, grain protein, fat, moisture, and other indicators. In recent years, domestic food equipment enterprises and universities have also developed NIR spectroscopy detector, which can realize quick and nondestructive detection of tea, grain, fruit, and other internal component indicators.

Electronic nose system is also widely used in the food detection. For example, PEN3 electronic nose from Germany AIRSENSE company, the Cyranose 320 sensors electronic nose system from the Intelligent Optical Systems company, and CMD-516 electronic nose from United States Sensigent company can be used to identify single or multiple gases from food with characters such as compact, fast, and efficient. That is to say, the portable electronic nose has been widely used in food rancid detection, molasses type and aroma characteristics analysis, meat freshness analysis, milk freshness discrimination, and other food quality and safety monitoring.

Furthermore, machine vision equipment combined with image processing technology can also be used for quick nondestructive detection of food quality indicators change such as appearance, size, color, and defects in the process of storage. Moreover, spectral imaging technology can quickly detect and analyze food micro-damage, hidden defects, component distribution, and other characteristics. Alternatively, vibration-acoustic detection technology can be used to analyze the structural characteristics of fruit hardness and maturity, egg crack, and eggshell hardness. With the accelerating pace of food purchase, storage, and transportation, NDD plays an increasingly important role in food quality and safety monitoring.

1.3.2   Intelligent Quality Monitoring Instrumentation in Food Processing

After the agricultural product is received and stored, it will normally experience commodity process or further processing by primary processing producing origins and then for sale. With the integration of information technology, automation technology, and intelligent technology with the field of circulation of agricultural products, raw agricultural products detection technology and complete classification equipment are growing mature, especially in the field of cleaning, grading, packaging, and drying of agricultural products, poultry slaughtering, and refrigerated storage. It determines the origin of the commercialization and intensive processing of agricultural products, which finally improves the product added value and the market competitiveness of agricultural products. In addition, in the process of food processing, some important physical and chemical indicators are continuously changing, which need to be monitored regularly. Moreover, a lot of process steps and parameters are also needed to be adjusted according to the real-time changes of these indicators. However, there is a difference between different batches of food products in the process of processing due to various impact factors, such as the variety of raw materials, processing season, environmental conditions, and many other factors. During the storage period, it is affected by the storage conditions such as container, temperature difference, humidity, ventilation, and storage time, which also cause some changes in its quality, like changes in volatile substances. Therefore, to ensure the consistency of production batches during food processing, it is necessary to monitor some important indicators in the process of food processing in real-time. Food nondestructive testing technology can monitor the change of color, flavor, size, chemistry, and other indexes in the process of food processing intelligently in real-time, so as to control the quality and safety of food and agricultural products.

As a typical example, Napel in Japan and the Netherlands Moba company explored eggs raw online classification system to meet the needs of quality eggs inside and outside. Many kinds of nondestructive testing equipment modules were adopted in the production line to ensure the quality of eggs and online classification. In detail, machine vision in production line equipment module can classify stained eggs, size, and appearance characteristics; the mechanical testing equipment module can orient the size of the egg head; and the acoustic vibration detection equipment module can detect and classify egg cracks. As another example, the Mafroda Company of France, the Autoline Company of the United States, and the Compac Company of New Zealand analysis the quality of fruits (cherry, tomato, strawberry, and winter jujube) through the detection of their shape, size, defect, and degree of maturity using photoelectric nondestructive detection and grading technology. In addition, tea production companies and universities in China also use vision detectors, NIR spectroscopy, and gas sensing equipment to detect and analyze changes in color, composition, and gas during solid-state fermentation of tea, and monitor the quality of tea aeration fermentation. In the process of rice wine fermentation, NIR spectrometer is used to monitor the changes of the contents of total sugar, alcohol, acidity, temperature, oxygen, and other environmental factors in the fermented mash of rice wine in real-time, and the dynamic changes of the quality of the whole fermentation process of rice wine can be clear at a glance. The odor imaging equipment can monitor the alcohol content in the solid fermentation process of vinegar in real-time, as well as the composition and content changes of various volatile gases, and monitor the fermentation quality of vinegar. With the high development of food industrialization, higher requirements are put forward for the intelligence and information of food detection. NDD and equipment are also playing an increasing importance in the intelligent monitoring of food quality and safety.

1.3.3   Development Trend of NDD Equipment

The development of science and technology has led to the modernization of food testing technology, which further raises food testing technology to a new level. The most prominent feature is to rely on high-tech, people-oriented, humanized design of instruments and equipment with an increasingly heavy weight. Food testing and analysis instruments are important carriers of food testing technology and pay more attention to practicality and accuracy with the development of science and technology, and the following trends are shown in the application of food quality and safety testing.

1. **Portable testing instruments**

   With the rapid development of the logistics industry, on-site monitoring of quality and safety in the process of food circulation has become an inevitable trend for future development. Traditional laboratory testing instruments are not convenient to carry and use because of their large size, so it is difficult to meet the needs of rapid on-site testing in the process of food circulation. Therefore, the development of miniaturized, portable testing equipment has become the inevitable development direction in the field of food quality and safety testing. The rapid advancement in microelectronic technology and micro-nano processing has provided developing possibilities in the miniaturization, and portability of the related equipment. At the same time, in the aspect of food quality and safety detection, with the continuous update of storage and transportation methods, higher requirements have been put forward for the on-site acquisition of food and the convenient and quick detection in the process of transit. Therefore, the miniaturization and portability of detection instruments is the inevitable trend of the development of food quality and safety detection instruments [28].

2. **Testing object specialization**

   At the end of the twentieth century, in order to obtain higher profits, instrument manufacturing companies vigorously promote the line of multifunctional products expecting to develop instruments that can cover a variety of test objects and multiple indicators that "One instrument multifunctional" is more economical. But it is proved by practice that the accuracy of the analysis instrument designed with multiple test objects can be reduced by the restriction of each function, and the reappearance of the function conversion device cannot be fully guaranteed after function conversion. In addition, the use of multifunctional analysis instruments on the operator's level of business is also very high. With the food quality and safety testing more stringent requirements, food quality and safety testing object and function of the specialization would be a trend in future.

3. **Intelligent modularization of testing process**

   With the diversification of material transportation forms, the rise of new technologies, such as artificial intelligence and the Internet of Things, intelligence is an inevitable trend in food quality and safety testing. Intelligent evaluation of food quality and safety includes both the intelligence of the sample detection itself, i.e., independent sampling and testing, and the transmission of samples at different testing processes, i.e., independent transmission and exchange of samples and data. Due to the highly complex and diverse nature of food materials with widely varying physical and chemical properties, it is necessary to modularization of the testing process for food quality and safety monitoring. The modularization of the testing process can ensure the high-efficient processing and testing of target molecules in samples, which reduces human labor intensity and the consequent uncertainty, contamination, and possibility of data intervention. In addition, the modularization processing of food quality and safety testing allows flexible multi-component integrated processing, object derivation, and intelligent solvent matching to improve the efficiency, reproducibility, and intelligence of food safety fast sample pretreatment and testing. In a word, modularization and intelligentization of food quality and safety are complementary and inseparable. Modularization processing can provide the foundation for intelligent detection, and intelligent management can help the modularization order management and intelligent circulation in the detection process, and provide the guarantee for modularization.

4. **Testing data informatization**

The intelligent process of modern NDD instruments is inevitably accompanied by information. Recently, since food safety incidents have happened frequently, the whole traceability including the construction of food quality and safety testing data traceability system, the realization of the whole chain of food quality and safety information data credible collection, credible storage, convenient sharing, is the trend of future development. When using modern NDD equipment to obtain food quality and safety information, the risk model of relevant information can be constructed, and the information can be actively associated with the cloud platform through the traceability QR code to build the risk database of the food quality and safety big data cloud platform. In the process of intelligent food quality and safety testing, information processing of data, the development of supporting traceability platform to achieve food quality and safety supply chain—acquisition—storage process of the whole chain information traceability, is the trend of future development.

References

1. Wolnik KA, Fricke FL, Capar SG et al (1983) Elements in major raw agricultural crops in the United States. 1. Cadmium and lead in lettuce, peanuts, potatoes, soybeans, sweet corn, and wheat. J Agric Food Chem 31:1240
2. Norris KH (1964) Simple spectroradiometer for 0.4 to 1.2-micron region. Trans ASAE 7:240–242
3. Ben-Gera I, Norris KH (1968) Direct spectrophotometric determination of fat and moisture in meat products. J Food Sci 33:64–67
4. Li X, Zhang Y, Wang D et al (2020) Review of NIR spectroscopy methods for nondestructive quality analysis of oilseeds and edible oils. Trends Food Sci Technol 101:172–181
5. Grassi S, Alamprese C (2018) Advances in NIR spectroscopy applied to process analytical technology in food industries. Curr Opin Food Sci 22:17–21

6. Czarnecki MA, Morisawa Y, Futami Y et al (2015) Advances in molecular structure and interaction studies using near-infrared spectroscopy. Chem Rev 115(18):9707–9744
7. Sun T, Huang K, Xu H et al (2010) Research advances in nondestructive determination of internal quality in watermelon/melon: a review. J Food Eng 100(4):569–577
8. Li X, Gao G, Niu L, Lin C, Qin Z, Liu J, Yao H (2012) Study on Raman imaging of living mouse ears. Anal Chem 40(10):1494–1499
9. Zhai C, Peng Y, Li Y, Zhao J (2017) Non-destructive testing of chemical additives in food based on Raman spectroscopy imaging. J Chem Coll Univ 38(03):369–375
10. Dachun J (2008) Image color eigenvalue extraction and matching based on C#. J Hebei Softw Vocat Techn Coll
11. Jiang L (2003) Adaptive median filtering under multiple windows in digital image processing. Comput Eng
12. Xu Y (2015) Study on nondestructive detection of peanut seeds quality based on hyperspectral imaging technology. Jiangsu University
13. Li Q, Xiao G, Xue Y, Zhang J (2008) Study of human blood cells based on microscopic hyperspectral imaging. Optoelectron Eng 5:98–101
14. Xin Z, Sun J, Tian Y, Lu B, Yingying H, Chen Q (2020) Hyperspectral technique combined with deep learning algorithm for detection of compound heavy metals in lettuce. Food Chem 321:126503
15. Zhu Y, Zou X, Shi J, Zhao J, Xu Y, Lin T (2014) Study on solid fermentation process of Zhenjiang vinegar based on hyperspectral image technology. Modern Food Sci Technol 30:119–125
16. Chen Q, Zhang Y, Wan X, Cai J, Zhao J (2010) Study on pork tenderness detection based on hyperspectral imaging technology. J Opt 30:2602–2607
17. Li H, Felix K, Zhao J, Chen Q (2016) Quantifying total viable count in pork meat using combined hyperspectral imaging and artificial olfaction techniques. Food Anal Methods 9:3015–3024
18. Zhenggan Z (2004) Defect extraction of X-ray images based on sub-region adaptive median filtering. Acta Aeronaut
19. Li PC, Wang QH, Zhang QL, Cao SW, Liu YL, Zhu T (2012) Non-destructive detection on the egg crack based on wavelet transform. IERI Proc 2:372–382
20. Liu Q, Zhao N, Zhou D et al (2018) Discrimination and growth tracking of fungi contamination in peaches using electronic nose. Food Chem 262:226–234
21. Majchrzak T, Wojnowski W, Dymerski T et al (2017) Electronic noses in classification and quality control of edible oils: a review. Food Chem 246:192–201
22. Ezhilan M, Nesakumar N, Babu KJ et al (2018) An electronic nose for royal delicious apple quality assessment – a tri-layer approach. Food Res Int 109:44–51
23. Chen Q, Sun C, Ouyang Q, Liu A, Li H, Zhao J (2014) Classification of vinegar with different marked ages using olfactory sensors and gustatory sensors. Anal Methods 6:9783–9790
24. Li H, Chen Q, Zhao J, Wu M (2015) Nondestructive detection of total volatile basic nitrogen (TVB-N) content in pork meat by integrating hyperspectral imaging and CSA combined with a nonlinear data fusion. LWT Food Sci Technol 63:268–274
25. Li H, Kutsanedzie F, Zhao J, Chen Q (2016) Quantifying total viable count in pork meat using combined hyperspectral imaging and artificial olfaction techniques. Food Anal Methods 9:3015–3024
26. Xu Y, Kutsanedzie FY, Sun H, Wang M, Chen Q, Guo Z, Wu J (2017) Rapid pseudomonas species identification from chicken by integrating CSAs with near-infrared spectroscopy. Food Anal Methods 11:1199–1208
27. Ouyang Q, Zhao J, Chen Q, Lin H (2013) Classification of rice wine according to different marked ages using a novel artificial olfactory technique based on CSA array. Food Chem 138:1320–1324
28. Sheng R, Cheng W, Li H, Ali S, Agyekum AA, Chen Q (2019) Model development for soluble solids and lycopene contents of cherry tomato at different temperatures using near-infrared spectroscopy. Postharvest Biol Technol 156:110952

Chapter 2
Near-Infrared Spectroscopy Technology in Food

What Is Near-Infrared Spectroscopy?

Near-infrared spectroscopy has the characteristics of higher spectral resolution, mature technology, nondestructive, and low sample consumption, which is suitable for real-time monitoring in production. In this technology, the reflective fiber is usually used to obtain the spectral information of the sample, and the external environmental factors have little influence on the spectral acquisition. This chapter mainly discusses the principle and application of near-infrared spectroscopy technology in the detection of food samples, as well as the processing and analysis of near-infrared spectroscopy data, as shown in Fig. 2.1.

Abbreviation	Full name
ACO	Ant colony optimization
ANN	Artificial neural network
BP	Back propagation
CARS	Competitive adaptive reweighted sampling
DPLS	Discriminant partial least squares
ELISA	Enzyme-linked immunosorbent assay
HCA	Hierarchical cluster analysis
HPLC	High-performance liquid chromatography
IR	Infrared spectroscopy
GA	Genetic algorithm
GC-MS	Gas chromatography mass spectrometry–mass spectrometry
KNN	K-nearest neighbor
LA	Local algorithm
LCMS-MS	Liquid chromatography mass spectrometry–mass spectrometry
LDA	Linear discriminant analysis
LDs	Linear discriminant
LOO-CV	Leave one out cross-validation
LVs	Latent variables
MC	Mean centering

(continued)

© The Author(s), under exclusive license to Springer Nature Singapore Pte Ltd. 2021
Q. Chen et al., *Advanced Nondestructive Detection Technologies in Food*,
https://doi.org/10.1007/978-981-16-3360-7_2

Abbreviation	Full name
MLR	Multiple linear regression
MLP	Multilayer linear perceptron
MRA	Multiple regression algorithm
NIR	Near-infrared
OSC	Orthogonal signal correction
PCs	Principal components
PCA	Principal component analysis
PCR	Principal component regression
PLS	Partial least squares
PLS-DA	Partial least squares discriminant analysis
PLSR	Partial least squares regression
RBFF	Radial basis function neural network
RF	Random forest
RMSEC	Root mean square error of calibration
RMSECV	Root means square error of cross-validation
RMSEP	Root mean square error of prediction
RPD	Residual predicted deviation
Si	Synergy interval
SPA	Successive projection algorithm
SNV	Standard normal variant
SVM	Support vector machine
WT	Wavelet transforms

2.1 Introduction

Food quality and safety are now progressively gaining public consideration as it has been deeply linked to public healthcare. The demand for high-quality food is on the rise in the development of economies and technological progress, requiring a timeless and objective quality assessment. Consequently, monitoring the toxic substances in food by fast detection systems is urgently needed. Near-infrared (NIR) spectroscopy is a fast, nondestructive approach and is being employed extensively during the past few years in food quality assessment [1]. The main region of NIR spectroscopy is 780–2500 nm. It deals with both electronic spectroscopy and vibrational spectroscopy because of NIR region bands, resulting in a combination of electronic transitions, furthermore owing combinations and overtones and diffuse-reflection of light. NIR spectroscopy involves emission, absorption, reflection, scattering, and diffuse reflection [2]. Traditional methods are considered time-consuming, laborious, and expensive as well. NIR spectroscopy has advantages of online monitoring feasibility that is considered more fruitful for the quality and safety monitoring of food and foodstuffs.

NIR spectroscopy is considered a versatile approach based on the very broad molecular overtone and band combinations used in the entire range that always lead

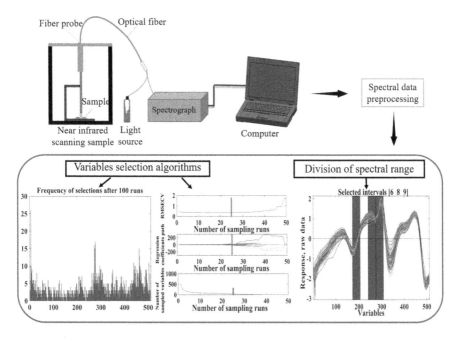

Fig. 2.1 The design of near-infrared spectroscopy

to complex spectral data. It can be difficult to assign particular features to specific food components; therefore, multivariate analysis plays a key role. NIR spectroscopy has a significant advantage and known as well matured technique coupled with chemometrics. The optimum model with better accuracy The optimum model with better accuracy yielded various kinds of pretreatments, different variables, and interval selection model to remove the noise and irrelevant information of various kinds of pretreatments, different variables, and interval selection methods [3, 4].

In this chapter, we have introduced state-of-the-art NIR spectroscopy. The working principle of NIR spectroscopy is discussed. The data mining, including various data analysis techniques such as qualitative and quantitative analysis, is presented. Furthermore, applications of NIR spectroscopy toward monitoring food quality and safety in the last years have been included. Significant research articles has already been published; however, the current book chapter reviewed limited literature based on the consideration of novelty.

2.2 Instrumentation

NIR spectroscopy light mainly covers the range of 4000–14,000 cm^{-1} during the spectra collection and complex structural information. In NIR spectroscopy, the sample absorbs in specific frequencies from the light source that lead to overtones and mixture bands of vibrational transformations of the molecule, which are primarily composed of CH, OH, CO, and NH groups [5]. These bonds are prone to vibratory energy shifts as NIR frequencies irradiate them, and there are two vibratory patterns, including stretch and bent vibration. In the NIR region, organic molecule energy absorption takes place as molecules vibrate or transformed into an absorption range inside the NIR spectrometer. Typically NIR works on the basic principle that by using NIR light, NIR spectroscopy first irradiates the sample and further the transmitted or reflected radiation [3]. Also, NIR detection complies with Lambert–Beer law; hence, the wavelength and quantity of light absorbed are determined by the unique chemical bond. There is a particular NIR range, and each material has infrared absorption. That is why the NIR spectrum can be used for qualitative and quantitative analyses [6–8].

The fundamental vibration frequency of the bond precisely corresponds to the minimum energy demands for inducing the connection vibration from its ground. Bonds at excitation frequencies (i.e., shorter wavelengths) often absorb integral values at necessary frequencies (i.e., double or thrice the basic energy level). Due to higher energy demand, comparatively fewer bonds are excited in the sample, and a smaller magnitude is formed in the absorbance spectra, rendering is much more difficult to classify and measure the bonds. Wavelengths below 1000 nm may usually lead to the weak third bond overtones, thus providing no useful details. Similarly, due to different vibration modes, a single type of bond may demonstrate absorption at various wavelengths. For instance, stretching vibrations exist when the two bound atoms have been in linear movement, and bending vibrations exist during the "scissoring" motion of both atoms bound to a central, common atom [3, 9, 10]. The electromagnetic radiation-enhanced food absorbs radiation at specific frequencies by bonds that are formed between certain atoms in the food. The chemical composition of an unknown mixture or food substance can be defined by detecting this absorption without specifying the precise molecular structure appropriate for the vibrational energy absorption. Thus, considering their obvious lack of observable and discriminatory data, NIR spectra are rich in chemical and physical details about organic molecules and therefore can provide valuable information on the composition of a particular food. At the same time, the overall sketch of the NIR spectroscopy has been presented in Fig. 2.2.

NIR spectroscopy is identical to the ultraviolet (UV)-visible and mid-IR instruments (Fig. 2.2). Generally, a source of light, detector, and dispersive element (such as prism or diffraction grids) are used to record the intensity at various wavelengths. The benchtop NIR spectroscopy instrumentation is developed from basic, filter-based, and dispersive equipment into the interferometric FT spectrometer, which currently directs most spectrometric technical conditions. A standard NIR spectrometer includes a source of radiation, a device for wavelength selection (i.e., monochromator), a holder for sample, a photoelectric detector for intensity assessment of observed light and to transform them into electric signals, and a computer system for the spectral data collection and processing (Fig. 2.1). Fourier transform technique can deliver the optimum spectral resolution (2 cm^{-1}), very high signal-to-noise ratio

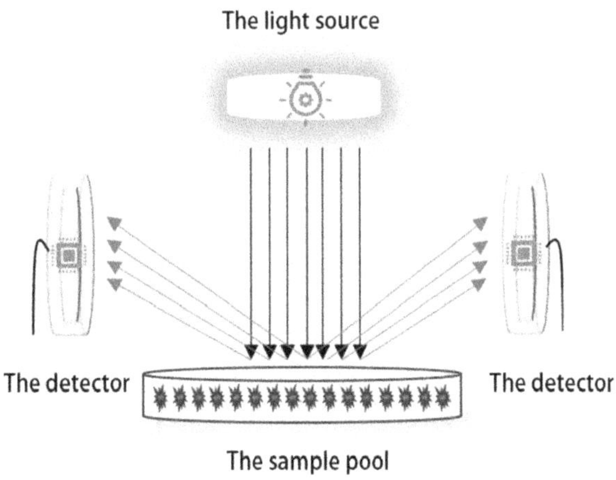

Fig. 2.2 Components of the
NIR spectroscopy [2]

The light source

The detector   The detector

The sample pool

(more than 10,000:1), quick spectral data acquisition (<1 s per scan, depending on resolution), and border spectral range of 800–2500 nm [11, 12].

2.2.1   Light/Radiation Source

Several sources of radiation are employed in NIR spectroscopy, including thermal and nonthermal sources. Thermal sources encompass radiant filaments, which produce radiation across a broad spectrum of NIR frequencies. A single polychromatic thermal source is mostly used for NIR spectroscopy. A typical example is an inert concrete, mainly tungsten (a tungsten halogen lamp), which is when electrically heated to 1500–2200 K irradiates uniformly in the IR spectral range. Tungsten halogen lamps are widely used as sources of radiation (as shown in Fig. 2.3).

In contrast, nonthermal radiation sources comprise discharge lamps, light source-emitting diodes (LEDs), laser diodes, or lasers and produce great slender bands of radiation compared to thermal. LEDs as NIR radiation sources must be increased if devices in the lower energy region of 1600–2500 nm can emit greater power at a lower cost. NIR source is based on the use of an LED framework. LEDs are quick-responding devices and are modulated to enhance NIR instruments' robustness [14].

2.2.2   Wavelength Selectors

NIR spectrometers are based on wavelength selection, separated, either continuous selection or discrete wavelength. While spectrophotometer assumed from discrete wavelength works by irritating samples with a minimum selected wavelength and utilizing LEDs or filters. Instruments that relays on filters are generally found

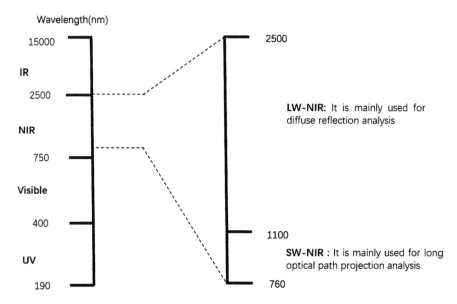

Fig. 2.3 Principal types of NIR absorption bands and their locations [13]

between 6 and 20 interference filters for the specific application in the particular spectral region to choose wavelength absorbed by targeted molecular species [15]. The use of cheap instruments is considered the most convenient way to determine analytes [13]. The NIR instruments for the continuous spectrum may comprise a diode array or diffraction grating, acousto-optical tunable filter, or FT-NIR type. These kinds of tools are considered even more flexible than the instruments based on the discrete wavelength and have a broader range of applications in various fields [16].

2.2.3   Sample Presentation Modes

Functional strengths of NIR spectroscopy are the vast variety of readily available options for the presentation of samples, enabling NIR spectroscopy applications in a wide range of types of modes (e.g., slurries, fluids, solid or powdered samples, and gasses). These sampling options typically require the use of independent, detachable accessories and encompass the possibility of using fiber optic for the spectral data collection from remote sides in a large industrial complex [17]. Three sampling modes were used for the sample data collection: transmittance mode, reflectance mode, and interactance mode (Fig. 2.4) [19]. The scattering of light that passes through a granulated material is triggered by interaction with a series of angular substances that emit light. The entire light may not contain any information about the sample composition.

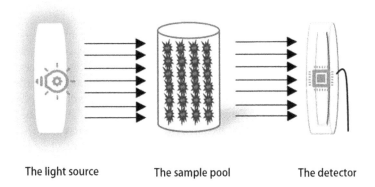

The light source   The sample pool   The detector

Fig. 2.4 Various data collection modes of NIR spectroscopy [18]

In the transmission measurement mode, incident light propagates through the sample and is measured as it exits the sample at a point directly opposite to the light source. Transmittance mode is normally used for the liquid samples but may also be used for the solid samples, such as cheese, meat, and whole grains form by NIR spectroscopy [13]. Transmittance mode determines the amount of transmitted light, which is probably small but may have more valuable information. NIR radiation may enter the sample and be absorbed in the form of either reflectance or transmittance. The Lambert–Beer law defines the proportionality between transmission, the concentration, and the sample path length of the absorbing component. This law may only be utilized for transparent, clear liquid samples only when no light is scattered. Beer's law stated that the transmittance of a solution remains constant if the product of concentration and path length stays constant [17]. These modes used wavelengths in this mode are generally found in the range between 700 and 1100 nm. Reflection usually is drastically reduced in transmittance measurements so that the radiation proportion enhanced by the sample can be determined as transmission.

Reflectance mode is mainly used for the sample in the form of solid or granular. Light dispersion is a significant complication in evaluating and interpreting diffuse reflectance measurements, which are considered NIR measurements. Scattered light has little or no information for the targeted sample chemical composition and effects, through intensity and baseline effects, with the collected sample information [20]. That is why chemometrics-based preprocessing techniques are applied to remove the irrelevant information from the sample. In reflectance mode, the radiation is transmitted through the sample, reflected from the ceramic then transmitted back through the sample before finally reaching the detector.

Interactance mode is also used for the spectral data collection by NIR. It is based on the combination of transmittance and reflectance mode, where the source of light and detector are in parallel positioned to each other on the similar side of the tested sample. This mode is usually suitable for whole fruits [21].

2.2.4   Spectrophotometer System

The spectrophotometer is used to measure sample in transmission or reflection modes by measuring light in absorbance/intensity at a precise wavelength. NIR spectrophotometer used for the measurements of those specific spectrums in the wavelengths of 4000–14,000 cm^{-1}. NIR spectroscopy instrument works based on the useful spectral range under particular wavelengths in either transmittance or reflectance mode. The range of the NIR instruments is usually divided into two categories depending on the wavelength range: the short-wave NIR range (700–1100 nm) and the long-wave NIR range (1100–2500 nm).

2.2.5   Detectors

Detectors are used in NIR spectroscopy and can be distinguished based on spectral response, speed of response, and the least amount of radiant power detected. Generally, two main types of IR detectors are used which vary because of their working principles, namely photon detectors and thermal detectors [17]. In NIR applications, devices used for the detection are mostly divided into single or multichannel detectors. Single-channel detectors consist of semiconductors with lead salt. Three different wavelength detectors are available: silicon detectors used in the range 400–1100 nm, epitaxially grown indium gallium arsenide in the range 800–1700 nm, and lead sulfide are used over the range 1100–2500 nm [15]. Some instruments have both lead sulfide and silicon detectors and ranged between 400 and 2500 nm. Multichannel detectors include diode arrays with multiple detector elements arranged in rows or charged-coupled devices organized in planes by multiple detection elements [22].

2.2.6   Data Mining Based on Chemometrics

Data mining based on chemometrics is a chemical discipline that uses statistical and mathematical design or selects optimal experimental producers to deliver the maximum targeted information by analyzing reference data and get knowledge about the used systems. Chemometrics is the science of relating measurements based on a chemical process or system via statistical or mathematical methods. Chemometrics does not consist of a single approach, but several techniques include signal processing, mathematical analysis, pre-processing, curve fitting, pattern recognition, calibration, testing, and validation sets. This way references data collection and applied instrumental data to build a prediction model for the targeted values. This is used to access the food products' authenticity, especially. Simultaneously, multivariate data analysis plays a significant role, using chemometrics in various applications for quality control, qualitative and quantitative measurement of chemical parameters studied previously. Data collected from the reference methods and data collected NIR spectroscopy are divided into two sets of calibration (training) and prediction (test) sets, where calibration set is used for building model while prediction set is used for testing the robustness of the model. Chemometrics is classified into qualitative pattern recognition and quantitative analysis, such as multivariate calibration.

2.3 Spectral Preprocessing and Data Analysis

The primary objective of the preprocessing of the spectral data is the transformation of the acquired data so that Beer's law, asserting the linear correlation of analyte concentration and absorbance, can be used [3]. Nevertheless, spectral data collected by NIR spectroscopy are often characterized by baseline shifts and unwanted spectral variations produced from a scattering of light in solid samples or opaque liquid samples, temperature variations, the particle size of samples, density, and noise of spectral data. Consequently, data pretreatment technique for reducing noise, correction of bassline, resolution enhancement, and normalization and centering of the spectra is considered necessary to build a robust and stable model. Different preprocessing techniques have been used to remove the unwanted information from the spectral data, such as orthogonal signal correction (OSC), multiplicative scatter correction (MSC), standard normal variate (SNV), direct orthogonal signal correction (DOSC), orthogonal wavelet correction (OWAVEC), improved version extended multiplicative signal correction (EMSC), de-trending, derivative techniques, smoothing, and mean centering. Some of the preprocessing techniques are presented in Fig. 2.5.

It is necessary to note that preprocessing cannot produce analytical information. A reliable preprocessing algorithm may help save the analytical information already present from undesired sources of variability in the spectral data collection, which may jeopardize the search for quantitative and qualitative information. The origins of this undesirable variability involve, among others, scattering, noise, essential shifts, and trends. The original data cannot be transformed to create information. However, the misuse will raise the probability that critical information may be lost [12]. Thus, the user must predict the nature of the data variation before implementing any preprocessing algorithm. The everyday use of preprocessing techniques for the collected data from NIR spectroscopy has been explored and critically assessed. The preprocessing choice is challenging to determine until model validation. Still, in particular, it is not appropriate to perform a series of multi-preprocessing as a minimal standard can maintain or minimize successful model specificity [23].

The SNV preprocessing technique is used to transform spectral data by deducting the spectral mean values from the spectral values of each individual collected spectra. The standard deviation of the spectrum then assessed these spectrums. SNV effectively eliminates particle size and multiplicative interferences of scattering to the specific NIR spectra [2]. The MSC is used to offset additive (baseline shift) and multiply (tilt) effects in spectral data provided by physical effects, including

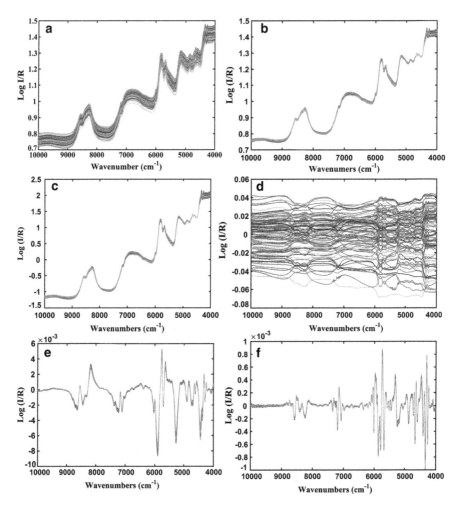

Fig. 2.5 Some of the preprocessing techniques used for the NIR spectral data: (**a**) Raw spectral data; (**b**) MSC; (**c**) SNV; (**d**) Centering; (**e**) First derivative; (**f**) Second derivative

nonuniform scattering in the spectrum. The degree of scattering relies on radiation, refractive index, and particle size. The approach aims to eradicate the impact of scattering by linearizing every spectrum to some "ideal" spectrum of the analyte, in practice the average spectrum of the study [24]. Additionally, the EMSC is an extension of the MSC preprocessing technique, which eliminates chemical-related interference consequences by combining recognized interferent spectral data and analytes. The EMSC allows separating and quantifying different chemical and physical sources of variability affecting the spectra set. Therefore, allowing qualitative and quantitative models to be interpreted more easily [25]. The OSC is a spectral preprocessing technique that attempts to delete the spectral data orthogonally to change sample properties or analyte concentrations. In the last decade, the application of OSC preprocessing indicates that the algorithm is not as appropriate as other preprocessing techniques. Likewise, the preprocessing technique DOSC, an improved form of OSC with better correction of spectral data by removing unwanted information, has also been used. Furthermore, mean centering may be considered the variance between the actual spectrum and an average spectrum, adapting the data set to the coordinating mechanism source by repositioning the data centroid. While preprocessing technique, OWAVEC's key goal is to precisely eliminate the scattering effects of the spectral data collected by NIR and take away systemic variations not relevant to the model response, which harm the efficiency of the subsequent calibration model [26].

Derivative techniques are considered as common preprocessing and used to improve minor spectral variations between samples and compensate for baseline changes induced by light scattering [27]. Normalization changes the collection of data by equalizing the sample magnitude. Normalization is conducted by centralizing the spectra (subtracting the mean absorbance from each spectrum) and then dividing the absorbance and further appropriate standard deviation for each wavelength [17].

2.3.1 Quantification/Regression Models for NIRS Data

NIR spectroscopy produces chemical information in a large quantity in the form of spectral data, and that data further can be used for the qualitative and quantitative measurement. In particular, several proposals for enhancing the accuracy of the predicted outcomes and the robustness of the predictive models were used to tackle the regression challenges. Some of the recently developed approaches also need further testing to show their efficiency in multiple data sets. Regrettably, most strategies with significant findings have not been introduced to user-friendly business applications to serve final consumers who are not chemometric experts.

1. **Partial least square (PLS)**

    The model PLS used for the regression of the collected data from NIRS considered a workhorse till now. PLS is a full spectral-based algorithm, used whole spectral data without interval or variable selection to build the final model through corresponding reference data. This model's precision and prediction efficiency are influenced by irrelevant information other than specific target variables in the sample spectra. PLS proved itself a very versatile algorithm for multivariate data analysis and proven good prediction accuracy for the single analyte. Furthermore, PLS as a supervised approach can guarantee that extracted principal components cover the maximum target values during the iteration process [28]. The PLS model worked with a combination of various intervals and variable selection models to improve model accuracy and stability.

2. **Interval-PLS (i-PLS)**

This model is initially introduced by Laars Nørgaard et al. [29]. The model i-PLS is a modified regression form of the PLS model, which is used for quantification purposes. Usually, this model works based on equidistant intervals, which are divided from the whole wavelength of NIR data. This allows the optimization of prediction models since subintervals without appropriate knowledge or intervention can be overlooked [30].

3. **Synergy interval-PLS (Si-PLS)**

The Si-PLS models are used to obtain better results compared to PLS because of the best variable selection from the full spectra for the target parameter (Fig. 2.6). Si-PLS select sthe best intervals from the spectral data by splitting them into various subintervals and then build the model based on the lowest RMSEC [31].

4. **Genetic algorithm-PLS (GA-PLS)**

The GA-PLS has the potential to search for the best variables from a large data for the selection of appropriate variables to the target parameter. Additionally, by removing the collinearity present in wavelet components, this algorithm enhances the final built model [32]. The GA works as follows: evaluation, selection, recombination, mutation, and reinsertion (Fig. 2.6). Until termination criteria are reached, these steps are repeated. Together GA and PLS can efficaciously solve the problem of multicollinearity and random correlation among the variables. The GA's primary function is to minimize the built model's error and make it more robust [33].

5. **Backward interval-PLS (Bi-PLS)**

Bi-PLS is commonly used to determine the number of intervals for a PLS model. The optimum number of principal components is determined by the lowest RMSECV [34]. This model also split the data into intervals as Si-PLS. The k intervals of the same width spectral region were divided and used to build the PLS models by removing the k intervals one by one [29]. Each interval was sequentially supplanted and removed to gain the k submodels built by $(k - 1)$ intervals based on the precision of the submodel's root mean square error of cross-validation.

6. **Competitive adaptive reweighted sampling-PLS (CARS-PLS)**

The CARS-PLS as a variable selection algorithm is used to improve the better robustness and better results than the PLS model. The CARS model works based on the "survival of the fittest" and used to select the best variable selection algorithm [35]. This algorithm worked by removing uninformative information from the data and selecting related information to build the final model. The variables selected by this algorithm are then used to construct a model on the target parameter. This algorithm worked based on the N sampling runs and N subsets of variable selection to get with lowest RMSECV the optimum subset variables [36].

7. **Ant colony optimization-PLS (ACO-PLS)**

The ACO-PLS was used as a variable selection algorithm which was developed from the inspiration of food foraging behavior of ants [37]. This path resemblance for food of ants from their nest to food source and back has been

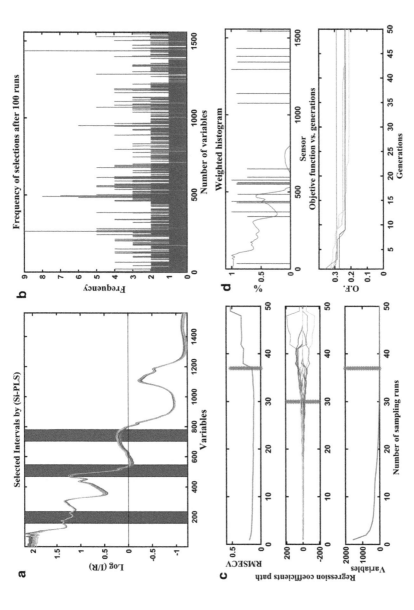

Fig. 2.6 Some of the models that are used for the prediction of target compounds in various food and food stuffs; (**a**) interval selection by Si-PLS, (**b**) GA runs performed to build the GA-PLS, (**c**) CARS sampling runs used for the CAR-PLS model, and (**d**) ACO-weighted histogram during ACO-PLS runs

performed into building the ACO, and it is written in mathematical form by *Shmygelska, A* [38]. This algorithm is used to select appropriate variables from the spectra with space dimension represented by code 1 or 0 when selected, and the Monte Carlo approach is used to select relevant spectra variables from the full spectral data.

8. **Multiple linear regression (MLR)**

   Multiple linear regression (MLR), also known as multiple regression, is a mathematical technique that predicts the outcome of a response variable using many explanatory variables. Multiple regression is a variant of linear regression that employs just one explanatory component. This regression method is simple and should promote the analysis of the model [39].

9. **Successive projections algorithm (SPA)**

   The SPA model is also used for the quantification and worked based on the forward selection method that utilizes the simple operations for variable minimization to build the model with better accuracy and stability. But the major drawback of this model is that the number of selected variables must be less than the number of samples used in the calibration model [40].

10. **Support vectors regression (SVR)**

    During the last decade, the model SVR has been used for the regression and has demonstrated excellent performance, generally against the outliers and tolerance to the noise in the spectral data collected by NIR spectroscopy. The SVR algorithm vastly improves on plain linear regression. It enables the development of nonlinear models and provides control over the stability vs. robustness of your models [41].

11. **Principal component regression (PCR)**

    The model PCR is also a regression model and is based on the PCA. In this model, PCs of the explanatory variables collected by PCA are used for the regression. This model is used to reduces the number of features of the mode, reduces the problem of overfitting, and considered as quite useful on datasets with highly correlated features, or even collinear features [42].

12. **Partial least squares regression (PLSR)**

    The PLSR is a regression model that generalizes and combines features from principal component analysis and multiple regression to build an optimum model. This model is useful when the high explanatory variable numbers are present, and explanatory variables are likely correlated [43].

13. **Back propagation artificial neural network (BPANN)**

    The model BPANN is also used for quantification or prediction purposes and is considered the most classical feed-forward multilayer (i.e., an output layer, an input layer, and one or more hidden layers) unidirectional connection (weights) from input to output. The model BPANN performance can be influenced by its various parameters such as its learning rate factors, hidden layer nodes, and initial weights and momentum factors [44].

Here we have just discussed the most commonly used classification and regression models for the NIR spectroscopic data, and the interested readers can go through these articles for more details and in-depth discussion about chemometrics [2, 12, 17, 45].

2.3.2 Qualitative/Discrimination of NIRS Data

Discrimination or authentication is mostly performed using classification or discrimination models based on NIR spectroscopic data. Presently, a wide variety of strategies for data reduction, classification, and regression are applicable.

1. **Principal Component Analysis (PCA)**
   The PCA model is mostly used as an unsupervised pattern recognition method. It is mainly used to verify or identify patterns in the data. PCA works based on converting primary variables into new independent variables called principal components PCs which are linear combinations of original variables. The PCA's objective is to extract latent variables from the data such that the collected latent variables are maximized [46].

2. **Linear Discriminant Analysis (LDA)**
   The LDA is used as a supervised classification algorithm. Among many other supervised algorithms, LDA is mostly used for classification. LDA splits the data into linear discriminant (LDs), which may be used as a linear classifier or distance reduction before discrimination. This LDA model is most popular for data analysis such as adulteration, identification, and authentication [47].

3. **K-Nearest Neighbor (KNN)**
   The KNN is a non-parametric and supervised method largely used to classify pattern recognition [48]. This algorithm's main working principle is that data points are measured as of the nearest K neighbors [49]. This algorithm stored all data attained on samples as a training set to classify the test set based on the sample's computation similarity within the training set. From the training set to test set, distance is calculated and the short distance is referred to as nearest neighbor. The optimum value of k has the lowest error rate used for the calibration process.

4. **Partial Least Squares Discriminant Analysis (PLS-DA)**
   The PLS-DA is a versatile algorithm that can be used for predictive and descriptive modelling as well as for discriminative variable selection [50]. First, PLS calculates the latent variables from the original data matrix so that these variables maximize the variance between groups associated with response vectors. Further, using the LDA approach, discriminant analysis is mostly used for each class [51].

5. **Soft Independent Modeling Class of Analogies (SIMCA)**
   The SIMCA model has been widely used for adulteration, discrimination, and authentication in foods and food products. One of the significant advantages of this model is that it can be used for low sample volumes. This model offers

interclass distances in between clusters, and better accuracy can be found if the interclass distance between two groups is large [52].

6. **Artificial Neural Network (ANN)**

   The artificial neural network modifies the configuration and operation of the human nervous system in order to assemble parallel, distributed, and adaptive information processing networks with a degree of intelligence. ANN combines the knowledge by capturing the relationship and patterns in data and learn via experience, not from programming [2]. The wide variety of applications and use of ANN is primarily attributed to its ability to work with dynamic functions while also allowing for the simulation of nonlinear relationships [53].

7. **Support Vector Machine (SVM)**

   SVM model is used for the classification and authentication, having the ability for both types such as linear and nonlinear. It is a pattern recognition method with the ability of supervised classification. SVM is considered helpful for linear and nonlinear classification [54]. SVM is considered a good classification algorithm because of its better performance even on small data sets. To deal with complex data, SVM is an excellent recognition method for categorical assignment [55]. The specific learning procedure in SVM is helpful in procurement of the global least of the error function and excellent generalization aptitude of the trained network. Also, based on nonlinearity, SVM is considered more useful than linear algorithms for qualitative and quantitative analysis of various foodstuff [56].

8. **Extreme Learning Machine (ELM)**

   The nonlinear ELM model is considered as predominant learning neural method with the ability of classification and regression. Due to the fast nature, intractable issues, and good computational scalability, ELM is more beneficial than other models. The ELM algorithm consists of randomly generated hidden layer node parameters and a linear output layer analytically determined by the output weights of single hidden layer input networks [57, 58].

2.3.3   Criteria for Built Model Evaluation

To build the model, generally spectral data is divided into calibration, prediction, and test sets. Evaluation of the final developed model is necessary and essential to search for an effective method to judge the predictive accuracy, reliability, and effectiveness for the model's practical use. The calibration set is used for training of the built model. At the same time, the prediction set is used to judge the stability and robustness of the built model. Leave one out cross-validation (LOO-CV) is used for the sample selection in model testing. LOO-CV working procedure is as follows: one sample in the calibration set is deleted, and the remaining one is used to build the model. LOOCV is a procedure used to estimate the performance of a machine learning algorithm when making predictions on data not used during the training of the model [59]. Then the final model performance was evaluated statistically.

Table 2.1 Guidelines for model performance indicator and their interpretation

Performance indicator	Value calculated	Interpretation
Calibration model	$R_C = 0$	No relationship between spectroscopic data and reference chemical data
	$R_C \leq 0.5$	Weak relationship between spectral data and data acquired from specific analyte
	$0.85 \leq R_C \leq 0.99$	Strong relationship between reference chemical data and acquired spectroscopic data
	$R_C = 1$	Perfect relationship between system generated data and reference chemical data
Prediction model	$R_C = 0$	No prediction ability from spectroscopic data
	$R_C \leq 0.5$	Low strength of prediction and weak relationship exist between system-acquired data and reference-generated data
	$0.85 \leq R_C \leq 0.99$	Higher prediction performance for future measurements
	$R_C = 1$	Perfect degree of prediction for future adoption from the system generated data
Calibration model error	RMSECV ≤ 0.5	Stronger the model and low variation or error in built model
	RMSECV ≥ 0.5	Low stability and higher error or variation associated to the model
Prediction model error	RMSEP ≤ 0.5	Low error associated with prediction model of future measurements
	RMSEP ≥ 0.5	High error or variation associated with built model of future measurements

Table 2.2 Statistical parameters and their interpretation

Parameters	Equation	Recommendation
Correlation coefficient	R_C and R_P	Should be close to 1 as possible
Precision	RMSECV	As small as possible
Tested precision	RMSEP	As small as possible
Trueness	Bias	Should be close to zero as possible
Ratio performance deviation	RPD	Above 2 is considered stable model

Likewise, the prediction set is used to assess the stability and robustness of the model. The lowest root mean square error of the calibration (RMSECV) and root mean square error of the prediction set (RMSEP) are generally used for model evaluation. Besides, the best model exhibits lower RECV and RMSEP and higher R_C and R_P values (Fig. 2.6). Model is more robust if the difference between R_C and R_P or RMSECV and RMSEP is small (Table 2.1). Residual predicted deviation (RPD) value is generally used to indicate the forecasting level of the model. Higher RPD shows the best model [60]. Table 2.2 summarizes the model performance that the built model is stable or weak.

$$\text{RMSEC/RMSEP} = \sqrt{\frac{\sum\limits_{i=1}^{n} (\widehat{y} - y_i)^2}{n}} \tag{2.1}$$

where $\widehat{y}_i$ is the prediction value and y_i is the measured value, and n shows the number of samples in Eq. (2.1). The R_C correlation coefficients for calibration and R_P are generally used to evaluate the correlation of the results.

$$R_C \text{ or } R_P = \sqrt{1 - \frac{\sum\limits_{i=1}^{n} (\widehat{y} - y_i)^2}{(\widehat{y} - \widehat{y}_i)^2}} \tag{2.2}$$

In Eq. (2.2), $\widehat{y}$ is the average measurement.

$$\text{Bias} = \sum\limits_{i=1}^{n} \frac{(\widehat{y}_i - y_i)}{n} \tag{2.3}$$

$\widehat{y}_i$ presents the predicted values and $\widehat{y}$ presents the reference values of the respective number of n samples.

$$\text{RPD} = \text{SD/RMSEP} \tag{2.4}$$

SD shows the standard deviation of prediction samples. If RPD is greater than 3, the model is the best, if less than 3 and greater than 2, the model is good, and if less than 2, the model is not good [21].

2.4   Applications in Food Quality and Safety Detection

2.4.1   Fruits and Vegetables

Recent advances have shown great potentials of NIR spectroscopy in real-time monitoring of fruits and vegetables to increase their effectiveness and final product quality. NIR spectroscopy has been increasingly used for the safety assessment and quality evaluation of food and agricultural products (Fig. 2.7) [61]. Research work focused on using NIR spectroscopy and reported the nondestructive measurement of apples and purees. A better classification rate based on varieties and storage was noted with 82% and 88% accuracy, respectively. While for the prediction based on PLS, better results were pointed out for both apple and apple puree samples [62].

Similarly, visible and near-infrared (Vis/NIR) spectroscopy is considered one of the most generally used nondestructive detection tools, which has been used to detect the internal defect of apples. The spectroscopic data were collected in the

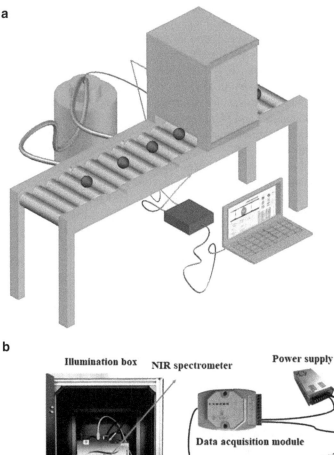

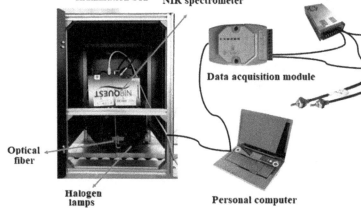

Fig. 2.7 Schematic diagram of the applied online NIR spectrometer system (**a**) and the sketch of main parts (**b**) [61]

range 550–1650 nm. The PLS-DA modeling approach was used for better classification. The results showed that the best PLS-DA models had minor misclassification rates between good apples (4.3%) and defective apples (10.0%). Besides, good classification results were noted for slightly defective apples using mean spectral data [63].

Likewise, another study recognized the Vis/NIR spectroscopy for forecasting apple fruitlet in response to the thinner application, optimum threshold values, and wavelength combinations were set on through the ROC (receiver operating characteristic) curve. The outcomes specify that a simplified portable device's improvement appears practicable [64]. Another study designed the nondestructive assessment of chilling injury in kiwifruit by Vis/NIR spectroscopy. NIR spectral data were collected in the wavelength range of 700–1000 nm. SNV was used as a preprocessing technique, and PCA was applied to build the classification model.

The overall results demonstrated that the incidence of water-soaked tissue might be less informative after SNV processing than corky and grainy tissue for the detection of chilling injury [65]. Likewise, another researcher used NIR spectroscopy for the quantitative detection of watercore degree and soluble solids content (SSC) in apple. In total, 663 samples were collected from various provinces, and spectral data were in the range 590–1200 nm. Various variable selection methods, including SPA, GA, and CARS, have been used to build the model.

The model CARS-PLS proved to be the best and have the ability for practical application, using the range of 600–1000 nm and the results with R_P, RMSEP, and RPD values for watercore degree (0.9562, 1.340%, and 3.720) and SSC (0.9808, 0.327°Bx, and 4.845), respectively [66]. Evidentially, an online application was performed in research work by using diffuse reflectance NIR spectroscopy for the tomato samples. The spectral data were recorded in diffuse reflectance mode with an entire wavelength of 900–1700 nm, and the model was built by using variable combination population analysis (VCPA), uninformative variable elimination (UVE), and CARS. Better results were noted with VCPA-PLS for the TSS attributes with $R_P = 0.9053$ and RMSEP $= 0.382$ in validation set [61]. A new low-cost handheld NIR spectrometer is used for the quality monitoring of umbu (*Spondias tuberosa Arruda*) fruit. The study's main objective was to evaluate the handheld NIR instrument's performance and have the wavelength range of 300–1150, 900–1700, and 740–1070 nm. Model PLSR is used for the determination of soluble solids (SS), dry matter (DM), skin color (SC), and flesh firmness (FF). Based on the model built, better RMSEP is obtained with the three spectrometers for dry matter (5.2 ± 0.9%), soluble solids (8.4 ± 1.5%), flesh firmness (27.6 ± 2.0%), and skin color (8.0 ± 0.6%), respectively, and indicates the acceptable accuracy of the used NIR spectrometers [67].

2.4.2   *Beverages*

NIR spectroscopy has been used for the quality and safety attributes of beverages, such as teas, beer, wines, distilled spirits, fruit juices, and soft drinks. Some of the recent applications of NIR spectroscopic data of beverages have been presented here. Tea is considered a widespread beverage worldwide and is of great interest due to its useful medicinal qualities. With the increasing consumption of tea, quality control of tea has become increasingly important nowadays. For the entire purpose, a study was conducted on white tea using NIR spectroscopy based on the discrimination. In total, 127 white tea samples of various maturity levels were collected and spectral data acquired in the wavelength range of 10,000–4000 cm^{-1}. For the classification, SVM, BPNN, and PSO models are used in single and combined forms. Better classification accuracy up to 93% is received by these models while optimum precision was noted with 98.92% by using SPO-SVM model. Overall results found that white tea leaves' maturity had potential correlations with exceptional sensory characteristics coupled with NIR spectroscopy [68]. In another research, Vis/NIR spectroscopy was used to determine the moisture content in tea leaves. Spectral data collected in the wavelength range of 220–1100 nm and PLSR was used to build the model. The overall findings proved that the suggested model transfer strategy by correcting the spectral data might be a powerful method for the quick detection of the water loss in different tea leaves [69]. Likewise, a handheld NIR spectrometer has assured the reliability and speed of analysis due to an accurate, compact, sensitive, and nondestructive analytical technique. It has been used to evaluate direct cup profiles in crushed and roasted coffee mixtures using handheld NIR spectroscopy. Spectral data were collected in the entire range of 906–1676 nm, and a total of 214 samples were used in this work. For the model building, PCA and PLS-DA were used, and better results were noted coupled with PLS-DA model, and accuracy was noted from 84% to 100% in the prediction set. Results revealed that the planned method is practical for the direct determination of cup profile at an industrial level; meanwhile, it is convenient, quick with specificity and sensitivity, and less costly compared to benchtops apparatus [70]. Another research work was conducted on the quality assessment of sugar beet juice through NIR spectroscopy coupled with the PLS model. The results illustrated that the PLS models were acceptable for predicting the pol, brix, and sucrose ($R^2 = 0.969, 0.984, 0.921$) in sugar beet juices samples, which found the convenient and rapid instrument for the assessment of quality parameters in sugar beet juice [71].

Based on health importance and commercial point of view, milk has its importance all around the globe. NIR spectroscopy was used for the quantification and adulteration of goat milk with cow milk. Spectral data collected in the range 4000–10,000 cm^{-1} for 36 samples. Results showed that preprocessed NIR spectra joined with interval selection by i-SPA-PLS gave the best outcomes for the determination of both adulteration and fat contents, while PLS algorithms provide good results for protein quantification [72]. Similarly, in another study, NIR and FTIRs were used for the determination of milk-surfactant adulteration coupled with PCA, PLS-DA, PLSR, and GA-PLS. Spectral data was collected for FTIR in the range 4000–600 cm^{-1} and NIR in the range 13,000–4000 cm^{-1}, while in total 55 samples were used in this work. The overall results showed apparent differences between pure and adulterated samples using PLS-DA, which was authenticated by the permutation test. The multivariate analysis makes it appropriate to control the quality of milk production [73]. Likewise, in another research work, near and mid-infrared transmission (NIR and MIR spectroscopy) was used for the chemical composition determination in raw and cow milk. In total, 150 cow milk samples and 217 camel milk samples were used for this work. NIR and MIR spectroscopy techniques both deliver valuable and comparable results for raw milk. While based on fat and protein concentration, clear differences have been noted by both techniques [74].

Data fusion strategies are also applied by some researchers to minimize the instruments' incompetence and the optimum results, for the entire purpose FT-NIR is combined with a computer vision system (CVS) for assessing the black tea fermentation degree. In total, 110 samples were collected for the entire work and PCA, LDA, and SVM were used to build the model. Data fusion strategy primary level and mid-level has been used for the classification and quantification. Feature level fusion gives up to 100% accuracy in training and test sets for the classification based on PCA [75].

2.4.3   Meat and Meat Products

Meat includes various types such as chicken, beef, pork, and their various products used worldwide, and most of them have a high production cost. Worldwide meat consumption is increasing day by day because of having more quantity of protein in its composition. The use of meat relates to the local food diet and price offered to consumers and production method. Due to the high consumption of meat and its products, consumers of today are thinking about the safety and quality concerns of meat. For the entire purpose, many kinds of research work have already been performed, such as a research work focused on the adulteration of beef meat with pork by using NIR spectroscopy. Spectral data was collected in the wavelength range 12,500–5400 cm^{-1} and a total of 64 samples were used for this study. The data analysis was performed using DA and PLS for quantitative and qualitative categorization of the binary adulteration (beef plus pork meat) and the ternary adulteration (pork plus duck meat). A better classification rate was achieved with optimum accuracy of 100% and 91.5% for binary and ternary systems, respectively. While better quantification is achieved by PLS with a correlation coefficient of prediction (R_P) (95.80–95.69) and RMSEP (7.27–9.27%) for the binary and ternary samples, respectively [76]. Likewise, research work focused on the pork meat quantification in other types of meats using FT-NIR spectroscopy. The spectral data collected in the range 10,000–4000 cm^{-1}, at a resolution of (2 cm^{-1}) and a total path length of (0.5 mm). PLSR was constructed to predict the pork meat contents in other meats. Better results were achieved with an R^2 value of 0.9774 and RMSECV value of 1.08% and also providing the lowest prediction error with RMSEP (1.84%) [77]. Another study was performed by using a portable NIR spectrometer for the adulteration in ground meat (beef, pork, and chicken) blends. Better results were noted based on both models PLS and SVR, but the SVR model was optimum. The values of R_C; R_P and RMSEC; RMSEP were noted as 0.99; 0.95% and 1.9 wt%; 3.9 wt%, respectively [78]. Similarly, Vis/NIR reflectance spectra at the wavelength 350–2500 nm were used for the rapid detection and adulteration in minced beef coupled with multivariate analysis. For the multivariate analysis, model SVM, RF, PCS, subwindow permutation analysis (SPA), locally linear embedding (LLE),

PLSR, and CARS were used to build the model. PCA gives better accuracy for the classification up to 99% for the adulteration samples, while better prediction for beef adulterated with the beef heart was obtained using PLSR and CARS models with $R_P = 0.960$ and RMSEP $= 2.758$ [79]. Evidentially, a research work performed on turkey meat by NIR spectroscopy and 400–2500 nm wavelength was used for the spectral data collection. PCA and LDA were used to obtain better classification, and the overall results were found acceptable for the classification of meat samples [80]. The freshness of the meat is considered important because of its great effect on the quality of meat and consumer acceptance. Keeping this in mind, a study was performed using NIR spectroscopy in the wavelength range of 1100–2200 nm coupled with multivariate analysis. Various preprocessing techniques such as MSC and SNV were used to preprocess spectral data for the prediction model SPA, MLR, and PLSR. Better results were noted by using model PLSR with $R_P = 0.83$ and RMSEP 0.068 mg MDA/kg meat [81]. The handheld NIR spectroscopy combined with machine learning was used for the quality monitoring of chicken. Spectral data was collected in the wavelength range 908–1676 nm with 125 variables. Data divided into calibration and prediction set with 70% and 30%, respectively, to build a classification model. Classification models PLS-DA, ANN, SVM, and random subspace discriminant ensemble (RSDE) were used to build the model. Optimum results were noted by the RSDE model compared with all other models with a classification accuracy of >95% [82]. Similarly, hand-held and portable NIR spectroscopy was used to differentiate game meat of various species. In total, 118 meat samples were collected for the entire study, and sample data were collected in the range 908–1700 nm. Overall better accuracies were noted up to 67–100%. Various preprocessing techniques were used to obtain an optimum classification model among the LDA, PLS-DA, and SIMCA. The outcomes indicate that NIRs can be used for the authentication of game meat, specifically impala, eland, and ostrich. Besides, it was convenient to differentiate species regardless of the muscle used than different muscles inside each species [83]. Likewise, game species discrimination was performed by NIR spectroscopy coupled with multivariate analysis. The best results were found by using the LDA model ranging from 68% to 100% and PLS-DA range 70–96% [84].

2.4.4 Cereal and Cereal Products

Cereals are stapled foods and are considered vital sources of vitamins, minerals, fiber, crude fats, proteins, and essentials fatty acids, which play significant roles in human beings' healthy lives. Cereals are considered usually grass members and based on grass family (known as *Gramineae*), a monocot family *Poaceae*, with the properties of thin, long stalks, e.g., rice, maize, wheat, millet, rye, sorghum, barley, and rye with starchy gain functions used in foodstuffs [85]. From a nutritional point of view, it has been proven that cereals and their products can reduce the prevalence of many diseases, such as obesity, diabetes, cancer, and cardiovascular diseases

[86, 87]. Food authenticity is currently a top priority for both consumers and suppliers, as food supply chains have become more global and dynamic. Fast, rapid, easy to use, and environment-friendly techniques are much needed today to compete with the requirements. NIR spectroscopy application in cereal and cereal products for the quality attributes considered an evergreen approach. A study was performed by using portable NIR spectroscopy for the wheat kernel's vigor assessment. Spectral data were recorded in the wavelength 1200–2400 nm, and SVM, ELM, RF, and AdaBoost were used to build the model. The optimum results were noted by the model used PCA-ELM and SPA-RF with classification accuracy up to 88.9% and 88.5%, respectively. A study showed that NIR spectroscopy successfully demonstrated distinct wheat kernel vigor discrimination for wheat quality assessors, plant breeders, and wheat processors [88]. Also, portable NIR spectroscopy was used to determine the wheat flour's fatty acid coupled with multivariate analysis. In total, 120 wheat samples' spectral data were collected in the wavelength range 899.22–1724 nm. Model variable combination population analysis (VCPA) and ELM were used for the prediction to build the model on the preprocessed data by SNV. Model ELM yielded better prediction accuracy with R_P over 0.96 in every mode to construct the model. The final results showed that the VCPA algorithm also has a useful application in optimizing NIR spectral characteristic wavelengths [89]. Evidentially, in another research work, FT-NIR spectroscopy was used to predict TPC coupled with multivariate analysis. A total of 107 wheat samples were used in this work, and further PCA and PLS were used to build a classification and prediction model. Better prediction results were noted for calibration and prediction with accuracy 0.92 and 0.90, respectively, while RPD was noted as 3.4 and considered model was robust and stable [90].

Similarly, the same wavelength NIR spectroscopy was used for adulteration in durum wheat pasta samples which were adulterated with simple wheat. The model PLS was applied to measure the mass fraction of simple wheat in durum wheat pasta. The overall results revealed that NIR screening was successfully used based on the PLS in the current work [91].

Consumers now demand nutritious food, and food safety is the primary concern of consumers, requiring safety between businesses and markets. In addition to considering the sources of food, including its practices and food hygiene, researchers are now using NIR spectroscopy for the safety-related problems in cereals and cereal goods for the whole purpose. Research works are focused on NIR for the detection of aflatoxin B1 with the safety purpose of maize. In total, 102 maize samples were used, and sample spectra were recorded in wavelength 400–2500 nm. Various preprocessing techniques were used to remove irrelevant information, noise, etc., while SVM, KNN, and extreme gradient boosting (XGBoost) were used for model building. Better results were noted with a root mean square error of prediction value of 3.57 μg/kg for the quantitative model, and the classification rate was 90.32% [92].

Similarly, hand-held portable NIR combined with a computer vision approach was used for the online detection of fungi (*Aspergillus* spp. and *Fusarium* spp.) in maize. Vis-NIR spectra were recorded in the wavelength range 600–1600 nm, and after that images of the samples were captured. Multivariate analysis, such as MSC,

SNV, and second derivatives, is used for the preprocessing while modeling PCA, LDA used for classification and CARS, PLSR for the quantification purpose. Better results were noted with 100% accuracy for discrimination, while the root mean square error of prediction value for the PLS model was noted as 25.0–17.4% [93]. Further, a research work demonstrated the determination of DON (deoxynivalenol) contamination using computer vision and Vis-NIR spectroscopy. With data fusion strategy, optimum results were archived for the classification rate up to 93.55%, and the purpose of the study was successfully demonstrated [94]. The same kind of approach was also used for the DON contamination detection based on the NIR spectroscopy. Models PLSDA and PLSR were used for multivariate analysis. By using both kinds of model classification and quantification, acceptable results were achieved. Accordingly, results described that NIR, coupled with a suitable chemometric technique, might be used as a green technique to detect DON contamination in barley [95].

2.4.5 Oils

Oilseeds and edible oils are important nutrient sources in the human diet, and the quality of oils has drawn increasing attention from industry and consumers. Therefore, it is crucial to evaluate the quality of oilseeds. Since the NIR spectroscopy is rapid, nondestructive, and cheap, it has a promising opportunity to detect quality factors of oilseeds, along with authenticity identification of oilseeds and edible oils. Moreover, the combination of NIR with chemometrics significantly broadens its application prospects. A quantitative detection NIR spectroscopy was recently performed for the oils coupled with chemometrics at the wavelength regions 9000–4500 cm^{-1}. PLS obtained best results after optimization with various spectral pretreatment approaches like comprising multiple scatter correction (CMSC), Savitzky–Golay smoothing (SGS), SNV, and normalization, with obtained coefficient R^2 greater than 0.995, RMSEC and RMSEP lesser than 6.79 and 4.98 consistently [96]. Similarly, another work based on NIRS was performed to quantify and identify adulterations of vegetable oils in extra virgin olive oil. Samples of extra virgin oil were blended with sunflower, soybean, canola, and corn oil. The overall results were found acceptable, and the study showed the successful application of NIR spectroscopy [97]. The FT-NIR spectroscopy coupled chemometrics for the detection of edible oil during storage times and applied an operative analysis technique for the detection of quality in stored edible oil. Spectral data pretreated with SNV and further PCA, SVM, KNN, and RF used for the qualitative identification pattern of edible oil in different storage periods. The results confirmed that the SVM model is significantly higher than KNN and RF models, which had a recognition rate of 100% in the prediction set when predicting independent samples [98]. Likewise, another work attempted with shortwave handheld NIR spectroscopy coupled with the multivariate algorithm to measure Sudan IV dye adulteration in palm oil samples. The KNN model was applied to differentiate among original palm oil samples and Sudan IV dye adulterated (0.10–0.002% w/w) ones. The incorporation of Sudan IV dye in original palm oil samples was quantified by PCR, PLSR, and SVMR algorithms. The best results were obtained by using SNV-PLSR with $R_C = 0.91$, $R_P = 0.90$, RMSEC $= 0.0841$, and RMSEP $= 0.0868$ [99]. Furthermore, a study based on spectroscopic techniques was performed to detect the total anthocyanin content (TAC), etc., in a single-seed state of soybean samples. The soybean seed samples were randomly selected from 70 different varieties, and the chemical components were evaluated with HPLC. The results found that FT-NIR spectroscopy spectra combined with PLSR models indicated R^2 of 0.88–0.90 and RMSEP of 9.4–19.5%, while FT-IR spectroscopy indicated an R^2 of 0.86–0.88 with RMSEP of 9.7–21.8%. The findings proved the potential of both spectroscopic techniques (FT-NIR and FT-IR) to predict nondestructively anthocyanin contents in soybean seed [100].

2.4.6   Other Foods

NIR spectroscopy has been recently applied in many studies that give chemical or molecular information about the sample. Recently, the NIR spectroscopy, coupled with chemometric analysis, was employed to detect the green pea and spinach adulteration in pistachio. NIR spectra were acquired at 908–1695 nm, and the model PCA was established effectively to differentiate original and adulterated samples. Furthermore, Model PLSR were used for data analysis and better results were noted as for green pea ($R^2 = 0.9957$ and RMSEP $= 7.87$) and for spinach ($R^2 = 0.9968$ and RMSEP $= 4.69$) [101]. Similarly, in another study, portable FT-MIR and FT-NIR spectroscopy were used to determine green pea and peanut adulteration in pistachio. The spectra were collected in the range of 3920–7400 cm^{-1} in absorbance mode for FT-NIRS and FT-MIRS 4000–650 cm^{-1} and then analyzed by model SIMCA and PLSR for classification and quantification. The overall results were found better with acceptable accuracy, and the use of both instruments was successfully demonstrated [102]. The Vis-NIR and Raman spectral data fusion approach was used with two PLS and SVM models for infant formula samples. The purpose was to observe the quality assessment of infant formula at different storage temperatures (20, 37 °C) and storage times (0–12 months). Overall results demonstrate that SVMR models established using medium-level data fusion had the best storage time predictions for infant formula samples stored at temperature 20 °C (RMSECV $= 0.7$, RMSEP $= 0.8$) and at 37 °C (RMSECV $= 0.7$, RMSEP $= 0.6$) [103]. Furthermore, the other study was based on comparing NIR and MIR spectroscopy for protein and glucose determination in various tuber and root flours, like arrowroot, *canna*, modified cassava, taro, and sweet potato (purple, yellow, and white). The spectra were collected for NIR over 10,000–4000 cm^{-1}, MIR over 4000–600 cm^{-1} wavelength ranges to build PCA and PLSR model. This study successfully presented NIR and MIR spectroscopy's potential for determining protein and glucose contents of root and tuber flours [104]. The chocolate confectionery product is considered very delicious around the globe in a young nation especially. A study reported using an NIR spectrometer to identify fat blooms in chocolate samples like milk chocolate, white chocolate, 40% cocoa chocolate, and 70% cocoa chocolate samples. The spectral data were recorded in the range 900–1700 nm, while further PCA PLS-DA was used to build the model. PLS-DA model obtained the best results, which provided sensibility, specificity, and accuracy values in the range of 80–100% [105]. Nowadays, a new trend is developed using many techniques together for the quality monitoring of food samples. A study reported CSA, NIR, and FT-IR spectroscopy (both separately and in the group) coupled with chemometrics for the quantitative investigation of volatile compounds in honey samples. Overall, the results demonstrated that combining CSA with NIR or FT-IR was feasible and could improve the prediction accuracy of the volatile compound in honey [106]. NIR spectroscopy has been widely used for qualitative and quantitative applications of various food stuffs, some of them also presented in Tables 2.3 and 2.4.

2.5 Conclusion and Future Trends

Although the NIR technique is quick, nondestructive, and inexpensive, it is a promising choice for identifying food and food quality parameters and for the validity and traceability of different foodstuffs. The potential feedback can, therefore, be extracted from the NIRS hyphenation of many other analytical technologies. Complementarily combining the analytical knowledge produced by NIRS with that given by other techniques will boost research findings for many areas of work. Portable, cost reduction, very compact, self-supporting, wireless spectrophotometers are now a reality. The combination of NIR and chemometrics also dramatically increases its scope for use in various foods.

Consequently, more robust calibrations for better sampling and enhancement of reference methods are also required. More attempts should also be made to incorporate more reliable algorithms, create more robust models, and minimize mistakes. A study must be carried out to unify spectrum pretreatment and refine quantitative and qualitative modeling and improve sensitivity to low concentrations.

In general, NIR spectroscopy can only obtain spectral information from solid and liquid samples but is limited to gaseous information. Gaseous information (flavor volatiles) plays a central role in monitoring food deterioration processes. Applications of NIR spectroscopy and chemodyes combined have been explored for volatiles detection. However, NIR spectroscopy can be used for online applications in various food industries in the future. But researchers must need to focus on the precision of technique and better results. Because of in an industrial environment, NIR measurements are sensitive to various external factors such as ambient temperature, spectrophotometer temperature, sample presentation to the NIR probes, wavelength shifts and others. Besides, sample appearance can become a crucial problem due to inhomogeneity for intact samples.

Table 2.3 Application of NIR spectroscopy for quantitative measurement in various food items

Food items	Mode	Type of study	References
Bayberry	Transmission	Titratable acidity, malic and citric acid	[107]
Coffee	Reflectance	Detection of addition of barley to coffee	[108]
Tomato	Absorption	Lycopene, total acid, sugar, phenols, and antioxidant properties	[109]
Radix	Reflectance	Free amino acid determination	[110]
Milk powder	Reflectance	Quantification of adulterants	[111]
Tea	Reflectance	Total polyphenol quantification	[112]
Honey	Reflectance	Antioxidant properties	[113]
Maize	Reflectance	Kernel rot and mycotoxins determination	[114]
Vinegar	Transmittance	Soluble solid contents and pH measurement	[115]
Tea	Reflectance	Quantification of caffeine and total polyphenol	[116]
Summer squash	Reflectance	Ascorbic acid, polyphenols, and chlorophyll	[117]
Olives oil	Reflectance	Total polyphenols	[118]
Sorghum grain	Reflectance	Polyphenols, condensed tannin, and 3-deoxyanthocyanidins	[119]
Karkade	Reflectance	Antioxidant properties	[120]
Goji berry	Reflectance	Chemical composition measurement	[121]
Goji berry	Reflectance	Antioxidant activity measurement	[122]
Black tea	Reflectance	Amino acid, water content, caffeine, and theaflavin	[123]
Cocoa beans	Transmittance	Quantification of total fungal content	[60]
Black tea	Reflectance	Prediction of fermentation quality	[124]
Coffee	Reflectance	Caffeine, caffeic acid, chlorogenic acid	[125]
Cocoa beans	Reflectance	Novel total fat content	[126]
Apple	Reflectance	Glucose, fructose, and sucrose	[127]
Bayberry	Transmittance	Glucose, fructose, and sucrose	[128]
Citrus	Transmittance	Quantification of sugars	[129]
Date fruit	Transmittance	Determination of sucrose	[130]
Fuji apple	Reflectance	Soluble solid content	[131]

Finally, it is incredible how NIRS technologies have extended its variety of uses over the last decade and how it is revived by easily integrating with other field technology. The ongoing effort needs to improve NIR spectroscopy coupled with fast data processing models for application in food and food materials. NIR spectroscopy is indeed a practical instrument with nondestructive nature and is used to inspect various food and food materials. NIR spectroscopy with the unparalleled combination of speed, precision, simplicity, and online applications has been used worldwide. A current qualitative and quantitative analysis technique by using NIR

Table 2.4 Application of NIR spectroscopy for qualitative measurement in various food items

Food items	Mode	Type of study	References
Bayberry	Reflectance	Varietal discrimination	[132]
Melon	Reflectance	Distinguishing genotypes	[133]
Paddy seeds	Reflectance	Storage age discrimination	[134]
Green tea	Reflectance	Origin discrimination	[59]
Coffee beans	Reflectance	Differentiation of varieties	[135]
Honey	Reflectance	Botanical discrimination	[113]
Tomato	Absorption	Production source discrimination	[109]
Roasted green tea	Reflectance	Discrimination of geographical origin	[59]
Nectarines	Reflectance	Postharvest shelf-life discrimination	[136]
Wheat	Reflectance	Discrimination based on ochratoxin A	[137]
Apple	Reflectance	Varietal discrimination	[138]
Persimmon fruit	Reflectance	Classification of origin	[139]
Asparagus	Reflectance	Harvest date discrimination	[140]
Strawberry	Reflectance	Discrimination of different production system	[141]
Olive fruit	Reflectance	Discrimination based on fly damage	[142]
Peer	Reflectance	Varieties discrimination	[143]
Chinese quince	Reflectance	Classification of varieties	[144]

spectroscopy will replace time-consuming chemical methods and be a valuable tool for food composition on larger scales. This chapter has shown NIR spectroscopy working principle coupled with chemometrics and further its application in various food and food products. Plenty of research studied have been published which examine the use of NIR spectroscopy for the quality and safety monitoring of food and food products. The importance of this approach for assessing multiple issues of authenticity is evident. It can serve with slight modification as new techniques for in-line, on-line, and at-line control in the future.

References

1. Cen H, He Y (2007) Theory and application of near infrared reflectance spectroscopy in determination of food quality. Trends Food Sci Technol 18(2):72–83
2. Zareef M, Chen Q, Hassan MM et al (2020) An overview on the applications of typical non-linear algorithms coupled with NIR spectroscopy in food analysis. Food Eng Rev 12:173–190
3. Porep JU, Kammerer DR, Carle R (2015) On-line application of near infrared (NIR) spectroscopy in food production. Trends Food Sci Technol 46(2):211–230
4. He Y, Bai X, Xaio Q et al (2020) Detection of adulteration in food based on nondestructive analysis techniques: a review. Crit Rev Food Sci Nutr:1–21
5. Li X, Zhang Y, Wang D et al (2020) Review of NIR spectroscopy methods for nondestructive quality analysis of oilseeds and edible oils. Trends Food Sci Technol 101:172–181
6. Grassi S, Alamprese C (2018) Advances in NIR spectroscopy applied to process analytical technology in food industries. Curr Opin Food Sci 22:17–21

7. Czarnecki MA, Morisawa Y, Futami Y et al (2015) Advances in molecular structure and interaction studies using near-infrared spectroscopy. Chem Rev 115(18):9707–9744
8. Sun T, Huang K, Xu H et al (2010) Research advances in nondestructive determination of internal quality in watermelon/melon: a review. J Food Eng 100(4):569–577
9. Huang H, Yu H, Xu H et al (2008) Near infrared spectroscopy for on/in-line monitoring of quality in foods and beverages: a review. J Food Eng 87(3):303–313
10. Tao F, Ngadi M (2017) Applications of spectroscopic techniques for fat and fatty acids analysis of dairy foods. Curr Opin Food Sci 17:100–112
11. Doddridge GD, Shi Z (2015) Multivariate figures of merit (FOM) investigation on the effect of instrument parameters on a Fourier transform-near infrared spectroscopy (FT-NIRS) based content uniformity method on core tablets. J Pharm Biomed Anal 102:535–543
12. Pasquini C (2018) Near infrared spectroscopy: a mature analytical technique with new perspectives–a review. Anal Chim Acta 1026:8–36
13. Osborne BG (2006) Near-infrared spectroscopy in food analysis. In: Encyclopedia of analytical chemistry: applications, theory and instrumentation
14. Giovenzana V, Civelli R, Oberti R et al (2015) Testing of a simplified LED based vis/NIR system for rapid ripeness evaluation of white grape (Vitis vinifera L.) for Franciacorta wine. Talanta 144:584–591
15. Blanco M, Villarroya I (2002) NIR spectroscopy: a rapid-response analytical tool. TrAC Trends Anal Chem 21(4):240–250
16. Barton FE (2001) Theory and principles of near infrared spectroscopy. In: Proceedings of the Korean Society of Near Infrared Spectroscopy Conference. The Korean Society of Near Infrared Spectroscopy
17. Manley M, Baeten V (2018) Spectroscopic technique: near infrared (NIR) spectroscopy. In: Modern techniques for food authentication, pp 51–102
18. Qu J, Liu D, Cheng J et al (2015) Applications of near-infrared spectroscopy in food safety evaluation and control: a review of recent research advances. Crit Rev Food Sci Nutr 55 (13):1939–1954
19. Wu D, Sun D (2013) Advanced applications of hyperspectral imaging technology for food quality and safety analysis and assessment: a review—Part I: Fundamentals. Innovative Food Sci Emerg Technol 19:1–14
20. Schaare P, Fraser D (2000) Comparison of reflectance, interactance and transmission modes of visible-near infrared spectroscopy for measuring internal properties of kiwifruit (Actinidia chinensis). Postharvest Biol Technol 20(2):175–184
21. Nicolai BM, Beullens K, Bobelyn E et al (2007) Nondestructive measurement of fruit and vegetable quality by means of NIR spectroscopy: a review. Postharvest Biol Technol 46 (2):99–118
22. Stchur P, Cleveland D, Zhou J et al (2002) A review of recent applications of near infrared spectroscopy, and of the characteristics of a novel PbS CCD array-based near-infrared spectrometer. Appl Spectrosc Rev 37(4):383–428
23. Rinnan A, Van Den Berg F, Engelsen SB (2009) Review of the most common pre-processing techniques for near-infrared spectra. TrAC Trends Anal Chem 28(10):1201–1222
24. Mishra P, Marini F, Biancolillo A et al (2020) Improved prediction of fuel properties with near-infrared spectroscopy using a complementary sequential fusion of scatter correction techniques. Talanta 223:121693
25. Afseth NK, Kohler A (2012) Extended multiplicative signal correction in vibrational spectroscopy, a tutorial. Chemom Intell Lab Syst 117:92–99
26. Esteban-Díez I, González-Sáiz J, Pizarro C (2005) Generalization of OWAVEC method for simultaneous noise suppression, data compression and orthogonal signal correction. Anal Chim Acta 544(1–2):89–99
27. Dotto AC, Dalmolin RSD, ten Caten A et al (2018) A systematic study on the application of scatter-corrective and spectral-derivative preprocessing for multivariate prediction of soil organic carbon by Vis-NIR spectra. Geoderma 314:262–274

28. Mehmood T, Liland KH, Snipen L et al (2012) A review of variable selection methods in partial least squares regression. Chemom Intell Lab Syst 118:62–69
29. Nørgaard L, Saudland A, Wagner J et al (2000) Interval partial least-squares regression (i PLS): a comparative chemometric study with an example from near-infrared spectroscopy. Appl Spectrosc 54(3):413–419
30. da Silva DJ, Wiebeck H (2017) Using PLS, iPLS and siPLS linear regressions to determine the composition of LDPE/HDPE blends: a comparison between confocal Raman and ATR-FTIR spectroscopies. Vib Spectrosc 92:259–266
31. Lin H, Duan Y, Yan S et al (2019) Quantitative analysis of volatile organic compound using novel chemoselective response dye based on Vis-NIRS coupled Si-PLS. Microchem J 145:1119–1128
32. Feng Y, Sun D (2013) Near-infrared hyperspectral imaging in tandem with partial least squares regression and genetic algorithm for non-destructive determination and visualization of Pseudomonas loads in chicken fillets. Talanta 109:74–83
33. Zareef M, Chen Q, Ouyang Q et al (2018) Prediction of amino acids, caffeine, theaflavins and water extract in black tea using FT-NIR spectroscopy coupled chemometrics algorithms. Anal Methods 10(25):3023–3031
34. Leardi R, Nørgaard L (2004) Sequential application of backward interval partial least squares and genetic algorithms for the selection of relevant spectral regions. J Chemom 18 (11):486–497
35. Talebi M, Schuster G, Shellie RA et al (2015) Performance comparison of partial least squares-related variable selection methods for quantitative structure retention relationships modelling of retention times in reversed-phase liquid chromatography. J Chromatogr A 1424:69–76
36. Han Y, Chen J, Pan T et al (2015) Determination of glycated hemoglobin using near-infrared spectroscopy combined with equidistant combination partial least squares. Chemom Intell Lab Syst 145:84–92
37. Allegrini F, Olivieri AC (2011) A new and efficient variable selection algorithm based on ant colony optimization. Applications to near infrared spectroscopy/partial least-squares analysis. Anal Chim Acta 699(1):18–25
38. Huang X, Zou X, Zhao J et al (2014) Measurement of total anthocyanins content in flowering tea using near infrared spectroscopy combined with ant colony optimization models. Food Chem 164:536–543
39. Liu K, Chen X, Li L et al (2015) A consensus successive projections algorithm–multiple linear regression method for analyzing near infrared spectra. Anal Chim Acta 858:16–23
40. Soares SFC, Gomes AA, Rodrigues Galvo Filho A et al (2013) The successive projections algorithm. TrAC Trends Anal Chem 42:84–98
41. Yuan Z, Huang B (2004) Prediction of protein accessible surface areas by support vector regression. Proteins Struct Funct Bioinf 57(3):558–564
42. Mevik BH, Cederkvist HR (2004) Mean squared error of prediction (MSEP) estimates for principal component regression (PCR) and partial least squares regression (PLSR). J Chemom 18(9):422–429
43. Cheng J, Sun D (2017) Partial least squares regression (PLSR) applied to NIR and HSI spectral data modeling to predict chemical properties of fish muscle. Food Eng Rev 9(1):36–49
44. Khulal U, Zhao J, Hu W et al (2016) Nondestructive quantifying total volatile basic nitrogen (TVB-N) content in chicken using hyperspectral imaging (HSI) technique combined with different data dimension reduction algorithms. Food Chem 197:1191–1199
45. Xu Y, Zhong P, Jiang A et al (2020) Raman spectroscopy coupled with chemometrics for food authentication: a review. TrAC Trends Anal Chem 131:116017
46. Elmqvist N, Fekete JD (2010) Hierarchical aggregation for information visualization: overview, techniques, and design guidelines. IEEE Trans Vis Comput Graph 16(3):439–454
47. Esteki M, Shahsavari Z, Simal-Gandara J (2018) Use of spectroscopic methods in combination with linear discriminant analysis for authentication of food products. Food Control 91:100–112

48. Miller J, Miller JC (2018) Statistics and chemometrics for analytical chemistry. Pearson Education, Prentice Hall, NJ

49. Gou J, Qiu W, Yi Z et al (2019) Locality constrained representation-based K-nearest neighbor classification. Knowl-Based Syst 167:38–52

50. Lee LC, Liong CY, Jemain AA (2018) Partial least squares-discriminant analysis (PLS-DA) for classification of high-dimensional (HD) data: a review of contemporary practice strategies and knowledge gaps. Analyst 143(15):3526–3539

51. Allen A, Williams MR, Sigman ME (2019) Application of likelihood ratios and optimal decision thresholds in fire debris analysis based on a partial least squares discriminant analysis (PLS-DA) model. Foren Chem 16:100188

52. Meza-Márquez OG, Gallardo-Velázquez T, Osorio-Revilla G (2010) Application of mid-infrared spectroscopy with multivariate analysis and soft independent modeling of class analogies (SIMCA) for the detection of adulterants in minced beef. Meat Sci 86(2):511–519

53. Agatonovic-Kustrin S, Beresford R (2000) Basic concepts of artificial neural network (ANN) modeling and its application in pharmaceutical research. J Pharm Biomed Anal 22(5):717–727

54. Nashat S, Abdullah M (2010) Multi-class colour inspection of baked foods featuring support vector machine and Wilk's λ analysis. J Food Eng 101(4):370–380

55. Huang C, Wang C (2006) A GA-based feature selection and parameters optimization for support vector machines. Expert Syst Appl 31(2):231–240

56. Pouladzadeh P, Shirmohammadi S, Bakirov A et al (2015) Cloud-based SVM for food categorization. Multimed Tools Appl 74(14):5243–5260

57. Chen H, Tan C, Lin Z (2020) Ensemble of extreme learning machines for multivariate calibration of near-infrared spectroscopy. Spectrochim Acta A Mol Biomol Spectrosc 229:117982

58. Li H, Hassan MM, Wang J et al (2020) Investigation of nonlinear relationship of surface enhanced Raman scattering signal for robust prediction of thiabendazole in apple. Food Chem 339:127843

59. Chen Q, Zhao J, Lin H (2009) Study on discrimination of roast green tea (Camellia sinensis L.) according to geographical origin by FT-NIR spectroscopy and supervised pattern recognition. Spectrochim Acta A Mol Biomol Spectrosc 72(4):845–850

60. Kutsanedzie FYH, Chen Q, Hassan MM et al (2018) Near infrared system coupled chemometric algorithms for enumeration of total fungi count in cocoa beans neat solution. Food Chem 240:231–238

61. Li H, Zhu J, Jiao T et al (2020) Development of a novel wavelength selection method VCPA-PLS for robust quantification of soluble solids in tomato by on-line diffuse reflectance NIR. Spectrochim Acta A Mol Biomol Spectrosc 243:118765

62. Lan W, Jaillais B, Leca A et al (2020) A new application of NIR spectroscopy to describe and predict purees quality from the non-destructive apple measurements. Food Chem 310:125944

63. Huang Y, Lu R, Chen K (2020) Detection of internal defect of apples by a multichannel Vis/NIR spectroscopic system. Postharvest Biol Technol 161:111065

64. Orlova Y, Linker R, Spektor B (2020) Selection of Vis-NIR wavebands for forecasting apple fruitlet drop in response to chemical treatment. Biosyst Eng 195:172–185

65. Wang Z, Künnemeyer R, Mcglone A et al (2020) Potential of Vis-NIR spectroscopy for detection of chilling injury in kiwifruit. Postharvest Biol Technol 164:111160

66. Guo Z, Wang M, Akomeah Agyekum A et al (2020) Quantitative detection of apple watercore and soluble solids content by near infrared transmittance spectroscopy. J Food Eng 279:109955

67. Marques EJN, de Freitas ST (2020) Performance of new low-cost handheld NIR spectrometers for nondestructive analysis of umbu (Spondias tuberosa Arruda) quality. Food Chem 323:126820

68. Li C, Zong B, Guo H et al (2020) Discrimination of white teas produced from fresh leaves with different maturity by near-infrared spectroscopy. Spectrochim Acta A Mol Biomol Spectrosc 227:117697

69. Huang Z, Sanaeifar A, Tian Y et al (2021) Improved generalization of spectral models associated with Vis-NIR spectroscopy for determining the moisture content of different tea leaves. J Food Eng 293:110374
70. Baqueta MR, Coqueiro A, Março P et al (2021) Multivariate classification for the direct determination of cup profile in coffee blends via handheld near-infrared spectroscopy. Talanta 222:121526
71. Bahrami ME, Honarvar M, Ansari K et al (2020) Measurement of quality parameters of sugar beet juices using near-infrared spectroscopy and chemometrics. J Food Eng 271:109775
72. Pereira EVS, Fernandes DDS et al (2020) Simultaneous determination of goat milk adulteration with cow milk and their fat and protein contents using NIR spectroscopy and PLS algorithms. LWT Food Sci Technol 127:109427
73. Hosseini E, Ghasemi JB, Daraei B et al (2021) Application of genetic algorithm and multivariate methods for the detection and measurement of milk-surfactant adulteration by attenuated total reflection and near-infrared spectroscopy. J Sci Food Agric 101:2696
74. Mohamed H, Nagy P, Agbaba J et al (2021) Use of near and mid infra-red spectroscopy for analysis of protein, fat, lactose and total solids in raw cow and camel milk. Food Chem 334:127436
75. Jin G, Wang Y, Li L et al (2020) Intelligent evaluation of black tea fermentation degree by FT-NIR and computer vision based on data fusion strategy. LWT Food Sci Technol 125:109216
76. Leng T, Li F, Xiong L et al (2020) Quantitative detection of binary and ternary adulteration of minced beef meat with pork and duck meat by NIR combined with chemometrics. Food Control 113:107203
77. Mabood F, Boqué R, Alkindi AY et al (2020) Fast detection and quantification of pork meat in other meats by reflectance FT-NIR spectroscopy and multivariate analysis. Meat Sci 163:108084
78. Silva LCR, Folli GS, Santos LP et al (2020) Quantification of beef, pork, and chicken in ground meat using a portable NIR spectrometer. Vib Spectrosc 111:103158
79. Weng S, Guo B, Tang P et al (2020) Rapid detection of adulteration of minced beef using Vis/NIR reflectance spectroscopy with multivariate methods. Spectrochim Acta A 230:118005
80. Barbin DF, Badaró AP, Honorato DCB et al (2020) Identification of turkey meat and processed products using near infrared spectroscopy. Food Control 107:106816
81. Kucha CT, Ngadi MO (2020) Rapid assessment of pork freshness using miniaturized NIR spectroscopy. J Food Measur Character 14(2):1105–1115
82. Parastar H, van Kollenburg G, Weesepoel Y et al (2020) Integration of handheld NIR and machine learning to "Measure & Monitor" chicken meat authenticity. Food Control 112:107149
83. Dumalisile P, Manley M, Hoffman L et al (2020) Discriminating muscle type of selected game species using near infrared (NIR) spectroscopy. Food Control 110:106981
84. Dumalisile P, Manley M, Hoffman L et al (2020) Near-infrared (NIR) spectroscopy to differentiate longissimus thoracis et lumborum (LTL) muscles of game species. Food Anal Methods 13(5):1220–1233
85. Shannon JC, Garwood DL, Boyer CD (2009) Genetics and physiology of starch development. In: Starch. Elsevier, Amsterdam, pp 23–82
86. Martinez CS, Ribotta PD, Leon AE et al (2007) Physical, sensory and chemical evaluation of cooked spaghetti. J Texture Stud 38(6):666–683
87. Czaja T, Kuzawińska E, Sobota A et al (2018) Determining moisture content in pasta by vibrational spectroscopy. Talanta 178:294–298
88. Fan Y, Ma S, Wu T (2020) Individual wheat kernels vigor assessment based on NIR spectroscopy coupled with machine learning methodologies. Infrared Phys Technol 105:103213

 89. Jiang H, Liu T, Chen Q (2020) Quantitative detection of fatty acid value during storage of wheat flour based on a portable near-infrared (NIR) spectroscopy system. Infrared Phys Technol 109:103423
 90. Tian W, Chen G, Zhang G et al (2021) Rapid determination of total phenolic content of whole wheat flour using near-infrared spectroscopy and chemometrics. Food Chem 344:128633
 91. De Girolamo A, Arroyo MC, Lippolis V et al (2020) A simple design for the validation of a FT-NIR screening method: application to the detection of durum wheat pasta adulteration. Food Chem 333:127449
 92. Zheng S, Wei Z et al (2020) Near-infrared reflectance spectroscopy-based fast versicolorin A detection in maize for early aflatoxin warning and safety sorting. Food Chem 332:127419
 93. Shen F, Huang Y, Jiang X et al (2020) On-line prediction of hazardous fungal contamination in stored maize by integrating Vis/NIR spectroscopy and computer vision. Spectrochim Acta A 229:118012
 94. Zhang B, Jiang X, Shen F et al (2021) Rapid screening of DON contamination in whole wheat meals by Vis/NIR spectroscopy and computer vision coupling technology. Int J Food Sci Technol 56:2588
 95. Caramês ET, Piacentini KC, Alves LT et al (2020) NIR spectroscopy and chemometric tools to identify high content of deoxynivalenol in barley. Food Addit Contam Part A 37 (9):1542–1552
 96. Du Q, Zhu M, Shi T et al (2021) Adulteration detection of corn oil, rapeseed oil and sunflower oil in camellia oil by in situ diffuse reflectance near-infrared spectroscopy and chemometrics. Food Control 121:107577
 97. Borghi FT, Santos PC, Santos FD et al (2020) Quantification and classification of vegetable oils in extra virgin olive oil samples using a portable near-infrared spectrometer associated with chemometrics. Microchem J 159:105544
 98. He Y, Jiang H, Chen Q (2020) High-precision identification of the actual storage periods of edible oil by FT-NIR spectroscopy combined with chemometric methods. Anal Methods 12 (29):3722–3728
 99. MacArthur RL, Teye E, Darkwa S (2020) Predicting adulteration of palm oil with Sudan IV dye using shortwave handheld spectroscopy and comparative analysis of models. Vib Spectrosc 110:103129
100. Amanah HZ, Joshi R, Masithoh RE et al (2020) Nondestructive measurement of anthocyanin in intact soybean seed using Fourier Transform Near-Infrared (FT-NIR) and Fourier Transform Infrared (FT-IR) spectroscopy. Infrared Phys Technol 111:103477
101. Genis HE, Durna S, Boyaci IH (2021) Determination of green pea and spinach adulteration in pistachio nuts using NIR spectroscopy. LWT Food Sci Technol 136:110008
102. Aykas DP, Menevseoglu A (2021) A rapid method to detect green pea and peanut adulteration in pistachio by using portable FT-MIR and FT-NIR spectroscopy combined with chemometrics. Food Control 121:107670
103. Wang X, Esquerre C, Downey G et al (2021) Development of chemometric models using Vis-NIR and Raman spectral data fusion for assessment of infant formula storage temperature and time. Innovative Food Sci Emerg Technol 67:102551
104. Masithoh RE, Amanah HZ, Yoon WS et al (2021) Determination of protein and glucose of tuber and root flours using NIR and MIR spectroscopy. Infrared Phys Technol 113:103577
105. Gatti RF, de Santana FB, Poppi RJ et al (2021) Portable NIR spectrometer for quick identification of fat bloom in chocolates. Food Chem 342:128267
106. Elrasheid Tahir H, Komla Mahunu G, Arslan M et al (2021) Feasibility study for the use of CSA arrays, NIR and FT-IR spectroscopy in the quantitative analysis of volatile components in honey. Microchem J 160:105730
107. Xie L, Ye X, Liu D et al (2011) Prediction of titratable acidity, malic acid, and citric acid in bayberry fruit by near-infrared spectroscopy. Food Res Int 44(7):2198–2204
108. Ebrahimi-Najafabadi H, Leardi R, Oliveri P et al (2012) Detection of addition of barley to coffee using near infrared spectroscopy and chemometric techniques. Talanta 99:175–179

109. Ding X, Guo Y, Ni Y et al (2016) A novel NIR spectroscopic method for rapid analyses of lycopene, total acid, sugar, phenols and antioxidant activity in dehydrated tomato samples. Vib Spectrosc 82:1–9

110. Lin H, Chen Q, Zhao J et al (2009) Determination of free amino acid content in Radix Pseudostellariae using near infrared (NIR) spectroscopy and different multivariate calibrations. J Pharm Biomed Anal 50(5):803–808

111. Borin A, Ferrao MF, Mello C et al (2006) Least-squares support vector machines and near infrared spectroscopy for quantification of common adulterants in powdered milk. Anal Chim Acta 579(1):25–32

112. Chen Q, Zhao J, Liu M et al (2008) Determination of total polyphenols content in green tea using FT-NIR spectroscopy and different PLS algorithms. J Pharm Biomed Anal 46 (3):568–573

113. Tahir HE, Zou X, Shen T et al (2016) Near-infrared (NIR) spectroscopy for rapid measurement of antioxidant properties and discrimination of Sudanese honeys from different botanical origin. Food Anal Methods 9(9):2631–2641

114. Berardo N, Pisacane V, Battilani P et al (2005) Rapid detection of kernel rots and mycotoxins in maize by near-infrared reflectance spectroscopy. J Agric Food Chem 53(21):8128–8134

115. Bao Y, Liu F, Kong W et al (2014) Measurement of soluble solid contents and pH of white vinegars using VIS/NIR spectroscopy and least squares support vector machine. Food Bioprocess Technol 7(1):54–61

116. Chen Q, Zhao J, Huang X et al (2006) Simultaneous determination of total polyphenols and caffeine contents of green tea by near-infrared reflectance spectroscopy. Microchem J 83 (1):42–47

117. Blanco-Díaz MT, Del Río-Celestino M, Martínez-Valdivieso D et al (2014) Use of visible and near-infrared spectroscopy for predicting antioxidant compounds in summer squash (Cucurbita pepo ssp pepo). Food Chem 164:301–308

118. Bellincontro A, Taticchi A, Servili M et al (2012) Feasible application of a portable NIR-AOTF tool for on-field prediction of phenolic compounds during the ripening of olives for oil production. J Agric Food Chem 60(10):2665–2673

119. Dykes L, Hoffmann L Jr, Portillo-Rodriguez O et al (2014) Prediction of total phenols, condensed tannins, and 3-deoxyanthocyanidins in sorghum grain using near-infrared (NIR) spectroscopy. J Cereal Sci 60(1):138–142

120. Tahir HE, Zou X, Shi J et al (2016) Rapid determination of antioxidant compounds and antioxidant activity of Sudanese Karkade (Hibiscus sabdariffa L.) using near infrared spectroscopy. Food Anal Methods 9(5):1228–1236

121. Arslan M, Zou X, Hu X et al (2018) Near infrared spectroscopy coupled with chemometric algorithms for predicting chemical components in black goji berries (Lycium ruthenicum Murr.). J Near Infrared Spectrosc 26(5):275–286

122. Arslan M, Zou X, Haroon Elrasheid T et al (2018) Near-infrared spectroscopy coupled chemometric algorithms for prediction of antioxidant activity of black goji berries (Lycium ruthenicum Murr.). J Food Measur Character 12(4):2366–2376

123. Zareef M, Chen Q, Ouyang Q et al (2018) Prediction of amino acids, caffeine, theaflavins and water extract in black tea by FT-NIR spectroscopy coupled chemometrics algorithms. Anal Methods 10:3023

124. Dong C, Zhu H, Wang J et al (2017) Prediction of black tea fermentation quality indices using NIRS and nonlinear tools. Food Sci Biotechnol 26(4):853–860

125. Magalhães LM, Machado S, Segundo MA et al (2016) Rapid assessment of bioactive phenolics and methylxanthines in spent coffee grounds by FT-NIR spectroscopy. Talanta 147:460–467

126. Teye E, Huang X (2015) Novel prediction of total fat content in cocoa beans by FT-NIR spectroscopy based on effective spectral selection multivariate regression. Food Anal Methods 8(4):945–953

127. Liu Y, Ying Y, Yu H et al (2006) Comparison of the HPLC method and FT-NIR analysis for quantification of glucose, fructose, and sucrose in intact apple fruits. J Agric Food Chem 54 (8):2810–2815
128. Xie L, Ye X, Liu D et al (2009) Quantification of glucose, fructose and sucrose in bayberry juice by NIR and PLS. Food Chem 114(3):1135–1140
129. Tewari JC, Dixit V, Cho B et al (2008) Determination of origin and sugars of citrus fruits using genetic algorithm, correspondence analysis and partial least square combined with fiber optic NIR spectroscopy. Spectrochim Acta A 71(3):1119–1127
130. Mabood F, Al-Harrasi A, Boqué R et al (2015) Determination of sucrose in date fruits (Phoenix dactylifera L.) growing in the Sultanate of Oman by NIR spectroscopy and multivariate calibration. Spectrochim Acta A 150:170–174
131. Zou X, Zhao J, Huang X et al (2007) Use of FT-NIR spectrometry in non-invasive measurements of soluble solid contents (SSC) of 'Fuji' apple based on different PLS models. Chemom Intell Lab Syst 87(1):43–51
132. Li X, He Y, Fang H (2007) Non-destructive discrimination of Chinese bayberry varieties using Vis/NIR spectroscopy. J Food Eng 81(2):357–363
133. Seregely Z, Deak T, Bisztray GD (2004) Distinguishing melon genotypes using NIR spectroscopy. Chemom Intell Lab Syst 72(2):195–203
134. Li X, He Y, Wu C (2008) Non-destructive discrimination of paddy seeds of different storage age based on Vis/NIR spectroscopy. J Stored Prod Res 44(3):264–268
135. Esteban-Diez I, González-Sáiz J, Sáenz-González C et al (2007) Coffee varietal differentiation based on near infrared spectroscopy. Talanta 71(1):221–229
136. Pérez-Marín D, Sánchez MT, Paz P et al (2011) Postharvest shelf-life discrimination of nectarines produced under different irrigation strategies using NIR-spectroscopy. LWT Food Sci Technol 44(6):1405–1414
137. De Girolamo A, von Holst C, Cortese M et al (2019) Rapid screening of ochratoxin A in wheat by infrared spectroscopy. Food Chem 282:95–100
138. He Y, Li X, Shao Y (2007) Fast discrimination of apple varieties using Vis/NIR spectroscopy. Int J Food Prop 10(1):9–18
139. Khanmohammadi M, Karami F, Mir-Marqués A et al (2014) Classification of persimmon fruit origin by near infrared spectrometry and least squares-support vector machines. J Food Eng 142:17–22
140. Jarén C, Arazuri S, García M et al (2006) White asparagus harvest date discrimination using NIRS technology. Int J Infrar Millim Wav 27(3):391–401
141. Amodio ML, Ceglie F, Chaudhry MMA et al (2017) Potential of NIR spectroscopy for predicting internal quality and discriminating among strawberry fruits from different production systems. Postharvest Biol Technol 125:112–121
142. Moscetti R, Haff RP, Stella E et al (2015) Feasibility of NIR spectroscopy to detect olive fruit infested by Bactrocera oleae. Postharvest Biol Technol 99:58–62
143. Fu X, Zhou Y, Ying Y et al (2007) Discrimination of pear varieties using three classification methods based on near-infrared spectroscopy. Trans ASABE 50(4):1355–1361
144. Shao W, Li Y, Diao S et al (2017) Rapid classification of Chinese quince (Chaenomeles speciosa Nakai) fruit provenance by near-infrared spectroscopy and multivariate calibration. Anal Bioanal Chem 409(1):115–120

Chapter 3
Bionic Sensors Technologies in Food

What Is Bionic Sensors Technologies?
Generally, people may use color, aroma, and taste to describe the quality of food.
"Color" refers to the vision seen by the eyes, "scent" refers to the smell of the nose,
and "taste" refers to the taste of the tongue. The quality evaluation of flavored food
usually involves the use of vision, smell, taste, and other sensory organs. Comprehensive sensory evaluation was also frequently used to analyze, evaluate, and judge
the color, aroma, taste, and style characteristics of food [1, 2]. Taste accounts for a
relatively high proportion, while color and other aspects also take up a certain
proportion. Although artificial sensory evaluation is the most commonly used
quality evaluation method for flavored foods such as wine and tea, it has many
limitations. First of all, even with well-trained senior evaluation experts, the sensory
organ's sensitivity may be easily affected by factors such as experience, gender, age,
mental state, physical condition, and geographical environment, thereby affecting
the accuracy and stability of sensory evaluation results. Second, with a certain degree
of subjectivity, artificial sensory organs are prone to fatigue after continuous stimulation for a period of time, and then become dull, and can only recover after a period
of rest [3–5]. Figure 3.1 illustrates three types of bionic sensor detection technology.

Abbreviation	Full name
AD	Analog-digital
BLE	Bluetooth low energy
DLP	Digital light processing
DMD	Digital micromirror devices
LDA	Linear discriminant analysis
MAE	Mean absolute error
MRE	Mean relative error
PLC	Programmable logic controller
R&D	Research and development
TVB-N	Total volatile basic nitrogen
Tablet PC	Tablet personal computer

(continued)

Abbreviation	Full name
UAV	Unmanned aerial vehicle
UV-vis	Ultraviolet and visible

3.1 Introduction

Bionics is a new comprehensive discipline that emerged internationally in the middle of the twentieth century, and it is an interdisciplinary combination of life sciences, mechanics, materials science, informatics, and computer science. Bionics studies the structure of an organ and its working principle [6, 7]. On this basis, instruments and devices with bionic functions are developed. Bionic sensors use sensitive units composed of physical and chemical materials with specific recognition capabilities to form a sensor array to simulate mammalian sense organs such as vision, smell, and taste [8]. They are cross-responsive, so they have unique advantages for the detection of complex systems with multiple components. The response signal generated by the biomimetic sensor array combined with the pattern recognition method can carry out qualitative and quantitative analysis of substances, and the overall identification of complex systems. It has the advantages of simple, fast, and objective operation without complicated sample pretreatment. Vision, smell, and taste sensors are designed to simulate human vision, smell, and taste functions. They are combined with signal preprocessing systems and pattern recognition to form electronic eyes (computer vision, color difference meter), electronic nose, and electronic tongue systems. The functions of perception, analysis, and recognition have been increasingly used in the quality analysis of food and agricultural products. With the advancement of life sciences and artificial intelligence, scientists have tried to imitate animals' and humans' smell and taste functions to develop artificial vision sensor

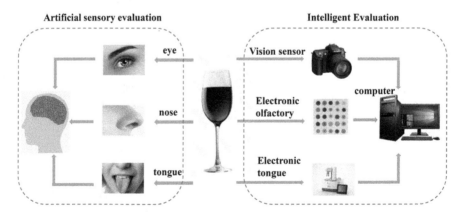

Fig. 3.1 Three types of bionic sensor detection technology (vision sensor, electronic olfactory, and electronic tongue)

systems—electronic eyes, olfactory sensor systems—electronic noses, and artificial taste sensor systems—electronic tongue [9–11]. The idea of an electronic tongue is gradually becoming a reality.

3.1.1 Brief Introduction of Bionic Sensor Detection Technology

1. **Vision Sensor**
   Vision sensor technology, also known as electronic eye technology, usually includes computer vision technology or color measurement technology. It is a visual function design that simulates living things and can be used for qualitative and quantitative food color, shape, texture, and other external qualities analysis. In bionic sensing, the main technology used in visual sensing is color measurement [12, 13]. Color measurement technology can be used to test the color characteristics of the surface of solid foods, while it can be used to characterize the color and clarity of liquid foods. The color measurement technology uses CIE Lab (L^* a^* b^* color space), which produces the same color difference at the same spatial distance, and is currently one of the most common color spaces for measuring object colors. L^* a^* b^* surface color system is a chromaticity space represented by L^* a^* b^* three color coordinates. It is a three-dimensional system, which is used to determine the three component values (L^*, a^*, b^*) of the color to be measured [12, 13].

2. **Electronic Olfactory**
   Electronic olfactory (commonly known as the electronic nose) technology simulates and detects the formation of biological smell. The electronic nose is artificial olfaction established by simulating the olfactory function of the mammalian nose. It consists of three parts: a gas sensor array, a signal processing system, and a pattern recognition system [14–16]. At work, gas molecules are adsorbed by the gas sensor to generate signals, and the signal processing system processes the generated signals. Finally, the pattern recognition system makes a comprehensive judgment on the results of the signal processing. The electronic nose system analyses of volatile components can be used for food quality testing, maturity, freshness, etc.

3. **Electronic Tongue**
   Taste is usually used to identify nonvolatile chemicals that enter the mouth and used to identify volatile substances. The receptors of the trigeminal nerve are distributed on the mucosa and skin. They react to both volatile and nonvolatile chemicals. The important thing is to be able to distinguish between the types of stimuli and chemical reactions. All three chemosensory systems are involved in fragrance perception, but the role of smell far exceeds the other two sensations. The basic ideas of artificial smell and artificial taste include the artificial reproduction of these reactions. Like the electronic olfactory, the electronic taste

(commonly known as the electronic tongue) is also composed of a sensor array, a signal processing system, and a pattern recognition system [17–20]. The sensor array responds to the liquid sample and outputs a signal. After the signal is processed by a computer and pattern recognition, a result reflecting the sample's taste characteristics is obtained.

4. **Features of Bionic Sensor Detection Technology**

Currently, the taste evaluation of food is mainly carried out by the sensory evaluation method, but the evaluation result of this method is determined by the experience of the taster. This method is highly subjective and has poor repeatability [21, 22]. Chemical methods to identify flavor components are generally laborious and time-consuming. For example, by measuring the optical density value of tea polyphenols and ferrous tartrate in tea soup and calculating the total flavor score value according to specific rules, the total score value is used to evaluate the flavor of tea soup. Although chemical methods' detection results are relatively objective and accurate, they have the disadvantages of complicated steps, time-consuming, and expensive. Relevant studies have shown a certain correlation between the flavor components of food and sensory evaluation, and a correlation model has been established [23].

Electronic olfactory and electronic taste both directly obtain the correlation data of flavor components through the sensor array, and through computer system processing and pattern recognition. They have the advantages of fast, accurate, and good repeatability. These two technologies are widely used in meat freshness detection, wine identification, and water pollution. Compared with artificial sensory organs, the color characterization value obtained by color measurement technology is more accurate, objective, and fast and can be used for online inspection of food and medicine quality in production practice. It can also base to promote the standardization and scientific management of industry production [24, 25].

3.2 Instrumentation

3.2.1 Electronic Vision Sensors

1. **Components of Electronic Vision Sensing Inspection System**

The working principle of the electronic vision system was briefly summarized as shown in Fig. 3.2. Mainly, the CCD detector will receive the signal about the information of samples, convert the received sample signal information into analog electrical signals, and the A/D converter will convert the analog electrical signals into digital signals and finally transmit them to the computer.

The spectroscopic system is the core component of the visible/near-infrared spectrum detection system. The spectroscopic system can be divided into fixed wavelength filters; fast Fourier transforms, acousto-optic tunable filters, and other types [26, 27]. The filter type is mainly used as a dedicated analysis instrument.

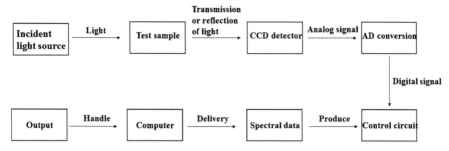

Fig. 3.2 Working principle diagram of a detection system

Due to the filter's limitation, only the spectral information under a limited number of wavelengths can be obtained, so it is not easy to analyze samples of complex systems. Fourier transform near-infrared spectrometer has high resolution and scanning speed, but there are mobile parts in the instrument, which are easily affected by the environment, so a stricter working environment is required. The acousto-optic tunable filter uses a birefringent crystal to adjust the scanning wavelength by changing the radiofrequency. The entire instrument system has no moving parts, and the scanning speed is fast. However, the resolution of this type of instrument is relatively low, and the price is high.

Due to the development and maturity of CCD array detector technology, fiber optic technology, and grating technology, the entire system structure has become more flexible and integrated [28]. Miniature fiber optic spectrometers have become common for on-site inspection and online inspection due to their compact size and portability. Compared with traditional spectroscopy technology, the miniature optical fiber spectrometer can fully adapt to the variability of the job site's spatial form in various fields and the variability of the production process conditions. The micro-fiber spectrometer uses grating as a light-splitting system, where the angular dispersion and linear dispersion parameters of the grating are critical indicators to measure the light-splitting effect of the grating. The greater the dispersion of the grating, the easier it is to separate two close spectral lines. After the light is split through the grating, the detector array of the miniature fiber spectrometer simultaneously receives the entire spectrum's energy distribution information. The linear CCD is the key component to receive the photoelectric conversion. The linear CCD used to measure the spectrum needs to have higher resolution and better spectral response, and at the same time, it needs to have a strong suppression of dark noise to improve the signal-to-noise ratio of the spectrum [29, 30]. The proper selection of CCD can improve the measurement accuracy of the entire spectrometer.

In terms of optical signal transmission, the electronic vision sensor detection system can use the characteristics of low optical fiber transmission loss and strong anti-interference ability to transmit the spectral signal to the spectrometer over a long distance. Simultaneously, due to the lightweight and small size of the optical fiber, the adjustment difficulty of the optical parts and the optical system of the

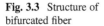

Fig. 3.3 Structure of bifurcated fiber

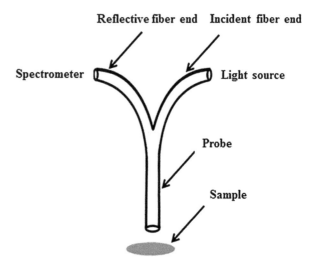

analytical instrument can be significantly reduced, and the miniaturization of the analytical instrument can be enhanced [31]. Figure 3.3 shows a Y-shaped bifurcated fiber structure. One end is the probe end, and the other end is bifurcated into the incident fiber end and the reflection fiber end.

2. **Electronic Vision Sensing Inspection System: Color Space Expression**

Electronic vision sensing technology usually adopts color measurement technology, namely CIE LAB ($L* a* b*$ color space). The color space produces the same color difference in the same space distance, which is one of the most common color spaces to measure objects' colors [32–34]. $L* a* b*$ surface color system is a chromaticity space represented by $L* a* b*$ three color coordinates. It is a three-dimensional system used to determine the three-component values ($L*$, $a*$, $b*$) of the color to be measured. $L*$ is the lightness; $a*$ is the degree of red and green, thus, red when it is positive and green when it is negative; $b*$ is the chromaticity of yellow and blue, which means the degree of yellow when it is positive and the degree of blue when it is negative. The color status of the object to be measured is expressed by the measured values of $L*$, $a*$, and $b*$. Compared with artificial sensory, the color characterization value obtained by color measurement technology is more accurate, objective, and rapid. It can be used for online detection of food and drug quality in production practice and promotes the standardization and scientific management level of industrial production.

Visual sensor data collection usually uses a colorimeter. In transmission mode, a quartz cuvette is used as a sample cell, and a standard D65 light source and a 10° viewing angle are used to measure the sample [35–38]. The CIE1976 color system is used to characterize the color information of the detected object, and measures the three components of the color—$L*$, $a*$, and $b*$ values, where $L*$ represents brightness, $a*$ represents red-greenness ($-a*$ represents green, $+a*$ means red), $b*$ represents yellow-blueness ($-b*$ means blue, $+b*$ means yellow).

The visible light wavelength range is 380–780 nm. Taking the spectral value at 380 nm as an example, first, one needs to take the two closest data points on the left and right sides of this wavelength and mark them as i and j, and then mark the corresponding wavelength and spectral intensity as λ_i, λ_j and I_i, I_j, respectively; Then, the spectral value I_{380} of this wavelength can be calculated using the following formula:

$$I_{380} = I_i - \frac{(I_j - I_i)(\lambda_{380} - \lambda_i)}{(\lambda_j - \lambda_i)} \tag{3.1}$$

Similarly, this method can be used to obtain the spectrum values of the remaining whole wavelengths in batches and record them as $I(\lambda)$. Second, calculate the reflectivity. The white correction value, W, and the dark correction value, D, saved before collecting the test object, using the same method as above to calculate the spectral value at 380 nm, respectively, are used to calculate the required white reference value $W(\lambda)$ and dark reference value D at the entire wavelength (λ). The calculation of the surface reflectance $R(\lambda)$ of fermented leaves can be obtained with the following formula:

$$R(\lambda) = \frac{I(\lambda) - D(\lambda)}{W(\lambda) - D(\lambda)} \tag{3.2}$$

Finally, on the basis of obtaining the reflectance, refer to the GB/T3979-2008 and CIE 1931 standard colorimetric system to calculate the XYZ value, and usually use the summation method instead of the integral. The processing formula is as follows:

$$X = K\sum_{\lambda} S(\lambda)\bar{x}(\lambda)R(\lambda)\Delta\lambda \tag{3.3}$$

$$Y = K\sum_{\lambda} S(\lambda)\bar{y}(\lambda)R(\lambda)\Delta\lambda \tag{3.4}$$

$$Z = K\sum_{\lambda} S(\lambda)\bar{z}(\lambda)R(\lambda)\Delta\lambda \tag{3.5}$$

where
K—normalization coefficient, $K = \frac{100}{\sum_{\lambda} S(\lambda)\bar{y}(\lambda)\Delta\lambda}$
$S(\lambda)$—relative spectral power distribution of CIE standard lighting
$\bar{x}(\lambda), \bar{y}(\lambda), \bar{z}(\lambda)$—CIE standard color matching degree of observer
$R(\lambda)$—spectral reflectance
$\Delta\lambda$—wavelength interval
The L^* a^* b^* color space is a standard formulated by CIE in 1976. Among them, the L^* value represents brightness, and the range is from 0 to 100; a^* value represents from green to red, and the range of variation is -128 to $+127$; b^* value represents from blue to yellow, and the range of variation is -128 to $+127$.

Generally, the RGB color space can represent 16.7 million colors, while the $L*$ $a*$ $b*$ can represent up to 27 million colors. Therefore, the $L*$ $a*$ $b*$ color space can more delicately describe the detection object's color change. The conversion formula is as follows:

$$L = 116 \times mX - 16 \tag{3.6}$$

$$*a = 500 \times (mX - mY) \tag{3.7}$$

$$*b = 200 \times (mY - mZ) \tag{3.8}$$

where

$$mX = \begin{cases} \left(\frac{X}{X_n}\right)^{1/3}, & \frac{X}{X_n} \geq 0.008856 \\ 7.787 \times \left(\frac{X}{X_n}\right) + \frac{16}{116}, & \frac{X}{X_n} < 0.008856 \end{cases} \tag{3.9}$$

$$mY = \begin{cases} \left(\frac{Y}{Y_n}\right)^{1/3}, & \frac{Y}{Y_n} \geq 0.008856 \\ 7.787 \times \left(\frac{Y}{Y_n}\right) + \frac{16}{116}, & \frac{Y}{Y_n} < 0.008856 \end{cases} \tag{3.10}$$

$$mZ = \begin{cases} \left(\frac{Z}{Z_n}\right)^{1/3}, & \frac{Z}{Z_n} \geq 0.008856 \\ 7.787 \times \left(\frac{Z}{Z_n}\right) + \frac{16}{116}, & \frac{Z}{Z_n} < 0.008856 \end{cases} \tag{3.11}$$

where
X, Y, Z—the three stimulation values of the test object
X_n, Y_n, Z_n—three stimulation values XYZ of total reflection diffuser (for 2° CIE observers) or $X_{10}Y_{10}Z_{10}$ (for 10° CIE observers)

3.2.2 Electronic Olfactory Sensing Technology

Artificial olfaction is a technology that simulates biological olfaction. Its principle is similar to the formation of biological olfaction. It uses sensors to respond to odor molecules, and the generated signals are processed and recognized to make a judgment on odors [39, 40].

To understand the working principle of electronic olfactory, we must first know how the biological smell is formed [41, 42]. The production of biological olfactory can be roughly divided into three stages: First, the signal generation stage. After the odor molecules diffuse through the air and reach the nasal cavity, they are adsorbed on the surface by the olfactory cells in the olfactory vesicles. Part of the charge on the surface of the negatively charged olfactory cells changes, generating an electric

Table 3.1 Comparison between human smell and electronic olfactory system

Human sense of smell	Electronic olfactory system
Primary olfactory neurons: olfactory cells and olfactory nerves	Gas sensor array
Secondary olfactory neurons: regulate and inhibit the signals from primary olfactory neurons	Operational amplifier, filtering, and other electronic circuits
Brain: process signals from secondary olfactory neurons and make judgments	Computer

current, making the nerve terminals stimulated and excited; it is the signal transmission and preprocessing stage. The excitement signal is sent to the brain after a series of processing and amplification in the olfactory bulb; finally, the brain recognition stage. The brain compares the input signal with experience and then makes a recognition judgment, whether this is the scent of milk, coffee, rose, or other scents. The brain's judgment and recognition function is formed by learning, memorizing, accumulating, and summarizing in the process of constant contact with the outside world since childhood.

Through experimental research on various neuroanatomy, neurophysiology, and neuro behavior levels, Freeman confirmed that every neuron in the olfactory neural network is involved in olfactory perception [43, 44] and believed that humans and animals would smell in the nasal cavity during inhalation. The olfactory cell array forms a specific spatial distribution, and then the olfactory system directly completes the classification in an abstract manner. When inhaling a familiar smell, the brain waves become more orderly than before, forming a unique spatial pattern. When an unfamiliar smell is inputted, the olfactory system's brain waves show a low-amplitude chaotic state, which is equivalent to a "I don't know" state.

The odor can be single or compound. A single odor is formed by a molecule of an odorous substance, while a compound odor is a mixture of many (possibly hundreds of) different odor molecules. In fact, the natural odors are all compound, and the single odor is artificial.

1. **Composition of an Electronic Olfactory Detection System**

The electronic olfactory system is a kind of bionic technology formed based on biological olfaction simulating. Table 3.1 lists the corresponding relationship between human smell and the electronic olfactory system. The human sense of smell needs to complete the signal reception, preprocessing, and recognition functions, so the signal generation, collection, and subsequent data processing of the electronic nose system are essential contents of the electronic nose system.

The electronic olfactory detection system is mainly composed of sensor array, data processing and analysis system, micro-processing and interface circuit, as shown in Fig. 3.4. The gas sensor array comprises multiple gas sensors with overlapping performances, which are functionally equivalent to human olfactory sensing cells that overlap each other. Compared with a single gas sensor, the gas sensor array has a wider detection range, and its sensitivity and reliability are greatly improved. The signal generated by the gas sensor array is transmitted to

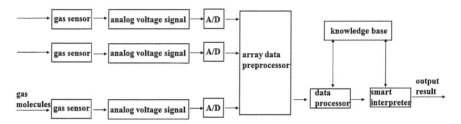

Fig. 3.4 Structure diagram of an electronic olfactory detection system

the data processing and analysis system, which is preprocessed (filtering, transformation, amplification, feature extraction, etc.), and then the gas composition analysis is realized through pattern recognition. The data processing and analysis system are equivalent to the second and third stages in forming the human sense of smell. They play the role of the middle meridian and brain of the human olfactory bulb and have the functions of analysis, judgment, and intelligent interpretation. The data processing analysis system is composed of A/D conversion data acquisition, array data preprocessor, data processor, intelligent interpreter, and knowledge base.

The intensity of the sense of smell measured can be represented by the absolute voltage, resistance, or conductance output of each sensor or represented by relative signal values such as normalized resistance or conductance. The sensor array's signal output is collected, processed by a special software, and compared with the known information obtained after "human learning and training." Finally, a quantitative quality factor is obtained, and the tested sample is judged by the quality factor based on the category, authenticity, pros and cons, qualified or not, etc.

The recognition methods used by the artificial olfactory system AOS (also for the artificial taste system ATS) mainly include statistical pattern recognition (such as linear classification, local minimum variance, principal element analysis, etc.) and artificial neural network pattern recognition. Statistical methods require known analytical expressions of response characteristics and often require linearization. Because the olfactory sensor array's response mechanism is relatively complex, it brings considerable difficulties to the approximation and linearization of the response characteristics, making it difficult to establish an accurate mathematical model thereby limiting its recognition accuracy. Artificial neural network has been widely used since the 1980s due to its ability to handle more complex nonlinear problems, suppress drift, and reduce errors.

2. **Electronic Olfactory Sensor Array**

An electronic olfactory sensor usually refers to a sensing device composed of gas sensors, circuits, and other components [45, 46]. A gas sensor refers to an element that can sense a specific gas (such as CO, CO_2, O_2, Cl_2) and its concentration in the environment. In practical applications, gas sensors should meet the following requirements:

Fig. 3.5 Metal oxide semiconductor sensor

(a) It has high sensitivity and a wide dynamic response range. When the measured gas concentration is low, there is a large enough response signal, and when the measured gas concentration is high, it has an excellent linear response value.
(b) The performance is stable, and the response of the sensor does not change with changes in ambient temperature and humidity.
(c) Fast response and good repeatability.
(d) Simple maintenance and low price. The semiconductor gas sensor developed at present cannot fully meet the above requirements, especially in terms of stability and selection range; there are still many problems that need to be further resolved.

The materials used as artificial olfactory gas sensors must meet two basic conditions:

(a) Respond to a variety of odors, that is, strong versatility, requiring the identification of thousands of different odors at the molecular level;
(b) The interaction or reaction with the smell molecule must be fast and reversible, without any "memory effect," that is, it has good reducibility.

According to different types of materials, existing sensors (referring to gas sensors, the same below) can be divided into metal oxide semiconductor sensors, organic conductive polymer sensors, quality sensors (including quartz crystal resonance sensors and surface acoustic wave sensors), and metal oxide semiconductor field effect tube sensors, infrared photoelectric sensors and metal gate MOS gas sensors, etc. The overview of these olfactory sensors are outlined below [47, 48]:

(a) Metal Oxide Semiconductor Sensor
The metal oxide semiconductor sensor (as shown in Fig. 3.5) is currently the world's most produced and most widely used gas sensor. It uses the detected odor molecules to be adsorbed on the sensitive film material,

resulting in the metal oxide semiconductor's resistance. Change in this characteristic results in detection. This type of sensor has low selectivity and prolonged recovery time, requires heating during operation, large, difficult to arrange when forming an array, and has a very narrow linear range of signal response. However, because of the low manufacturing cost of this type of sensor, simple method of signal detection, good working stability, and high detection sensitivity, it is currently the most widely used and most practical type of gas sensor. The primary measurement objects are various reducing gases, such as CO, H_2, ethanol, and methanol.

(b) Organic Electro Polymer Sensor

The working principle of the organic conductive polymer sensor is that the conductivity of the heterocyclic molecular coating on the surface of the working electrode changes after the gas molecules are absorbed and released. The conductive polymer material is an organic sensitive film material, such as pyrrole, aniline, and thiophene. This kind of sensor is characterized by small size, low energy consumption, no heating during work, good stability, fast adsorption and release, the concentration of the measured object, and the sensor's response are almost linear in an extensive range, which can be used for data processing. It is very convenient to use. The application of this type of sensor array has increased in recent years.

(c) Grease-Coated Sensor

Grease-coated sensors are also called mass sensors. Typical grease-coated sensors include surface acoustic wave type and quartz crystal resonance type. Surface acoustic wave (SAW) gas sensors have been developed for more than 20 years. In 1979, Wohhjen and Dessy successfully used SAW elements coated with organic polymers as a gas chromatograph's detectors unveiling the first page of the SAW gas sensor [49]. The surface acoustic wave sensor's working principle is to coat a layer of gas-sensitive material on the piezoelectric crystal. When the measured gas is adsorbed on the sensitive film during the flow process, the quality of the piezoelectric crystal substrate changes. The load effect causes a corresponding change in the substrate's oscillation frequency, thereby stimulating the detection of the measured gas. Although SAW can also detect some inorganic gases, the primary measurement objects are various organic gases. Its gas sensitivity and selectivity depend on the gas-sensitive membrane material on the surface of the element. It is generally used to detect multiple gases with similar chemical properties simultaneously. It is not suitable for detecting a single gas component in an unknown gas component. The quartz crystal resonator sensor's working principle is to coat a sensitive film (such as lipids, serotonin) on the quartz oscillator. When the sensitive film adsorbs molecules, the oscillation frequency of the oscillator is proportional to the mass load effect. The change triggers the detection of the measured gas. The performance of the sensor is different if the sensitive film material coated on the resonator is different.

Table 3.2 Commonly used olfactory sensor arrays and characteristics

Gas sensitive materials	Sensor type	Number of sensors	Typical measured object
Metal oxide	Chemical resistance	6, 8, 12	Flammable gas
Organic polymer	Chemical resistance	12, 20, 24, 32	NH_3, NO, H_2, alcohol
Grease coating	Surface acoustic wave, the piezoelectric material	6, 8, 12	Organic matter
Infrared	Light energy absorption	20, 22, 36	CH_4, COX, NOX, SO_2

(d) Infrared Photoelectric Sensor

The working principle of the infrared photoelectric sensor is that, in a given optical path, after the infrared passes through a different medium (a gas), the light intensity and the position and shape of the spectral peak will all change. The composition and concentration of the measured objects can then be analyzed. Its characteristic is that sensor's output is basically linear with the concentration of the gas to be measured within a specific range. However, the large size, high price, and harsh use conditions of this device restrict its application range.

These sensors' common feature is their strong sensitivity to temperature and humidity, so the influence of temperature and humidity must be strictly controlled during testing. As mentioned earlier, the performance (such as sensitivity and sensing range) of a single olfactory receptor cell in the biological olfactory system is not high, but biological olfactory system's overall performance is terrific. Similarly, we should not deliberately pursue the higher performance of a single gas sensor but combine multiple gas sensors with overlapping performance to form an olfactory sensor array. As mentioned in the first section, compared with a single gas sensor, the olfactory sensor array does not only have a broader detection range but also its sensitivity and reliability are greatly improved. Therefore, when detecting gas or odor recently, most people tend to use olfactory sensor array devices. The development trend of olfactory sensor array devices is its integration, wide monitoring range, and convenient portability. Table 3.2 lists commonly used olfactory sensor array devices and related characteristics.

3.2.3 Electronic Tongue Sensing Technology

Electronic tongue is also called artificial taste. Its working principle is based on the process of simulating the formation of biological taste. The sensation caused by chemical substances is not inherent in the chemical substance itself but occurs after the chemical substance reacts with the sensory organs. For example, the sense of taste can be regarded as caused by the physical and chemical reaction between the taste substance and the taste bud's receptive membrane. The basic idea of artificial taste includes the artificial reproduction of these reactions [50–52].

Similar to the formation of olfactory sensation, the formation process of taste sensation can also be divided into three stages: (1) The biofilm of taste cells on the surface of the tongue senses taste substances and forms bioelectric signals; (2) The bioelectric signals are transmitted to the brain through nerve fibers; (3) Brain recognition. Numerous flavors are composed of four basic tastes: sweet, salty, sour, and bitter. Some scholars abroad define the basic tastes as sweet, salty, sour, bitter, and fresh. In recent years, astringent and spicy tastes have been introduced. Usually, sourness is caused by hydrogen ions, such as hydrochloric acid, amino acids, citric acid; NaCl mainly causes salty taste; sweet taste is mainly caused by sucrose, glucose, etc.; bitter taste is caused by quinine, caffeine, etc. The umami taste is caused by monosodium glutamate (MSG) in seaweed, disodium inosinate (IMP) in fish and meat, disodium guanylate (GMP) in mushrooms. The taste of different substances is related to their molecular structure. For example, H^+ in inorganic acids is a critical factor in causing sourness, but the taste of organic acids is also related to their negatively charged acid roots; sweetness is related to the main structure of glucose; and the structure of quinine and some poisonous plant alkaloids can cause the typical bitter taste. Taste stimulating substances must have a certain degree of water solubility, be adsorbed on the surface of taste cell membranes, and react with the biofilm of taste cells to produce taste. The biofilm's main components are lipids, proteins, and inorganic ions, and a small amount of sugar and nucleic acids. For different tastes, the components involved in the reaction in the biofilm are different. Experiments have shown that when producing acid, salty, and bitter tastes, the components involved in the reaction in the biofilm of taste cells are all lipids; the proteins in the biofilm of taste cells may participate in the reaction that produces bitter taste. In the case of sweet and umami taste, only proteins are involved in the reaction in the biofilm of taste cells.

1. **Composition of the Electronic Tongue System**

    Electronic tongue is composed of three parts: taste sensor array, signal collector, and pattern recognition system. The taste sensor array is equivalent to the tongue in the biological system, feeling different components in the measured solution. The signal collector is like a neurosensory system which collects excited signal and transmits it to the computer [53]. The computer plays the role of the brain in the biological system. The software processes and analyzes the data and distinguishes different substances to obtain sensory information of different substances. Each independent sensor in the sensor array is like a taste bud on the tongue, with interactive sensitivity. Thus, an independent sensor senses not only a chemical substance but a type of chemical substance and a particular type of specific chemical substance and, at the same time, also feels a part of other chemical substances.

    In 1985, *Kiyoshi Toko and his colleagues* developed the world first electronic tongue capable of distinguishing five basic tastes. In 1995, *Andrey V. Legin et al.* constructed a new electronic tongue system with nonspecific sensors as a sensor array, and for the first time, officially proposed the name electronic tongue. In 1997, *Ingemar Lundstrom et al.* successfully developed a new electronic tongue system based on bare metal electrodes and voltammetry [54]. The voltammetry electronic tongue sensor is composed of a set of precious metal working electrode arrays instead of a single working electrode in the traditional method. In recent years, research scholars have developed a sensor array composed of eight different metal materials with overlapping properties of tin, niobium, palladium, titanium, iron, tungsten, platinum, and aluminum. A constant potential meter picks up the sensors of different metal electrodes. The taste signal is converted into a potential signal and combined with the BP algorithm's neural network pattern recognition tool, and it can identify several different fruit juice drinks, dry red wine, etc.

2. **Electronic Tongue Sensor Array**

Just as people do not need to distinguish each chemical substance when expressing taste, what artificial taste sensor measures is not the qualitative and quantitative result of a specific chemical component, but the overall information of the whole measured substance's taste. On the other hand, there are roughly more than 1000 chemical substances in food, and there are still interactions between taste substances, so it is impractical to use so many chemical sensors. As mentioned earlier, the most effective and most researched one for achieving artificial taste is the multi-channel lipid membrane taste sensor array, which can partially reproduce the response of human taste to taste substances. Among the four basic tastes (sweet, salty, sour, and bitter), bitterness is the most difficult to detect, so here we focus on the multi-channel lipid membrane taste sensor array's detection mechanism [55, 56].

Research results show that bitter substances can increase the resistance of phospholipid membranes. According to food chemistry, there are many bitter substances, mainly organic bitter substances such as quinine, brucine or nicotine, and inorganic bitter substances containing alkaline earth metal ions (Ca^{2+}, Mg^{2+}) such as halogen salts. Although they have different molecular properties, they can all cause an increase in the phospholipid membrane's resistance. For example, quinine, brucine, or nicotine are highly resistant to water. They pass through the hydrocarbon-based chain layer entering the membrane and occupy the small pores in the membrane. This compresses the lipid membrane, thereby increasing the resistance of the membrane, while phospholipid molecules easily bind the bitter substances containing alkaline earth metal ions due to alkaline earth metal ions such as Ca^{2+} and Mg^{2+}.

On the one hand, the bitter substances are in the intermolecular space of the phospholipid membrane, the lipid molecules are compressed in the narrow groove to increase the resistance of the membrane. On the other hand, the ion exchange between the bitter substance and the lipid molecules increases the membrane's resistance. Therefore, it can be considered that the increase in the impedance of the phospholipid membrane can mimic the process of bitterness in the biological, physiological system. However, the bitter taste sensing system based on phospholipid membrane impedance measurement still has the following challenges:

(a) Some taste substances that do not produce bitterness, such as sucrose and sodium glutamate (MSG), can also increase the phospholipid membrane's resistance. They may have a high affinity for the phospholipid membrane and can be adsorbed on the membrane surface. Therefore, the current sensing system cannot distinguish bitter substances from highly absorbent substances.
(b) Bitter substances with relatively low toxicity, such as caffeine, theobromine, and L-amino acids, cause a smaller increase in membrane resistance than those of highly toxic substances. The sensitivity of current sensing systems is not sufficient enough to detect low-toxic bitter substances.
(c) Although the impedance changes caused by some bitter substances are relatively large, the impedance changes caused by them are discontinuous at a specific concentration point. We call this discontinuous change a "transition," that is, the concentration and membrane impedance change show extreme strong nonlinearity. The discontinuity of this change makes it challenging to detect bitterness.
(d) $CaCl_2$ and $MgSO_4$ both contain alkaline earth metal ions, and the concentration that they cause to transition is lower than the threshold concentration for the human body to feel bitter. Except for $CaCl_2$ and $MgSO_4$, the concentration of bitter substances that cause membrane resistance transition is higher than the corresponding threshold concentration in the human body. Therefore, it is necessary to clarify the mechanism and find the innate response of bitter substances. Based on the current research situation, the multi-channel lipid membrane taste sensor array needs further research.

3.3 Signal Processing and Recognition of Imitation Sensor Signals

Pattern recognition is to identify and classify the researched objects according to their common characteristics or attributes. The characteristics of a pattern category should belong to the common attributes of the pattern. Such attributes are called intra-class characteristics and represent the characteristics of different pattern categories [57, 58]. It is an interclass feature. Pattern recognition has many branches, such as cluster analysis, neural network, and related algorithms contained therein, such as KNN, SIMCA algorithm, and SVM.

3.3.1 Response Expression of Bionic Sensing

The mathematical model of bionic sensing is very complicated. At present, only the mathematical model of artificial olfaction is expressed relatively clearly, so only the response expression of artificial olfaction is introduced here. Suppose that m gas sensors form an array, and the detection object is a mixed gas composed of h gases with different compositions and different concentrations.

The mathematical expression of the relationship between the sensitivity k_i of the ith ($i = 1, 2, \ldots, m$) gas sensor and a single chemical component j ($j = 1, 2, \ldots, h$) (concentration b_j) is:

$$k_i = G_{ij}^{W}/G_{ij}^{S} = a_{ij}\left(b_j + p_{ij}\right)^{t_j} \tag{3.12}$$

where G_{ij}^{S}, G_{ij}^{W} —The conductivity of the sensor in the standard state and working state

t_j—A constant between 0 and 1

a_{ij}—To be determined coefficient

The expression of the relationship between the total response q_i of the ith gas sensor to the mixed gas composed of various component concentrations b_j ($j = 1, 2, \ldots, h$) is:

$$q_i = a_{i0} + a_{i1}b_1^{t_1} + \cdots + a_{ij}b_j^{t_j} + \cdots a_{ih}b_h^{t_h} \tag{3.13}$$

In the formula: a_{i0}—constant term.

An array composed of m gas sensors measures a particular gas mixture to obtain a numerical vector:

$$\begin{pmatrix} q_1 \\ q_2 \\ \vdots \\ q_m \end{pmatrix} = \begin{bmatrix} a_{11} & a_{12} & \cdots & a_{1h} \\ a_{21} & a_{22} & \cdots & a_{2h} \\ \cdots & \cdots & \cdots \\ a_{m1} & a_{m2} & \cdots & a_{mh} \end{bmatrix} \begin{pmatrix} b_1^{t_1} \\ b_2^{t_2} \\ \vdots \\ b_h^{t_h} \end{pmatrix} + \begin{pmatrix} a_{10} \\ a_{20} \\ \vdots \\ a_{m0} \end{pmatrix} \tag{3.14}$$

Equation (3.13) can be considered a gas sensor response model to gas mixture; Eq. (3.14) describes the gas sensor array response model. This shows that the relationship between the measured value of the sensor and the gas concentration is very complicated, and there are many undetermined coefficients, and it is difficult to find the parameters in Formula (3.14) using conventional data processing methods. For a highly selective gas sensor array, the coefficient matrix can be simplified to a diagonal matrix, and when $h < m$, the equation has a unique solution. The system can accurately perform component analysis but requires that the mixed gas composition does not exceed m species. In fact, due to the low selectivity of the sensors that make up the sensor array and overlap in performance, the coefficient matrix is often not diagonal; that is, there are many nonzero nondiagonal elements. This allows the system to monitor different kinds of gases, but the accuracy will be reduced. The actual sensor characteristics are also affected by environmental factors such as temperature (these changes are reflected in the undetermined coefficient), which makes the response model of the gas sensor array to the gas mixture more complicated. This is also the reason why the neural network is used for data processing in the future.

3.3.2 Feature Extraction, Selection, and Normalization

In pattern recognition, feature extraction is an important issue. If all the features that can distinguish different categories are obtained from the input data, pattern recognition and classification are not difficult. But in fact, you only need to extract the most important features for distinguishing different categories to classify and calculate effectively [59, 60]. This is called feature selection. They can be divided into three types: physical features, structural features, and mathematical features. The first two features are detected by contact, visual observation, or other sensory organs. Mathematical features such as statistical mean, correlation coefficient, eigenvalues, and eigenvectors of the covariance matrix are often used for machine recognition.

1. **Feature Extraction**

   The sampled sensor output is a time series, and its steady-state response value and transient response values are the basis for extracting features. Common feature extraction methods are shown in Table 3.3. Experiments show that the relative method and the difference quotient method help compensate sensitive devices' temperature sensitivity. A logarithmic analysis is often used for concentration determination, which can linearize highly nonlinear concentration-response values. In the table, x_{ij} is the response characteristic value of the i-th sensor to the j-th gas, $V_{ij}^{\max}$ is the maximum response value of the i-th sensor to the j-th gas, and $V_{ij}^{\min}$ is the minimum response value of the i-th sensor to the j-th gas.

   The data difference between the different characteristic values of different sensors in the sensor array may be huge, sometimes by several orders of magnitude. Therefore, based on extracting sensor characteristics, the sensor response value must be normalized; that is, the sensor response characteristic value is between [0, 1]. Several common normalization methods are shown in Table 3.4. Where y_{ij} is the normalized characteristic value, $x_{ij}^{\max}$ is the maximum characteristic value of the response of the ith sensor to the jth gas, and $x_{ij}^{\min}$ is the minimum characteristic value of the response of the ith sensor to the jth gas. $\bar{x}_{ij}$, σ_{ij} are the

Table 3.3 Common feature extraction methods for gas sensor response

Method	Formula	Sensor type
Difference	$x_{ij} = \left(V_{ij}^{\max} - V_{ij}^{\min} \right)$	Metal oxide chemical resistance, SAW
Relative method	$x_{ij} = \left(V_{ij}^{\max} / V_{ij}^{\min} \right)$	Metal oxide chemical resistance, SAW
Difference quotient	$x_{ij} = \left(V_{ij}^{\max} - V_{ij}^{\min} \right) / V_{ij}^{\min}$	Metal oxide resistor, conductive polymer
Logarithm	$x_{ij} = \log \left(V_{ij}^{\max} - V_{ij}^{\min} \right)$	Metal oxide resistance

Table 3.4 Normalization method of eigenvalues

Method	Formula
General normalization (sensor normalization) expression	$y_{ij} = \frac{x_{ij} - x_{ij}^{min}}{x_{ij}^{max} - x_{ij}^{min}}$
Vector normalization (vector array normalization) expression	$y_{ij} = \frac{x_{ij}}{\sqrt{x_{1j}^2 + x_{2j}^2 + \cdots + x_{nj}^2}}$
Self-contained (auto scaling) expression	$y_{ij} = \frac{x_{ij} - \overline{x_{ij}}}{\sigma_{ij}}$

average and variance of multiple response characteristic values of the ith sensor in the jth gas.

2. **Principal Component Analysis**

Principal component analysis (PCA) is the most used method for statistical processing of artificial smell and taste data at home and abroad. This statistical method transforms the original multiple indicators into a few new uncorrelated or independent comprehensive indicators. It can simplify data and reveal the relationship between variables. The relationship and purpose of statistical interpretation provide some important information for further analysis of the nature of the population and the data's statistical characteristics. For the overall $X = (x_1, \ldots, x_p)'$, the principle of comprehensive index $(k \leq p)$ of X is proposed:

(a) $(i = 1, \ldots, k)$ is a linear function of X.
(b) The variance of the new feature value y_i $(i = 1, \ldots, k)$ is required to be as large as possible, that is, to reflect the information of the original data as much as possible.
(c) It is required that y_i $(i = 1, \ldots, k)$ are not related to each other or that there should be no duplicate information between $y_1, \ldots, y_k$ as much as possible. In this way, $y_1, \ldots, y_k$ is called the principal component of X.

The primary goal of PCA is to reduce the dimensionality of the data. There are many uses for dimensionality reduction. First, the calculation of the subsequent processing steps is reduced. Second, the noise can be reduced because the data that is not included in the first few principal components may be mostly noise, which is conducive to optimizing feature data. Third, projection to one very low-dimensional (such as two-dimensional) subspaces are useful for data visualization. It is a precise method because it has the above functions and has been widely used in simplifying artificial smell and artificial taste characteristic values.

3.3.3   Pattern Recognition Method

1. **SIMCA Method**

SIMCA (soft independent modeling of class analogy or statistical isolinear multicategory analogy) belongs to the class model method, which constructs a principal component regression mathematical model for each class, and classifies the samples [61]. The SIMCA is a pattern recognition method based on principal

Fig. 3.6 Schematic
diagram of the principle of
K-nearest neighbor method

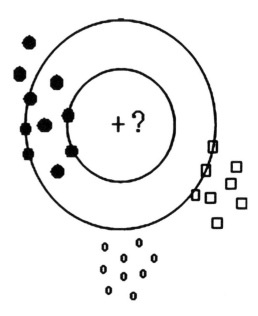

component analysis. The SIMCA pattern recognition method first conducts
principal component analysis for each type of sample's spectral data matrix,
establishes a principal component regression model, and then classifies the
unknown sample according to the model, that is, to test the unknown sample
and the class model of each sample. Fitting is done to determine the unknown
sample category.

2. ***K*-Nearest Neighbor Method**

K-nearest neighbors (KNN) is a classification method based directly on the
basic assumption of pattern recognition—similar samples are close to each other
in the pattern space. It calculates the distance between the K known samples in the
nearest neighbor and the unknown sample to be judged. This method is still
applicable even if the system under study is linearly inseparable. The KNN
method is very intuitive in terms of algorithm. In this method, all the sample
data of the training set are stored in the computer. For each unknown sample to be
discriminated, it is calculated one by one with each training sample. The nearest
K among them is then found for discrimination. The schematic diagram of its
basic principle is given in Fig. 3.6. There are three different types of samples in
the figure, which are represented by solid dots, hollow dots, and square dots. One
of the samples represented by "+" needs to be distinguished from others. In the
figure, this sample is the center of the circle, and two circles are drawn. If a small
circle is used as the boundary, there are only two solid points in this circle.
According to the principle of *K*-nearest neighbor method, it should be marked as
the solid point class. If you take the great circle as the boundary, there are
11 points (including boundary points) in this circle, of which eight are solid
point types and three are box point types. According to the *K*-nearest neighbor

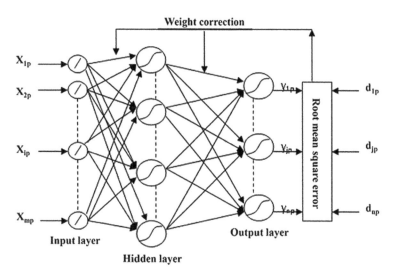

Fig. 3.7 The topology of a forward three-layer neural network

method principle, they should also be classified as solid. Click this category. Obviously, the value of K in the KNN method has a specific impact on the discrimination result. Generally, it is determined by experience. The K value can also be optimized through the method of interactive verification.

3. **Artificial Neural Networks**

Artificial neural network (ANN) works by simulating the human brain. The rise of artificial neural network information processing technology has injected vitality into the development of artificial tongue and artificial olfactory detection technology [62]. Problems such as serious linearity can suppress sensor drift or noise to a certain extent, which helps to improve the accuracy of gas detection. The current problems faced by artificial neural networks for artificial olfactory information processing are: there is still a lack of certain guidance in network construction, the training time of the network is long, especially when the characteristics of the receptor pieces are not stable enough or fatigue often cannot meet the requirements, etc. In the future, artificial smell and artificial taste will be a large-scale integration of sensor arrays and processing circuits. The hardware implementation of neural networks will be one of the first choices. Therefore, new neural network algorithms be developed and combined with other pattern recognition and signal processing methods. Overcoming the practical problems of ANN in applying artificial smell and artificial taste has become another hot spot in this field. In artificial neural networks, forward multi-layer neural networks (MLNNs) with back propagation (BP) algorithm as the mathematical model are used in pattern recognition, classification, and nonlinear mapping. They have been successfully applied in many fields such as feature extraction.

Figure 3.7 is a schematic diagram of a forward three-layer neural network's topology, which consists of an input layer, a hidden layer, and an output layer. Of course, there can be more than one hidden layer. There is no interconnection between units in the same layer, and weights connect units in adjacent layers. Assuming that the dimension of an input pattern is m, the number of nodes in the network's input layer is m. The number of output layer nodes is related to the research object. If the network is used for classification, the number of output nodes is generally equal to the number of known pattern categories. The number of hidden layer nodes can be selected as required. An input layer unit is a linear unit; that is, the neuron output of this layer is directly equal to the input. The commonly used transfer function of each unit of the hidden layer and the output layer is the Sigmoid function; that is, if the network input of the unit is x, the output is the error back propagation algorithm. The working principle of the forward multilayer neural network is to set the training set mode. The number is N, and the subscript of a certain mode is p, that is, $p = 1, 2, \ldots, N$. When each unit of the input layer receives a specific input mode $X_p = (x_{1p}, x_{2p}, \ldots, x_{ip}, \ldots, x_{mp})$, it directly outputs without any processing, and the output variables are weighted and sent to the hidden layer. Each unit of the hidden layer will output the received information after being processed by the transfer function, then sent to the output units after weighting processing, and finally, produce an actual output vector after the output units are processed. This layer-by-layer update process is called the forward process. If the error between the network's actual output and the expected target output does not meet the specified requirements, the error is transmitted layer by layer in the reverse direction, and the connection weights between the layers are corrected. This is called the error direction propagation process. For a set of training patterns, the network is continuously trained with one input pattern, and the forward process and error back propagation processes are repeated. When the error between the actual output of the network and the expected target output meets the specified requirements for the entire input training set, we say that the network has been learned or trained. Since the input of each unit of the previous layer of this network is the linear weighted sum of the outputs of all the units of the latter layer, it is also called linear basic function (Linear-Basis Function, LBF) neural network.

4. **Vector Support Machine**

Vector support machine (SVM) is developed from the optimal classification surface in linear divisibility [63]. The basic idea is illustrated in the two-dimensional situations in Fig. 3.8.

The square and circle in the figure represent class A samples and class B samples, respectively, $H((w \cdot x) + b = 0)$ is the classification line, $H_1((w \cdot x) + b = +1)$ and $H_2((w \cdot x) + b = -1)$ are respectively called the classification edge lines, the distance between them is the classification interval, and the training sample points on H_1 and H_2 are called support vector. For two types of classification problems, let $X = \{x_i\} \in R^n$ be the input space and $Y \in \{-1, 1\}$ the output domain. For the training set composed of n samples, it can be expressed as $S = ((x_1, y_1), \ldots, (x_n, y_n)) \subseteq (X \times Y)^n$.

According to the principle of minimum structural risk, an objective function is constructed to find a segmentation hyperplane that meets the requirements, and

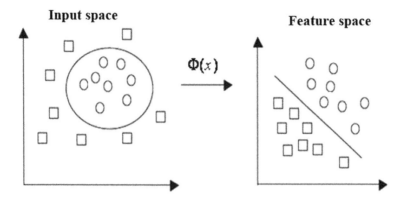

Fig. 3.8 Nonlinear separable input space and feature space

the points in the training set are as far as possible from the segmentation hyperplane, even if the blank areas on both sides are separated by the largest distance $margin = 2/\|w\|$. It is equivalent to minimizing $\|w\|$. Therefore, the SVM method transforms the unsolved pattern classification problem into a secondary planning optimization problem. In a nutshell, the support vector machine first transforms the input space into a high-dimensional space through a nonlinear transformation defined by the inner product function, and finds the (generalized) optimal classification surface in this space [64].

Different inner product kernel functions in SVM will form different algorithms. At present, the most studied kernel functions mainly include polynomial (Poly) kernel functions, radial basis function (RBF) kernel functions, and Sigmoid kernel functions. The first kind of kernel function obtains a p-order polynomial classifier; the second kind of kernel function obtains a classifier based on the radial basis function (RBF) kernel function, which is significantly different from the traditional RBF artificial neural network method [65]. The center of each basis function corresponds to a support vector, and the algorithm automatically determines their output weights; the third kernel function implements a multilayer perceptron with a hidden layer, and the number of hidden layer nodes is automatically determined by the algorithm. Certainly, the algorithm does not have the local minimum problem that plagues neural network methods.

3.4 Application in Food Quality and Safety Inspection

Bionics is the science of building technological devices that mimic living things. It is an emerging fringe discipline that emerged only in the mid-twentieth century. Bionics studies the working principles of the structure or function of living organisms (including animals and plants) and applies these principles to artificial technologies to invent new instruments, devices, and machines and create new technologies [66]. Bionics is an interdisciplinary discipline that combines life sciences and engineering disciplines such as mechanics, materials, and information. In nature, various organisms have grown and evolved over a long period of time and are highly adaptable to their survival environment. Organisms have extreme strong environmental adaptability in information processing, control and regulation, energy conversion, direction discrimination, navigation, and detection. The functions of organisms such as perception, command, decision-making, feedback, movement, and external organ structure provide inspiration for the optimal design of new products and inventions, and provide imitable objects for human beings to solve scientific and technological problems, thus creating new theories and technologies and promoting the development of bionics [67]. Modern bionics involves a variety of disciplines, such as physics, chemistry, medicine, mathematics, materials science, mechanics, dynamics, cybernetics; at the same time, the research and development of bionics has a positive role in promoting the development of mechanical engineering, medical engineering, biotechnology, tissue engineering, and other related engineering disciplines. In short, bionics provides new ideas, new theories and new methods for scientific and technological innovation. The main applications of bionic sensing technology in food inspection are vision sensor, electronic olfactory, and electronic tongue.

3.4.1   Application of Vision Sensing in Food Quality and Safety Inspection

Vision sensor uses camera means to acquire images of objects, converts analog images into digital images, and uses image recognition technology to simulate human discriminative guidelines to understand and recognize images for the purpose of analyzing images and making conclusions. Vision sensor can detect agricultural and food products' size, shape, color, surface cracks, and surface defects and damage. Its advantages are: it is fast and informative and can detect multiple quality indicators at once [68]. The use of vision sensor technology can overcome human eye discrepancies and visual fatigue and can also enable nondestructive testing.

Jianrong Cai et al. used vision sensor to develop a tobacco quality sorting system, which calibrated the collection system and controlled the sensitivity as shown in [69]. A total of 180 feature parameters were extracted and selected to form a feature vector to remove odd samples from the standard samples. An artificial neural network was used to learn and classify tobacco leaves from several regions with a detection accuracy of over 80%. Among the features of tobacco leaf size and shape are mainly area, length, residual injury, and distribution features after unfolding the contour line of the tobacco leaf.

Color is one of the most important indicators for determining the quality of agricultural and food products. Rice variety classification is currently detected manually, which is a huge workload, and the results are highly subjective and poorly consistent. *Xing Yi Huang et al.* studied the use of visual sensing technology to replace the human eye for rice variety recognition [70]. The combination of color features and shape features of images was carried out, and the recognition classifier was established by Bayesian decision method, and the correct rate of rice variety recognition reached over 88.3%.

3.4.2 Application of Electronic Nose Technology in Food Quality and Safety Inspection

The electronic nose consists of functional devices such as data acquisition processors, gas-sensitive sensor arrays, and pattern recognition systems, which perceive, analyze, and judge odors by simulating the human olfactory organs. The gas-sensitive sensor converts chemical signals from chemical reactions with different odorant substance molecules into electrical signals [71, 72]. The instrument further processes the electrical signal, extracts its features, classifies the odor information obtained by the odor sensor array through multivariate statistical methods, and finally realizes the differentiated discrimination of the detected samples. The electronic nose has the advantages of fast detection speed, good ease of use, and low detection cost.

Guo et al. applied the electronic nose system and other methods to detect and predict the effectiveness of spoilage or infestation of *Penicillium expansum* on apple storage [73]. Based on the acquisition of electronic nose signals, selected sensitive feature sensors of spoilage apple and all sensors were analyzed and compared by the recognition effect. PCA, PCA-DA, LDA, PLS-DA, and KNN were used to establish the classification model of apple with different degrees of corruption. PCA-DA has the best prediction, the accuracy of training set and prediction set was 100% and 97.22%, respectively. Synergy interval (SI), GA, and CARS are three selection methods used to accurately and quickly extract appropriate feature variables, while constructing a PLS model to predict plaque area. Among them, the PLS model with unique variables was optimized by CARS method, and the best prediction result of the area of the rotten apple was obtained. The best results are as follows: $R_C = 0.953$, root mean square error of calibration (RMSEC) $= 1.28$, $R_P = 0.972$, root mean square error of prediction (RMSEP) $= 1.01$ (Fig. 3.9).

3.4.3 Application of Electronic Tongue in Food Quality and Safety Inspection

In the taste inspection of food, the current main method is to rely on professional and technical personnel for taste assessment. The judging results contain human factors, which are extremely difficult to grasp and not reproducible, making it difficult to meet the requirements of the trend of high-volume and automated production in the

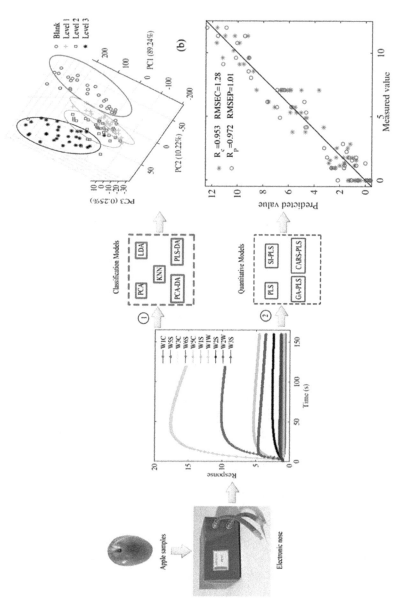

Fig. 3.9 Schematic procedures of classification and prediction of defects of *Penicillium expansum* in apples by electronic nose combined with chemometrics

food industry [74]. Although sensitive testing of taste can be achieved by some instruments, such as pH meter for acidity detection, conductivity meter for salinity detection, specific gravity meter or refractometer for sweetness detection, but because there are interactions between various odors, such as the contrast of taste, change of tone, phase into and phase kill phenomena, so these measured physical and chemical characteristic parameters are not true and comprehensively reflect the taste characteristics of the sample.

Electronic tongue technology is a kind of electronic sensor for analyzing and recognizing liquid composition, which mainly consists of sensor array and pattern recognition system [75]. The sensor array responds to the liquid sample and outputs a signal, which is processed by a computer system to obtain a result that reflects the taste characteristics of the sample. This technology is also known as taste sensor technology or artificial taste recognition technology. This is different from the common chemical analysis method in that the sensor does not output the results of the composition of the sample, but rather outputs different signals to distinguish between the different compositions, and the output signals rely on computer processing.

Qin Ouyang et al. investigated the potentiometric determination of free amino acids in black tea by combining cyclic voltammetry electron tongue (CVET) with chemometrics as shown in Fig. 3.10 [76]. Two working electrodes—glassy carbon electrode and platinum electrode—were comparative and syncretic in the prediction of total FAA content. Si-PLS combined with variable combination population analysis (VCPA), namely Si-VCPA-PLS, was creatively employed for searching characteristic variables selection from the whole cyclic voltammetry data set. Cyclic voltammetry signals coupled with Si-VCPA-PLS model results represented its feasibility in estimating total FAA content in black tea, achieved with $R_P = 0.8185$ for the glassy carbon electrode; $R_P = 0.8366$ for the platinum electrode; and $R_P = 0.8414$ for their data fusion. The extracted characteristic variables were only occupying about 0.2% of the origin, which confirmed that the cyclic voltammetry electronic tongue could be as a fast, low-cost, efficient, and complementary approach for predicting total FAA content in black tea.

Chen Quansheng et al. attempted to detect the total content of theaflavin in black tea using a portable electronic tongue with integrated chemical measurement algorithm [77]. Glassy carbon as a working electrode was used to collect the cyclic voltammetry current signals from the black tea samples. The Si-CARS-PLS was attempted to optimize and select the most informative current signal variables at special potentials for the prediction of the total theaflavins content in black tea. Models were optimized via cross-validation. Compared with other characteristic variable selection methods, Si-CARS-PLS showed the best performance, employing only 13 variables (0.23% of the total variables), to achieve $R_P = 0.8302$ and RMSEP $= 0.257$ in the prediction set. The portable electronic tongue based on glassy carbon electrode and cyclic voltammetry combined with variable selection algorithm Si-CARS-PLS proved a promising, rapid, and cost-effective method to measure the total theaflavins content in black tea (Fig. 3.11).

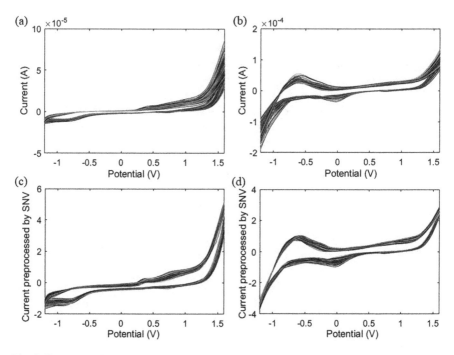

Fig. 3.10 Raw cyclic voltammograms from glassy carbon (**a**) and platinum electrode (**b**) of the black tea samples; and SNV preprocessed cyclic voltammograms from glassy carbon (**c**) and platinum electrode (**d**) of the black tea samples

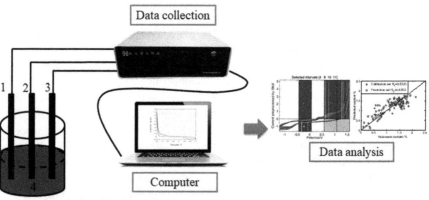

1. The working electrode(i.e., glassy carbon)
2. The Ag/AgCl reference electrode
3. The platinum wire counter electrode
4. The black tea sample infusion

Fig. 3.11 The scheme of detecting the total content of theaflavin in black tea using a portable electronic tongue with integrated chemical measurement algorithm

3.5 Conclusion and Future Trends

Bionic sensors technology does not require complex preprocessing of samples and can detect different kinds of food products, while avoiding human judgment errors and good repeatability, enabling simple, fast, and nondestructive online detection of the overall characteristics of unknown liquid samples and precise differentiation of samples in a short time. In addition, with the development of modern technology, the bionic sensors intelligent detection system will play an increasingly important role in many fields of the food industry in combination with biochips and nanomaterials. Although some research results have been achieved, there is still a lot of work to be done in depth. (1) The current equipment also has to operate under the control of a microcomputer with a cumbersome structure. The miniaturization of equipment is the key to transforming research results into productivity. (2) Artificial neural network is a complex algorithm where the number of intermediate layers, the number of neurons in the intermediate layers, the functions on the neurons, the learning training function of the network, and the performance function of the network all have an impact on the operation of the artificial neural network. In this study, it was found that changing different network training functions, i.e., using different iterative algorithms, can have an impact on both the training process and the results. A faster training speed (fewer training steps) does not necessarily mean better network performance. (3) The number of sensors that make up the sensor array is also important. Obviously, the more sensors that make up the array, the larger the amount of data, the more accurate the results will be; at the same time, the increase in the amount of data will make computing difficult and reduce efficiency.

References

1. Ai Y, Lou Z, Chen S et al (2017) All rGO-on-PVDF-nanofibers based self-powered electronic skins. Nano Energy 35:121–127
2. Cai G, Wang J, Lin M et al (2017) A semitransparent snake-like tactile and olfactory bionic sensor with reversibly stretchable properties. NPG Asia Mater 9(10):e437–e437
3. Cao A, Wang J, Pang H et al (2018) Design and fabrication of a multifocal bionic compound eye for imaging. Bioinspir Biomim 13(2):026012
4. Chang L, Liu Y, Yang Q et al (2018) Ionic electroactive polymers used in bionic robots: a review. J Bion Eng 15(5):765–782
5. Chang Z, Sun Y, Zhang Y et al (2018) Bionic optimization design of electronic nose chamber for oil and gas detection. J Bion Eng 15(3):533–544
6. Chen BD, Tang W, He C et al (2018) Water wave energy harvesting and self-powered liquid-surface fluctuation sensing based on bionic-jellyfish triboelectric nanogenerator. Mater Today 21(1):88–97
7. Chen C, Wang C, Wang Y et al (2017) Fuzzy logic controller design for intelligent robots. Math Probl Eng 2017:1–12
8. Chen J, Chen B, Han K et al (2019) A triboelectric nanogenerator as a self-powered sensor for a soft–rigid hybrid actuator. Adv Mater Technol 4(9):1900337

9. Chen T, Wei S, Cheng Z et al (2020) Specific detection of monosaccharide by dual-channel sensing platform based on dual catalytic system constructed by bio-enzyme and bionic enzyme using molecular imprinting polymers. Sensors Actuators B Chem 320:128430

10. Chen X, Lin X, Mo D et al (2020) High-sensitivity, fast-response flexible pressure sensor for electronic skin using direct writing printing. RSC Adv 10(44):26188–26196

11. Deng H, Zhong G, Li X et al (2017) Slippage and deformation preventive control of bionic prosthetic hands. IEEE/ASME Trans Mechatron 22(2):888–897

12. Ding L, Wang Y, Sun C et al (2020) Three-dimensional structured dual-mode flexible sensors for highly sensitive tactile perception and noncontact sensing. ACS Appl Mater Interfaces 12 (18):20955–20964

13. Du T, Li X, Wang Y et al (2019) Multiple disturbance analysis and calibration of an inspired polarization sensor. IEEE Access 7:58507–58518

14. Du W, Yang Y, Liu L (2020) Research on the recognition performance of bionic sensors based on active electrolocation for different materials. Sensors (Basel) 20(16):4608

15. Ferri S, Sode K (2012) Biomolecular engineering of biosensing molecules—the challenges in creating sensing molecules for glycated protein biosensing—. Electrochemistry 80(5):293–298

16. Guo SZ, Qiu K, Meng F et al (2017) 3D printed stretchable tactile sensors. Adv Mater 29 (27):1701218

17. Guo X, Yang J, Du T et al (2019) An autonomous navigation system integrated with air data and bionic polarization information. Trans Inst Meas Control 41(13):3679–3687

18. Guo Z, Guo C, Chen Q et al (2020) Classification for penicillium expansum spoilage and defect in apples by electronic nose combined with chemometrics. Sensors (Basel) 20(7):2130

19. Gupta S, Kumar A (2018) Bionic functionality of prosthetic hand, intelligent communication. In: Control devices, pp 1177–1190

20. Han S, Zhao J, Wang D et al (2017) Bionic ion channel and single-ion conductor design for artificial skin sensors. J Mater Chem B 5(34):7126–7132

21. Hao Q, Wang Z, Cao J et al (2018) A hybrid bionic image sensor achieving FOV extension and foveated imaging. Sensors (Basel) 18(4):1042

22. Haruyama T (2006) Cellular biosensing: chemical and genetic approaches. Anal Chim Acta 568 (1–2):211–216

23. Hu G-D, He C-L, Li H-Y et al (2018) Test on stress distribution of bionic C-leg wheel-soil interaction with its data processing. IOP Conf Ser Mater Sci Eng 428:012010

24. Hu Y, Li Z, Li G et al (2016) Development of sensory-motor fusion-based manipulation and grasping control for a robotic hand-eye system. IEEE Trans Syst Man Cybernet Syst 47:1169–1180

25. Huang H, Han L, Fu X et al (2020) Multiple stimuli responsive and identifiable zwitterionic ionic conductive hydrogel for bionic electronic skin. Adv Electron Mater 6(7):2000239

26. Jiang B (2020) Research on wireless sensor location technology for biologic signal measuring based on intelligent bionic algorithm. Peer Peer Netw Appl

27. Jiang C, Li Q, Sun N et al (2020) A high-performance bionic pressure memory device based on piezo-OLED and piezo-memristor as luminescence-fish neuromorphic tactile system. Nano Energy 77:105120

28. Jiang J, Sun Z, Duan F et al (2018) Disguised bionic sonar signal waveform design with its possible camouflage application strategy for underwater sensor platforms. IEEE Sensors J 18 (20):8436–8449

29. Li H, Ren Y, Zhang G et al (2019) Design of a high SNR electronic heart sound sensor based on a MEMS bionic hydrophone. AIP Adv 9(1):015005

30. Li J, Zhou X, Liu Z (2020) Recent advances in photoactuators and their applications in intelligent bionic movements. Adv Opt Mater 8(18):2000886

31. Li X, Jiang C, Zhao F et al (2020) A self-charging device with bionic self-cleaning interface for energy harvesting. Nano Energy 73:104738

32. Liao X, Song W, Zhang X et al (2020) A bioinspired analogous nerve towards artificial intelligence. Nat Commun 11(1):268

33. Liao X, Wang W, Wang L et al (2021) A highly stretchable and deformation-insensitive bionic electronic exteroceptive neural sensor for human-machine interfaces. Nano Energy 80:105548
34. Lindau ST, Bensmaia SJ (2020) Using bionics to restore sensation to reconstructed breasts. Front Neurorobot 14:24
35. Liu F, Wang K, Liu Y et al (2019) A bionic vibration source localization device inspired by the hunting localization mechanism of scorpions. J Bion Eng 16(6):1019–1029
36. Liu G, Wang M, Xu L et al (2020) A new bionic lateral line system applied to pitch motion parameters perception for autonomous underwater vehicles. Appl Ocean Res 99:102142
37. Liu J, Cui Y, Chen Y et al (2018) Evaluation of food fineness by the bionic tongue distributed mechanical testing device. Sensors (Basel) 18(12):4250
38. Liu Y, Zhu Y, Liu J et al (2018) Design of bionic cochlear basilar membrane acoustic sensor for frequency selectivity based on film triboelectric nanogenerator. Nanoscale Res Lett 13(1):191
39. Lü X-Z (2017) A multi-scale flexible tactile-pressure sensor. In: Wearable sensors and robots, pp 49–54
40. Magee EG, Ourselin S, Nikitichev D et al (2017) The bionic clicker Mark I & II. J Vis Exp (126):55705
41. Miao Y, Xu M, Yu J et al (2021) Conductive cold-resistant and elastic hydrogel: a potential bionic skin for human-machine interaction control over artificial limbs. Sensors Actuators B Chem 327:128916
42. Michelini E, Cevenini L, Mezzanotte L et al (2010) Bioluminescent genetically engineered cells for biosensing applications. J Biotechnol 150:194–194
43. Moore P (2017) Review. Dawn Nafus (ed.), Quantified: biosensing technologies in everyday life. Theor Cult Soc 34(7–8):269–275
44. Nie P, Wang R, Xu X et al (2017) High-performance piezoresistive electronic skin with bionic hierarchical microstructure and microcracks. ACS Appl Mater Interfaces 9(17):14911–14919
45. Ouyang Q, Yang Y, Wu J et al (2020) Measurement of total free amino acids content in black tea using electronic tongue technology coupled with chemometrics. LWT Food Sci Technol 118:108768
46. Ouyang Q, Yang Y, Wu J et al (2019) Rapid sensing of total theaflavins content in black tea using a portable electronic tongue system coupled to efficient variables selection algorithms. J Food Compos Anal 75:43–48
47. Pagaduan JV, Bhatta A, Romer LH et al (2018) 3D hybrid small scale devices. Small 14(27): e1702497
48. Pan JY, Lin CC, Wu CJ et al (2016) Early and intermediate-term results of the extracardiac conduit total cavopulmonary connection for functional single-ventricle hearts. J Formos Med Assoc 115(5):318–324
49. Patil AC, Xiong Z, Thakor NV et al (2020) Toward nontransient silk bioelectronics: engineering silk fibroin for bionic links. Small Methods 4(10):2000274
50. Peng S, Wu S, Yu Y et al (2020) Multimodal capacitive and piezoresistive sensor for simultaneous measurement of multiple forces. ACS Appl Mater Interfaces 12(19):22179–22190
51. Preechaburana P, Suska A, Filippini D (2014) Biosensing with cell phones. Trends Biotechnol 32(7):351–355
52. Qian C, Li X, Zhu J et al (2019) A bionic manipulator based on multi-sensor data fusion. Integr Ferroelectr 192(1):10–15
53. Su K, Qiu X, Fang J et al (2017) An improved efficient biochemical detection method to marine toxins with a smartphone-based portable system—bionic e-eye. Sensors Actuators B Chem 238:1165–1172
54. Syed Mubarak Ali SAA, Ahmad NS, Goh P (2019) Flex sensor compensator via Hammerstein-Wiener modeling approach for improved dynamic goniometry and constrained control of a bionic hand. Sensors (Basel) 19(18):3896
55. Syu Y-C, Hsu W-E, Lin C-T (2018) Review—field-effect transistor biosensing: devices and clinical applications, ECS. J Solid State Sci Technol 7(7):Q3196–Q3207

56. Tang W, Zhang K, Jiang D (2017) Physarum-inspired routing protocol for energy harvesting wireless sensor networks. Telecommun Syst 67(4):745–762
57. Tavakoli M, Lopes P, Lourenco J et al (2017) Autonomous selection of closing posture of a robotic hand through embodied soft matter capacitive sensors. IEEE Sensors J 17 (17):5669–5677
58. Ulusan H, Muhtaroglu A, Kulah H (2019) A sub-500 μW interface electronics for bionic ears. IEEE Access 7:132140–132152
59. Wan Y, Han Z, Zhong J et al (2018) Pattern recognition and bionic manipulator driving by surface electromyography signals using convolutional neural network. Int J Adv Robot Syst 15 (5)
60. Wang C, Sun Z, Long J et al (2020) Development of a novel massage platform for medical training. Technol Health Care 28(S1):89–101
61. Wang F, Ren Z, Nie J et al (2019) Self-powered sensor based on bionic antennae arrays and triboelectric nanogenerator for identifying noncontact motions. Adv Mater Technol 5 (1):1900789
62. Wang J, Suzuki R, Shao M et al (2019) Capacitive pressure sensor with wide-range, bendable, and high sensitivity based on the bionic Komochi Konbu structure and Cu/Ni nanofiber network. ACS Appl Mater Interfaces 11(12):11928–11935
63. Wang J, Tenjimbayashi M, Tokura Y et al (2018) Bionic fish-scale surface structures fabricated via air/water interface for flexible and ultrasensitive pressure sensors. ACS Appl Mater Interfaces 10(36):30689–30697
64. Wang Q, Yu P, Bai L et al (2017) Self-assembled nano-leaf/vein bionic structure of TiO$_2$/MoS$_2$ composites for photoelectric sensors. Nanoscale 9(46):18194–18201
65. Wang S, Qu C, Liu L et al (2019) Rhinophore bio-inspired stretchable and programmable electrochemical sensor. Biosens Bioelectron 142:111519
66. Wang X, Wang H, Cai Y et al (2017) Research on a beef tenderness detection method using a bionic mastication system based on a pressure sensor. Anal Methods 9(32):4695–4701
67. Wang Y, Wang H, Wang H et al (2019) Calcium gluconate derived carbon nanosheet intrinsically decorated with nanopapillae for multifunctional printed flexible electronics. ACS Appl Mater Interfaces 11(22):20272–20280
68. Wei Y, Li S, Zhang X et al (2020) Smart devices based on the soft actuator with nafion-polypropylene-PDMS/graphite multilayer structure. Appl Sci 10(5):1829
69. Wen H-Y, Liu Y-C, Chiang C-C et al (2020) The use of doped conductive bionic muscle nanofibers in a tennis racket–shaped optical fiber humidity sensor. Sensors Actuators B Chem 320:128340
70. Yang B, Zhang T, Liang Z et al (2019) Research on an artificial lateral line system based on a bionic hair sensor with resonant readout. Micromachines (Basel) 10(11):736
71. Yang J, Du T, Niu B et al (2018) A bionic polarization navigation sensor based on polarizing beam splitter. IEEE Access 6:11472–11481
72. Yue Z, Ye X, Liu S et al (2019) Towards ultra-wide operation range and high sensitivity: graphene film based pressure sensors for fingertips. Biosens Bioelectron 139:111296
73. Zhang C (2019) Simulation analysis of bionic robot fish based on MFC materials. Math Probl Eng 2019:1–9
74. Zhang C, Dai K, Liu D et al (2020) Ultralow quiescent power-consumption wake-up technology based on the bionic triboelectric nanogenerator. Adv Sci (Weinh) 7(12):2000254
75. Zhang X, Xu Q, Zhang G et al (2018) Design and analysis of a multiple sensor units vector hydrophone. AIP Adv 8(8):085124
76. Zhou L, He X, He D et al (2011) Biosensing technologies for Mycobacterium tuberculosis detection: status and new developments. Clin Dev Immunol 2011:193963
77. Zou Y, Tan P, Shi B et al (2019) A bionic stretchable nanogenerator for underwater sensing and energy harvesting. Nat Commun 10(1):2695

Chapter 4
Computer Vision Technology in Food

What Is Computer Vision Technology?

Computer vision technology is an explicit and meaningful description of an image from a physical object in visible light. It is the main task of computer vision technology to acquire images employing physical image sensors and special computer software and hardware and to analyze and process images with purpose. Appearance feature evaluation is a commonly used evaluation method of food quality, including food and agricultural products' appearance, size, color evaluation, texture feature analysis, defect and damage analysis, and the analysis in the composition. In the 1960s, there was an increase in computer vision uses in a variety of fields: medical diagnostic imaging, factory automation, remote sensing, automated vehicle, and robot guidance. Color imaging is an explicit and meaningful description of images from physical objects within the visible light range. Computer vision technology's main task is to obtain images through physical image sensors and special computer software and hardware and analyze and process images with purpose. As expectations of food quality and safety standards rise, so does the need for accurate, rapid, and objective measurements of these characteristics in food. Here we will focus the measurements by computer vision overview and basic principle, and computer image processing method. Published scientific research application cases to the reader in the use of color measurements by computer vision detection of food quality and safety of ideas and reference for the scientific research and practical application. The technologies and applications in this chapter are shown in Fig. 4.1.

Abbreviation	Full name
ALU	Arithmetic/logic unit
BP	Back propagation
CCD	Charge-coupled device
CMOS	Complementary metal oxide semiconductor
LUT	Look-up-table

(continued)

Abbreviation	Full name
LDA	Linear discriminant analysis
PCA	Principal component analysis

4.1 Introduction

4.1.1 Brief Introduction of Computer Vision for Food Quality Detection

The technology of computer vision obtain the images of objects in the wavelength range of visible light to analyze their appearance and color characteristics. Computer vision technology provides an automated, nondestructive, low-cost, and effective alternative to these requirements. This detection method based on image analysis and processing has been widely used in the food industry. Appearance size is an essential basis for the classification of food and agricultural products. Computer vision technology combined with the image processing method could determine measured objects' information by detecting the contour dimension parameters such as perimeter, area, socket orientation pore volume ratio. Besides, product shape is also an essential aspect of appearance size detection, especially for fruits, whose shape is an essential indicator of classification. Color is an essential feature in the quality evaluation of food and agricultural products. Using visible light imaging system to evaluate its color could overcome the fatigue and difference of human eyes. The defects of fresh peach could be identified by the difference in color characteristics, mildew soybean, etc. Texture feature could also be used to analyze and evaluate the quality of food, and the texture of images can be used to evaluate the quality of puffed food. There is a strong correspondence between the texture feature and the tenderness of beef. The quality of beef could be analyzed by combining the

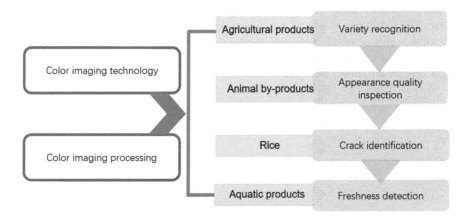

Fig. 4.1 The related technologies and applications

conventional color imaging technology with the texture analysis method. The detection of surface defects and damage has always been problematic for food and agricultural product classification. By combining computer vision technology with corresponding image processing methods, these features could be rapidly detected. For example, computer vision technology could also detect the damage and defects in fruit, the bursting waist rate of rice particles, and the appearance of worm-eaten defects. Computer vision technology also has many applications in detecting food and agricultural products, such as determining the lysine content based on the color characteristics of digital images at the bottom of the pizza. The digital image processing technology has been used to analyze the characteristics of bread cross-section size uniformity and the thickness of wall and brightness of bread core and to judge the role of emulsifier in bread baking. Visible light imaging could also be used to evaluate cheese melting and browning. The new contactless method is used to analyze the properties of cheddar and mozzarella during cooking. The results show that this method could provide an objective and straightforward way to explore cheese functional properties.

After more than 30 years of development, computer vision has developed from a simple visual simulation to replace and interpret human visual information. At the same time, due to the rapid growth of sensing technology and people's in-depth understanding of food materials, infrared and near-infrared imaging technology and its image processing research, the computer vision technology from the appearance of vision to the internal properties of food materials, components, and other directions of the research. Besides, with the help of 3D visualization technology, the appearance, nutritional, and cooking quality of rice could be observed and measured more intuitively and objectively [1, 2]. Especially in the evaluation of nutrient distribution density, the analysis of white rice internal structure, and the observation of the change of white rice structure during cooking, the results that are difficult to obtain by traditional research methods have been obtained.

4.1.2   Image Processing in Computer Vision

As the main body of computer vision technology, computer image processing characteristics represent the main peculiarity of computer vision technology. Image processing could be divided into two classifications: analog image processing and digital image processing. Analog image processing including optical lens processing, photography, broadcast television production, and so on, all belong to real-time image processing. The processing speed is fast, but the accuracy is low, the flexibility is poor, and there is no judgment and nonlinear processing ability. Digital image processing is mostly realized by computer software or use real-time hardware processing. Digital image processing could be complicated nonlinear processing and thus has the flexible ability. The disadvantage is that the processing speed is slow, and the image processing precision is higher; the more complex the processing speed is slower. Digital image processing technology is one of the most effective methods to realize real-time monitoring, detection, and product classification. With the further development of computer hardware and software, scientific research on thinking, pattern recognition, machine vision system, and other related technologies, this technology's existing problems are gradually overcome. The following mainly discusses the characteristics of digital image processing technology [3]. Digital image processing technology mainly shows the following characteristics: (1) digital image contains a large amount of data, digital image data occupies a large storage capacity. In digital imaging, transmission, storage, processing, display, and other links occupy a wide frequency band. For black and white images in digital imaging processing, 1 byte could store 8 pixels; for 16 grayscale images, 1 byte could store 2-pixel points; for a 256 grayscale image, 1 byte could store 1-pixel point. In the present image processing applications, most of the 256 grayscale images are used. A 256 grayscale image with a resolution of 600×800 requires 480 kB of storage. (2) Image processing technology integrated robust digital image processing is a multidisciplinary technical discipline in optics, electronics, computer science, artificial intelligence, pattern recognition, and photography. Its theoretical and technical system is extensive and complex.

4.2 Instrumentation

The standard color optical imaging system comprises three parts: image acquisition component, image processing component (computer), and image output component. Figure 4.2 is a typical visible light imaging system schematic diagram.

4.2.1 Image Acquisition

The image acquisition component is also the image digital input equipment. The image acquisition component converts the three-dimensional solid objects into a pair of the two-dimensional plane image and converts it into suitable for input computer or digital equipment's digital information. This process includes photoelectric conversion, image uptake, and image digitization, and other steps.

The digital camera is usually used to obtain digital image information in nondestructive testing of food. A camera that converts an optical image into electronic data by a photosensitive CCD or CMOS. CCD technology has developed early and matured. As only one or a few signal output nodes are read out uniformly, the signal output has a good consistency and low noise. However, the signal bandwidth of the output amplifier must be broad, and the power consumption and speed are slow. In the CMOS chip, each pixel has its signal amplifier, which makes the signal output consistency low, so the image has a high noise point and low power consumption, fast reading speed, and other advantages. CCD is also sensitive to near-infrared, the response spectrum extends to about 1.0 μm, and the imaging sensitivity of CCD is

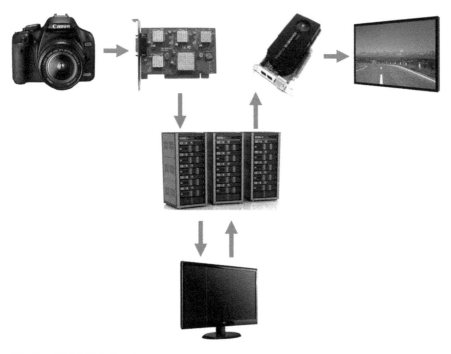

Fig. 4.2 Visible light imaging system

also excellent. High-resolution CCD cameras are very mature, and 600-line cameras are usually used in image processing. Rapid storage technology has been developing rapidly in recent years, and a large amount of image information could be accessed quickly [4].

4.2.2   Image Processing

Digital image processing is a huge amount of information and requires high processing speed of the image processing system's hardware. For point processing, a fast hardware pipeline processor is adopted, composed of an ALU and an LUT, and is used for real-time addition, subtraction, multiplication, division, logical operation, and gray transformation. For domain processing, a fast real-time small core volume machine, consisting of multiplier, accumulator, shift register, and lookup table, is applied for real-time convolution filtering, noise removal, enhancement, smoothing, and edge extraction [5]. For large-domain processing, a fast array machine, equipped with a digital signal processor could be used for fast Fourier transform, matrix operation, and vector operation [6].

4.2.3 Image Output

Image processing ultimate goal is to provide a more easily interpreted and recognized image for humans or machines. Therefore, the image output is also an essential part of image processing. There are two kinds of image output hard to copy, the other is a soft copy. Its resolution has been improved with the development of science and technology, so far, it has 2048 × 2048 high-resolution display equipment. The common hard copy methods include photography, laser copy, color inkjet printing, and so on. Soft copy methods include CRT display, liquid crystal display, field luminescence display, and so on [7].

4.3 Image Processing and Data Analysis

Image processing technology refers to acquiring and analyzing image information by computer to achieve the desired results. The complete image processing technology could be divided into image digitization, digital image storage, color image processing, image filtering, digital image segmentation, binary image morphological analysis, image shape feature analysis, image texture feature analysis, and other aspects.

4.3.1 Representation and Processing of Digital Images

1. **Color Model of Digital Images**
   In the image analysis of food and agricultural products, color features are often used to judge food quality. Color information is the most commonly used feature in the appearance quality inspection of food and agricultural products, such as the rot of fruit, surface stains, maturity, yellowing of grain food, quality control of baked food processing. In digital image processing of color features, the photometry and chromaticity knowledge foundation are required. In chromaticity, color is usually defined as a sensory image transmitted through the eye, namely vision. The three-color models commonly used in image processing in food engineering are RGB, CMY, and HIS.

   (a) RGB Model
       The RGB color model is based on the principle of the three primary colors. The RGB color model will be red, green, blue, three primary colors in different proportions, to generate a color perception of color images the model as shown in the picture. The RGB color model is a three rectangular coordinate color system with a cube unit, on the main diagonal of the cube, the amount of the primary colors is equal, from dark to bright white, gray. (0,0,0) is black, (1,1,1) is white, and the coordinates of the other six angles of

Fig. 4.3 RGB color model

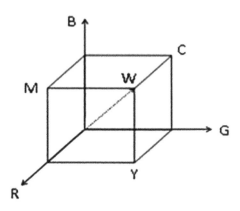

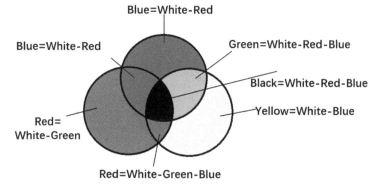

Fig. 4.4 CMY color model

the cube represent red, yellow, green, cyan, blue, and magenta, respectively, as shown in Fig. 4.3.

(b) CMY Model

CMY color model is a color model composed of complementary green, magenta, and yellow as the primary colors of red, green, and blue. It is often used to filter specific colors from white light, so it is called the subtractive primary color space, as shown in Fig. 4.4. The rectangular coordinate corresponding to the CMY color model is almost identical to that corresponding to the RGB color model. The subtracting effect of CMY in primary color is shown in the image. The CYM color model could be transformed by RGB color model, and the transformation formula is as follows:

$$C = 255 - R; \quad M = 255 - G; \quad Y = 255 - B \tag{4.1}$$

(c) HSI Model

The H stands for hue in the HIS model, S stands for saturation, and I stands for light intensity. The Hue is a color attribute that describes a solid color (yellow, orange, or red), while saturation provides a measure of how diluted the solid color is by white light. The importance of HSI color lies in two aspects. First, the relationship between strength component I and color information in the image is removed. Second, hue and saturation are similar to how the human eye perceives color. These features make the HSI model an ideal tool to study the algorithm of image processing. Therefore, the HSI model is often used in computer vision systems in food engineering, and the same HSI model could also be obtained through RGB model transformation [8].

2. **Image Filtering**

Image filtering suppresses image noise under preserving as much detail as possible and plays an essential important role in image processing. The processing effect will directly affect the effectiveness and reliability of subsequent image processing and analysis [9].

Linear filtering is the most basic image processing method, and it could allow us to image processing and produce a lot of different effects. First, there needs to be a two-dimensional filter matrix (convolution kernel) and a two-dimensional image to be processed. Then, for each pixel point of the image, its neighborhood pixel and the corresponding element of the filter matrix are calculated. Then, it adds to the pixel position's value.

The operation of multiplying images and filtering matrices element by element and summation is equivalent to moving a two-dimensional function to all locations of another two-dimensional part. This operation is called convolution or correlation. Convolution and correlation's distinction are that convolution requires the filtering matrix to be flipped 180 times. Still, if the matrix is symmetric, then there is no difference.

Correlation and convolution are the most basic operations in image processing but are very useful. The two functions have two very critical characteristics: they are linear and have translational invariant shift-invariant. Translation invariance means that we perform the same operation at every position of the image. Linearity implies that the procedure is linear. That is when replacing the pixel with a linear combination of the neighborhoods of each pixel. These two properties make this operation very simple because linear functions are the simplest.

There are related non-hard constraints for filters: the filter size should be technical, such as 3×3, 5×5, or 7×7 and so on, to ensure that the filter has a center. If the sum of all the filter matrix elements is greater than 1, then the filtered image will be brighter than the original image. If the sum is 0, the image will become very dark. The filtered structure values range from 0 to 255. By changing the size and weight of the filter matrix, the final filter image could be varied. For example, the following image processing could be carried out:

(a) Sharpening (as shown in the convolution kernel below)

$$\begin{pmatrix} -1 & -1 & -1 \\ -1 & 9 & -1 \\ -1 & -1 & -1 \end{pmatrix}$$

The simple linear filter that makes the image look sharper by enhancing the image's edges. Similarly, the convolution kernel could be magnified to achieve a finer sharpening effect.

(b) Edge detection

$$\begin{pmatrix} -1 & -1 & -1 \\ -1 & 7 & -1 \\ -1 & -1 & -1 \end{pmatrix}$$

The sum of the convolution kernel matrix elements is -1, so the resulting image becomes very dark and highlights the edges.

3. **Segmentation of Digital Images**

An image could be divided into areas based on specific colors, geometric shapes, textures, and other features. The image must be decomposed into a series of nonoverlapping areas for image analysis convenience. This operation is called image segmentation. Image segmentation is one of the early processing in the process of image analysis and understanding. In digital image processing, image segmentation is defined as separating a target from an image. Sometimes, segmentation is also called target isolation. Although image segmentation has nothing in common with human visual experience, it plays an essential role in digital image analysis.

(a) Threshold segmentation of image

Using threshold image segmentation is a technique of region segmentation. The threshold method is especially useful for distinguishing target and background. Assuming that the target is placed on a background with a clear contrast and using the threshold rule, the gray value of each pixel could be obtained comparing with the threshold 'T'. The area with pixel value smaller than or equal to the threshold value 'T' would be considered as the background, and the area with pixel value larger than the threshold value 'T' could be considered as the target. The selection of optimal threshold is an essential and challenging task in image analysis, and the histogram is the basis for selecting an optimal point. In the histogram, the image of the gray value of frequency is calculated and drawn into the figure, background and target have obvious gray difference of image, histogram twin peaks shaped, screening optimization threshold's job is to choose between peak a gray value, so that the target area and background could be separated by thresholding. Otherwise, it will cause error in the subsequent measurement and analysis of images.

If the object of interest has a uniform and consistent gray value within it and is distributed on a uniform background with another gray matter, the threshold method works well. If the object differs from some property background rather than grayscale value (such as texture, etc.), first, the property could be converted to grayscale, and then the image to be processed could be segmented by the grayscale thresholding technique.

Usually, two techniques are used to screen optimization thresholds: automatic selection and manual selection. In automatic selection, the choice of entries is based on mathematical and statistical methods. The selection process is conducted without human intervention. In the manual selection technique, the operator observes the histogram's distribution with the eye through trial and error and selects a gray value with a better segmentation effect as the threshold value. Due to the simplicity of the manual selection threshold method, it is widely used in practice.

- Global threshold

  The simplest way to determine the boundary by threshold is to set the gray threshold value to a constant throughout the image. If the background's gray value is reasonably constant throughout the image, all objects and the background have almost the same contrast. In that cost, as long as the correct threshold is chosen, the use of a fixed global threshold is generally better.

- Adaptive threshold

  In many cases, the background's gray value is not constant, and the contrast between the object and the background varies in the image. In this case, a threshold that works well in one area of the image may not work well in other areas. In this case, it is appropriate to take the gray threshold as a function value that changes slowly with image's position.

- Selection of optimal thresholding

  Unless the object in the image has a steep edge, the gray threshold value greatly impacts the boundary location and overall size of the extracted object. This means that measurements of size (especially area) are sensitive to the selection of gray thresholds. For this reason, we need an optimal or at least consistent way to determine the threshold. The commonly used optimal threshold segmentation methods include histogram threshold and automatic threshold determination. The histogram threshold method is used to construct a histogram containing pixels with large gradient amplitude. The average gradient value of each gray-level pixel is used to divide the histogram to enhance the concave valley. The gray average value of high gradient pixels is used to determine the threshold value. Automatically choosing the threshold value method is to identify the basis of analysis. Its characteristic does not require any prior information of threshold value. First, calculate the image histogram, and the image pixel could be divided into two categories according to threshold "T," background, and target. The pixel in the background area is composed of $0-T$, and the target refers to the gray value of $T + 1$–255 area. The probability distribution calculation of background and target intra-domain variance, inter-domain variance, total variance. "T" could be selected and confirmed according to the following different measurements. After the threshold is determined, the image could be segmented and processed.

(b) Gradient-based image segmentation method

The threshold segmentation method uses a threshold value to achieve segmentation. In contrast, the boundary method uses the property of boundary with a high gradient value to find the boundary instantly. Here are three such approaches.

- The boundary tracking

   Supposing we start with a gradient amplitude image, which is calculated from an image of a single object in a background that contrasts to the object. Since the point with the highest gradient value in the image must be on the boundary, we could use as the starting point of the boundary tracking process. Then, the 3×3 neighborhood with the boundary starting point as the center is searched, and the neighborhood point with the maximum gray level is found as the second boundary point. If two neighborhood points have the same top gray level as the puffy image, choose one of them. Starting from this point, we create an iterative process of finding the next boundary point given the current and previous boundary points. In the 3×3 neighborhood centered on the current boundary point, we investigate the adjacent point relative to the last point boundary and the two points on both sides of the adjacent point (Fig. 4.5). The next boundary point is the one with the highest gray level of the above three points. If all three or two adjacent boundary points have the same highest gray level, we select the middle point. If two nonadjacent points have the same highest gray level, you could choose either one.

- Gradient image binarization

   If we binarize a gradient image with a moderate threshold, we will find that the background points are below the threshold, while most of the edge points are above it (see Fig. 4.6). Kirsch's segmentation method exploits this phenomenon. This technique first binarizes the gradient image with a low to a medium gray threshold to detect the object and the background, separated by the boundary points above the threshold. As the threshold increases, both the object and the background increase simultaneously. When the object and the background area are almost in contact without merging, the contact point could define the boundary. This is the application of the watershed algorithm in gradient image.

   Although the Kirsch method has a larger calculation than binarization, it could produce the maximum gradient boundary. For an image that contains multiple objects, if the initial binarization is properly segmented, the segmentation could be ensured correct. Smoothing the gradient image in advance produces a smoother boundary.

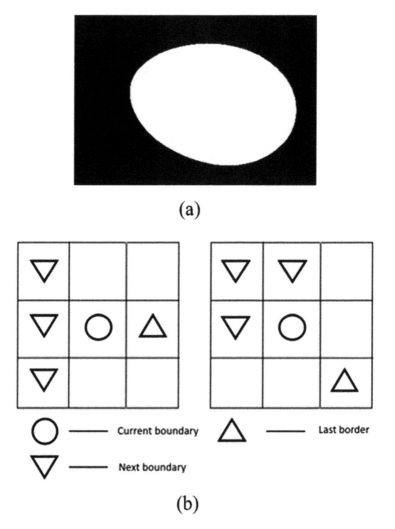

Fig. 4.5 (a) Egg image and (b) boundary tracking

- Laplace edge detection

    Laplace operator is a scalar operator of the second derivative of a function that operates on two dimensions. It defines

$$\nabla^2 f(x,y) = \frac{\delta}{\delta x^2}f(x,y) + \frac{\delta}{\delta y^2}f(x,y) \qquad (4.2)$$

    It could usually be implemented digitally with one of the convolution kernels (templates) shown in Fig. 4.7.

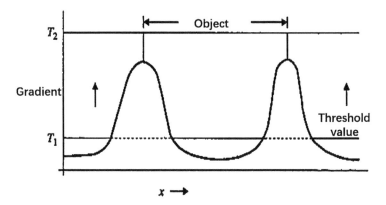

Fig. 4.6 Kirsch's segmentation method

0	-1	0
-1	4	-1
0	-1	0

-1	-1	-1
-1	8	-1
-1	-1	-1

Fig. 4.7 Laplace convolution kernel

If a noiseless image has steep edges, Laplacian operators could be used to find them. The image filtered by the Laplace operator is binarized with zero gray value, which will produce a closed and connected contour and eliminate all the internal points. As it is a second-order differential operator, it is more sensitive to noise, so the image with noise needs to be filtered before using the Laplace operator. It is appropriate to select a low pass filter for smoothing in advance. According to the associative law of convolution, Laplace operator and Gaussian impulse response could be combined into a single Gaussian Laplace kernel:

$$-\nabla^2 \frac{1}{2\pi\sigma^2} e^{\frac{x^2+y^2}{2\sigma^2}} = \frac{1}{\pi e^4} \left[1 - \frac{x^2+y^2}{2\sigma^2} \right] e^{\frac{x^2+y^2}{2\sigma^2}} \tag{4.3}$$

This impulse response is separable for x and y and therefore could be implemented effectively.

4. **Morphological Analysis of Binary Digital Images**

Binomial morphology is to measure and extract the corresponding shape of the image by the structural elements with a certain shape to achieve the purpose of image analysis and recognition. Binomial morphology has some advantages. The edge information extraction and processing are not as sensitive to noise as the differential algorithm. Simultaneously, the extracted edge is relatively smooth, and the image skeleton extracted by the mathematical morphology method is relatively continuous with fewer breakpoints.

(a) Corrosion and expansion

Expansion: Add pixels to the border of the object in the image. The output pixel value is the maximum value of all pixels in the input image's corresponding pixel field.

Corrosion: Delete the border element of the image. The output pixel value is the minimum value of all pixels in the input image's corresponding pixel field.

(b) Open operation and closed operation

Corrosion and expansion are not inverse operations. The combination of corrosion and expansion produces two operations: open and closed.

The open operation generally makes the outline of the object smooth, widening the gap of the narrow port and eliminating the thin protrusion. The closing operation also makes the outline smoother. Still contrary to open operation, it usually eliminates the narrow gap and the long gap, eliminates the small hole, and fills the fracture in the outline.

Open operation: Perform corrosion operation on the image first, and then perform expansion operation on the corrosion results, you could eliminate the small size of the details. Closed operation: The image expansion operation, and then the expansion of the result of the corrosion operation, you could eliminate the size of the small dark details.

4.3.2   Feature Analysis in Digital Images

Generally, there are two kinds of representation methods for shape features of images: contour features, and the other is regional features. The contour feature of the image is mainly aimed at the outer boundary of the object, while the regional feature of the image is related to the whole shape area.

1. **Typical Description Methods of Features**

(a) Boundary feature method

The shape parameters of the image are obtained by describing the boundary features. Hough transform uses the global feature of the image to connect the edge pixels to form the closed region's boundary. It is the most classical method. The basic idea of Hough transform is the duality of point and line. The boundary direction histogram method first differentiates the image to obtain the image edge and then makes the histogram about the edge's size and direction. The usual method is to construct the image gray gradient direction matrix.

(b) Fourier shape descriptor

   Fourier shape descriptor's basic idea is to use the Fourier transform of the object boundary as the description of the shape and transforms the two-dimensional problem into a one-dimensional problem using the closure and periodicity of the regional boundary. Three shape expressions are derived from boundary points: curvature function, centroid distance, and complex coordinate function.

(c) Geometric parameter method

   More simple regional feature description methods are adopted for shape expression and matching, such as shape factor, which is related to shape quantitative measures (such as moment, area, circumference, etc.). In the QBIC system, shape feature-based image retrieval is carried out using geometric parameters such as roundness, eccentricity, principal axis, and algebraic invariant moments.

(d) Shape invariant moment method

   The rectangle of the target area is used as the shape description parameter.

(e) Others

   Finite element method (FEM), transform function (turning function), wavelet descriptor, etc. It should be noted that the extraction of shape parameters must be based on image processing and image segmentation, and the effect of image segmentation has a significant impact on the accuracy of the extraction parameters. Just extracting the shape of the object is not the difficulty of image processing. The problem lies in how to use these features.

2. **Commonly Used Shape Parameters**

(a) Rectangularity

   The parameter that reflects the rectangularity of an object is the rectangle fitting factor.

$$R = \frac{A_O}{A_R} \tag{4.4}$$

A_O is the area of the object; A_R is the area of its MER.

   R reflects how full an object is to its MER. For the rectangular object, the maximum value is 1.0; For a circular object, the value is $\frac{\pi}{4}$; Get smaller for thin, curved objects. The value of the rectangle fitting factor is limited to 0–1.

   Another characteristic aspect ratio is relating to shape

$$A = \frac{W}{L} \tag{4.5}$$

(b) Circularity

   It is the ratio of the width to the length of the MER (minimum bounding rectangle). This feature could distinguish thinner objects from square or round things.

There is a set of shape features called roundness because they take the minimum value when calculating a circular shape. Their amplitude values reflect the complexity of the measured boundary. The most commonly used indicator of roundness is

$$C = \frac{P^2}{A} \tag{4.6}$$

P is the perimeter; A is the area. This feature takes a minimum value of 4π for a circular shape. If the shape is more complex, the value is more considerable. The roundness index C is related to boundary complexity.

(c) Moments

Moments of functions are often used in probability theory, and several expected values derived from moments are also applicable to shape analysis. The moment set defining a bounded function $f(x, y)$ with two variables is defined as $M_{jk} = \int\limits_0^\infty \int\limits_0^\infty x^j y^k f(x,y) \mathrm{d}x\mathrm{d}y$ in the formula, j and k could take all the nonnegative integer values. They produce an infinite set of moments. And the set could identify the $f(x, y)$. In other words, the collection $\{M_{jk}\}$ is unique for the function $f(x, y)$. Only $f(x, y)$ has this particular set of moments. To describe the shape, assuming that $f(x, y)$ takes 1 in this object and 0 in everything else. The silhouette function only reflects the shape of the object and ignores the grayscale details inside. Each particular shape has a particular outline and a particular set of moments.

3. **Image Texture Feature Analysis**

The texture is also a significant global property of the image, which describes the surface property of the object corresponding to the image. Generally speaking, the texture shows some regularity in the grayscale and color distribution in image. Textures fall into two broad categories: One is regular texture, and the other is near-regular texture. Even the near-regular texture could show certain statistical characteristics on the whole.

(a) Texture roughness analysis

Its roughness often describes texture. The size of roughness is related to the spatial repetition period of local structure, the texture with large period is coarse, and the texture with a small period is fine. This sense of roughness is not enough to be used as a quantitative texture measure, but it could at least explain the variation tendency of texture measure. The small value texture measure represents fine texture, while the large value measure represents coarse texture.

The method of using spatial autocorrelation function as texture measure is as follows: Assuming the image is $f(m, n)$, and the autocorrelation function could be defined as

$$C(\varepsilon,\eta,j,k) = \frac{\displaystyle\sum_{m=j-\bar{\omega}}^{j+\bar{\omega}}\sum_{n=k-\bar{\omega}}^{k+\bar{\omega}} f(m,n)f(m-\eta,n-\eta)}{\displaystyle\sum_{m=j-\omega}^{j+\bar{\omega}}\sum_{n=k-\bar{\omega}}^{k+\bar{\omega}} [f(m,n)]^2} \tag{4.7}$$

It calculates the correlation value between $(2\bar{\omega}+1) \times (2\bar{\omega}+1)$ each image point (j, k) and the pixel whose deviation value is $\varepsilon,\eta = 0, \pm1, \pm2, \ldots, \pm T$. The correlation of a coarse-grained region with a given deviation (ε, η) is higher than that of a fine-grained region, so the roughness of the texture should be proportional to the expansion of the autocorrelation function. One measure of the extension of the autocorrelation function is the second moment, i.e.

$$T(j,k) = \sum_{\varepsilon=-T}^{j}\sum_{\eta=-T}^{k} \varepsilon^2\eta^2 C(\varepsilon,\eta,j,k) \tag{4.8}$$

If the texture roughness is greater, the T is greater. Therefore, T could be conveniently used as a parameter to measure the roughness.

(b) Fourier power spectrum

To calculate a texture, select a window. Only one point is texture-free, so the texture is two-dimensional. Let the texture image be $f(x, y)$, and its Fourier transform could be expressed by the following formula.

$$F(u,v) = \int_{-\infty}^{+\infty}\int_{-\infty}^{+\infty} f(x,y)\exp\{-j2\pi(ux+vy)\}dxdy \tag{4.9}$$

The power spectrum of the two-dimensional Fourier transform is defined as follows:

$$|F|^2 = FF^* \tag{4.10}$$

where F^* is conjugate of F.

The power spectrum $|F|^2$ reflects the property of the whole image. If you write the Fourier transform in polar coordinates, you have $F(r, \theta)$. As shown in Fig. 4.8, the energy on a circle with distance r from the origin is considered to be

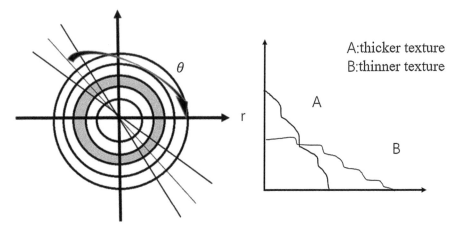

Fig. 4.8 Energy on a circle and energy versus radius curve

$$\Phi_r = \int_0^{2\pi} [F(r,\theta)]^2 d\theta \qquad (4.11)$$

Thus, the change curve of energy with radius r could be obtained as shown in Fig. 4.8. The study on actual texture images shows that in the thicker texture, the energy is mostly concentrated in the range close to the origin, like curve A in Fig. 4.8. In comparison, in the case of thinner texture, the energy is dispersed in the range far from the origin, as shown in Curve B in Fig. 4.8. Thus, the following analysis rules could be summarized: if r is small or Φ_r is large, or r is large but Φ_r is small, then the texture is rough; On the contrary, if the effect of r change is not very great, then the texture is relatively fine.

In addition, when a texture image has a large number of lines and edges along the θ direction. In addition, when there are a large number of lines and edges along the θ direction of a texture image, the frequency domain of the power spectrum concentrated in perpendicular to the θ direction. Otherwise, the power spectrum does not show obvious direction. Therefore, $|F|^2$ values could reflect the orientation of a texture.

(c) Joint probability matrix

The joint probability matrix method is a statistical survey of all pixels in an image to describe their gray distribution. Take any point (x, y) in the image and another point deviating from it $(x + a, y + b)$, and set the gray value of this point pair as (g_1, g_2).

If points (x, y) are moved across the whole screen, (g_1, g_2) values will be obtained. If the grayscale value is set as k, there are k_2 combinations of g_1 and g_2. For the whole screen, count the occurrence times of each (g_1, g_2), arrange them into a square matrix, and then use the total number of (g_1, g_2) to normalize them into the occurrence probability $P(g_1, g_2)$, and call such square joint probability matrix.

Set (x, y) as the image of a point, the grayscale difference value between (x, y) and $(x + \Delta x, y + \Delta y)$ which has only small distance with (x, y) is

$$g_\Delta(x, y) = g(x, y) - g(x + \Delta x, y + \Delta y)g_\Delta \qquad (4.12)$$

called gray-level difference. Suppose that all possible values of the gray difference value have m level, let the point (x, y) move on the whole picture, count the times of taking each value, and then make the histogram. The histogram shows the probability $p_\Delta(i)$.

When the probability of a smaller i value is large, the texture is rough. When the possibility is flat, the texture is thinner. The characteristics of texture images are generally described by the parameters such as contrast, second moment of Angle direction, entropy, and average value.

If the gray value of the point (x, y) is g, the gray value of its adjacent point may also be g. The probability that n consecutive points along the direction of had a gray value g from any point is calculated, denoted as $P(g, n)$. The number of pixels with the same gray value in one direction is called the run length. From $P(g, n)$, some parameters could be derived to describe the texture image variation characteristics better. The commonly used statistical methods of stroke length include the term stroke weighting method, gray value distribution, stroke length distribution, stroke ratio, etc.

4.4 Application in Food Quality Detection

Size, shape, color, and surface defects are important indexes for the variety identification of tea. The external characteristics of food with different grades, textures, maturities could be presented in images obtained from computer vision.

4.4.1 Application in Agricultural Products Quality Identification

1. **Application in Tea Variety Identification**
   Color, shape, and other external quality characteristics are important indexes for the variety identification of tea. The external characteristics of tea leaves from different characteristics could be presented in images obtained from computer vision.

   In the work presented by Chen et al. [10], the varieties of "Biluochun," "Tunchaoqing," "Maofeng," "Queshe," and "MaoshanChangqing" were classified by using computer vision. Figure 4.9 showed the original images of tea with

(1): Biluochun;   (2): Tunchaoqing;   (3): Maofeng   (4): Queshe;   (5): Maoshanchangqing

Fig. 4.9 Original images of tea with different grades

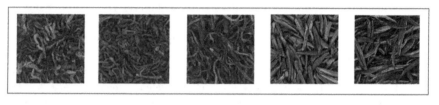

(1): Biluochun;   (2): Tunchaoqing;   (3): Maofeng;   (4): Queshe;   (5): Maoshanchangqing

Fig. 4.10 Preprocessed images of tea

different varieties. Before features extraction, the images were first preprocessed. The central location of the original image was found, and then 400 × 400 neighborhood around the center was selected as target areas, and the processed images are shown in Fig. 4.10.

Different varieties of tea leaves could vary greatly in their color characteristic parameters. Generally, the color features extracted from single color space could not well express the appearance information. Therefore, HIS and RGB are used to extract the color feature variables for identifying tea leaves.

A total number of 120 tea samples were used for the experiment, including 20 samples for each variety. Six texture feature parameters of statistical moments and six texture feature parameters of spectrum characteristics were extracted, so 12 variables were selected for further analysis. Since these 12 variables have different dimensions and orders of magnitude, to make these variables comparable, all sample data involved in the analysis are standardized for further analysis.

The correlation between the extracted color feature variables and texture feature variables may result in certain information redundancy. The interference of redundant information could reduce the predictive performance during modeling. PCA is a statistical method to convert multiple indicators into several comprehensive indicators. It projects from multidimensional spectral data space to low-dimensional data space along the maximum direction of covariance. Therefore, PCA was carried out on these characteristic variables, and extracted vectors of PCA were used as input of the pattern recognition model. The LDA method was used to establish the classification model with 60 samples as the training set, and 60 samples as the prediction set.

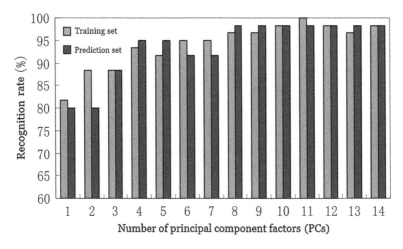

Fig. 4.11 LDA model training and prediction results under different principal components

Figure 4.11 shows the results of model training and prediction under different principal component factors. It could be seen from Fig. 4.11, when PC numbers were 11, the recognition rate reached 100% in the training set. In the prediction set, when the PC number was 8, the recognition rate reached a maximum of 98.33%. The work showed that the computer vision combined with image processing could be used for tea variety classification through analysis of external quality characteristics.

2. **Application in the Fruit Defects Detection**

(a) Defect detection of apples

The defect is an important indicator for apple quality detection. Machine vision is widely used in the rapid and nondestructive grading of apple quality. Generally, it is difficult to fully express the defect information of apple images only using a signal camera, which may result in information omission. Aiming at overcoming the limitation of image information detection by a single camera, [11] proposed an apple defect detection method based on tri-CCD cameras computer vision system [11].

The tri-CCD camera system comprised three parts: conveyor line, lighting room, and camera image acquisition and processing system. The conveying line adopted chain drive mode, and the chain was embedded with a roller. The apple was rolled forward along with the horizontal conveying production line. When the fruit passed through the lighting room, the image acquisition and processing system were triggered to shoot and process the apple images. The system consisted of three cameras, which were placed on the top and two sides around the analyte. Three computers form a small local area network, which could exchange data in real-time [12].

In this work, apple image edge detection was first used to get the outline of objects. The convolution operation of the apple image was performed using

Table 4.1 Detection effects of different differential operators

Edge detection operator	The original image before detection	The result image after detection
Roberts operator	Defects before detection Fruit peduncle before detection	Defects detected Fruit peduncle detected
Laplace operator	Defects before detection Fruit peduncle before detection	Defects detected
Sobel operator	Fruit peduncle before detection Defects before detection	Fruit peduncle detected Defects detected
Canny operator	Defects before detection Fruit peduncle before detection	Defects detected Fruit peduncle detected

an edge detection operator by setting a threshold. Repeated tests were carried out to determine the threshold value according to the actual effect of edge detection. The setting of the threshold value could make the edge of the suspicious area be preserved as far as possible, and at the same time making noise signals to be removed. Table 4.1 compared the detection effects of several edge detection operators. Compared with other methods, using the method of Sobel may obtain a better performance, so the Sobel edge extraction operator was used as an apple defect feature extraction method.

(b) Defect detection of carrots

In the process of carrot growth, harvest, and transportation, it is inevitable to produce defects on the surface. The removal of defective carrots is necessary before carrot marketing. At present, defective carrots mainly rely on manual sorting, which has the disadvantages of unstable sorting standards,

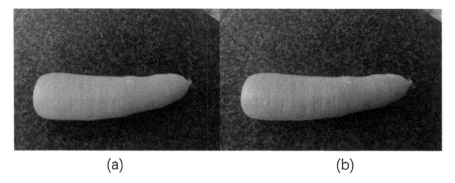

(a)   (b)

Fig. 4.12 Image preprocessing. (**a**) The original image. (**b**) The gray level image

high labor intensity, and high cost. To detect faulty carrots quickly, accurately, and nondestructive, computer vision was introduced into the carrot sorting process to improve the sorting accuracy and efficiency. The surface defects of carrots include green heads, bending, breaking, bifurcating, cracking, and so on.

Image preprocessing extracts the carrot region image and deletes other irrelevant regions. First, the carrot was segmented from the background according to the gray-level image of the carrot (Fig. 4.12b). The binary image was used as a mask to perform morphologically and operations with the original image to obtain the color carrot image with the background removed.

The crack detection is realized by using the texture difference between normal and cracked carrot areas. The crack area was extracted by the Sobel horizontal edge detection operator and the Canny edge detection operator combined with the morphological operation. The results showed that 94.91% of defects were correctly identified [13].

3. **Application on Fruit Ripeness Evaluation**

(a) Strawberry ripeness evaluation

Appearance characteristics were greatly related to the quality of strawberries. Strawberries could be graded according to their images obtained from computer vision, which has the advantage of simple calculation and high real-time performance. Lv et al. (2009) presented a work of online ripeness grading of strawberries based on their appearance characteristics. Strawberries with different ripeness were picked from a strawberry garden in Dazhou Rock Town in China. Image processing of strawberries included denoising, gray transformation, binarization, edge detection, and feature extraction. Figure 4.13 shows the strawberries dealt by median filtering, and the image background noise was almost removed.

The processing of strawberries was as follows. The pixel value of the grayscale strawberry image mainly concentrated between 29 and 120, and the

(a)   (b)

Fig. 4.13 Processing of strawberry image. (**a**) Image of original strawberry. (**b**) After median Filtering

contrast was not apparent. It was necessary to enhance the image grayscale, which was operated by binarization. The threshold value is automatically obtained by the OTSU method for the strawberry image after gray enhancement. Sobel operator, Prewitt operator, Canny operator, Roberts operator, and Log operator were used for edge detection and comparative analysis of strawberry images [14–16]. After image processing, the size of the strawberries could be detected.

The color feature of the strawberry image is an important visual property that could be used to identify the ripeness of the strawberry. HSV model could reflect the visual characteristics of color as human being. In the work, the HSV model of the strawberry image converted from RGB model was used to extract the color features of the strawberry image. The results showed that strawberries could be segmented using white as the strawberry image. The median filter algorithm may remove pepper and salt noise in the process of strawberries image acquisition. All the five classical edge detection methods can clearly and continuously segment the contour of strawberries. To evaluate the size and quality grade of strawberry by using the area, circumference, and color characteristics of strawberry fruit can be an important basis for developing automatic strawberry grading equipment [17].

(b) Mango ripeness identification

The traditional grading of mangoes method mainly relied on manual observation and chemical analysis, which does not fit the industry's development. Xin Huajian designed a mango ripeness detection method based on computer vision.

After obtaining images of mangoes, the wiener filtering method was used to remove noise. Then, an image operation was used to obtain the outline. The background of the preprocessed image was black, and the distinction between the mango and the background was obvious, and the outline of the fruit was clear. The mango surface presents multiple colors, so HSL and RGB models were used to express the color of mangoes. The edge detection of mango

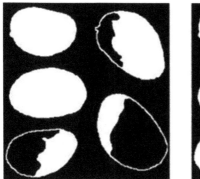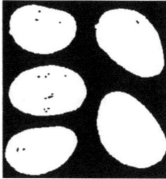

Fig. 4.14 Extraction of appearance character (**a**: color; **b**: surface defect)

images was based on the adaptive Canny algorithm. The adaptive Canny algorithm defined the point where the vector's maximum value appears in a specific direction as the edge of the image and introduced information entropy to adapt to the ratio of the high and low thresholds of the Canny operator. The median filter of the image R component in the RGB model was used to calculate the segmentation threshold through the OTSU algorithm. The red part was extracted from the image area of single fruit and the area of fruit could be calculated, as shown in Fig. 4.14a. The surface defects of mango were black damage caused by diseases and collisions. Therefore, by setting the threshold value of the L component in HSL color space, the black areas of the surface defects were extracted. The area ratio of the surface defects in the image of single fruit could be calculated, as shown in Fig. 4.14b. The experimental results showed that the accuracy of mango quality grading by using computer vision exceeded 93% [18].

4.4.2   Application in Animal Products

1. **Application in Egg Crack Identification**
   In commercial egg production, one of the main physical quality defects is the presence of cracks in the eggshell. The crack characteristics of eggs could be shown in the transmitted images, and egg cracks could be detected by computer vision combined with the image processing [19].
   In the work presented by Lin et al. [20], cracks of eggs were detected by computer vision. The measured egg was placed at the top of the lighting obscuration, the light source is emitted from the halogen lamp, and the transmitted image of the egg was collected by the CCD camera. The corresponding filtering and denoising were carried out. Images of eggs were collected in BMP format, and the size of the RGB color image is 640×480 [20].

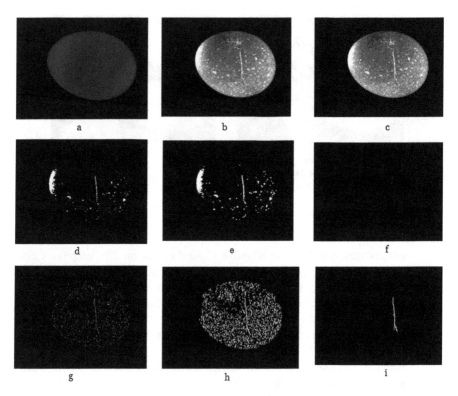

Fig. 4.15 Cracked egg processing image. (**a**) Original image of egg. (**b**) Egg grayscale. (**c**) After median filtering image. (**d**) Segmentation based on color threshold. (**e**) Expansion connection threshold. (**f**). Top-Hat transform. (**g**) Automatic threshold segmentation of images. (**h**) After secondary expansion. (**i**) Image segmentation based on threshold value of area

Figure 4.15a shows the original transmission image of cracked egg. This image usually has a dark background, and the main area is red. The cracked area's transmit light was higher than that of the noncracked ones, showing a narrow bright line or a bright area. In order to be convenient for processing, the original RGB color image was transferred into grayscale, where a crack area was relatively obvious. Furthermore, median filtering as nonlinear filtering was used to process the noise in the image. Figure 4.15c shows the cracked egg image processed by median filtering with 3×3 rectangular windows. Then, the target area of the egg extracted by the color-based threshold segmentation method was adopted. After segmentation, the crack and noise areas were presented in white with black background (Fig. 4.15d). After threshold segmentation, large noise interference areas may be generated and crack areas could not be well extracted. Therefore, corrosion in morphology was used to join the white gray area (Fig. 4.15e). According to the egg crack morphology characteristics, select the appropriate structural elements and after Top-Hat transform (Fig. 4.15f), and then treated with threshold binarization (Fig. 4.15g). The denoising processing, the

Table 4.2 Identification results of egg cracks

| Egg type | Sample size | Identification | | Total recognition rate |
		Intact eggs	Crack eggs	
Intact eggs	88	80	8	84.1%
Crack eggs	88	20	68	

effect of crack lines appeared many breakpoints, could be connected by morphological dilation, longitudinal structural elements are selected for processing (Fig. 4.15h). The crack area was presented in a line shape, and their ratios of length to width were bigger than noise areas. After demarcating all regions with 8-connectivity, the area of each region was calculated, the largest crack area was extracted and segmented as shown in Fig. 4.15i. The discriminant results of 88 intact eggs and 88 cracked eggs by the above methods are shown in Table 4.2. Among the 88 intact eggs, 8 were misjudged as cracked eggs, while 20 eggs were misjudged as entire eggs. The overall recognition rate of the test eggs was 84.1%.

2. **Application in Egg Freshness Measurement**

Freshness is an important standard for internal egg quality, and consumers usually regard the decline of freshness as the main sign of egg quality decline. First, freshness changes of eggs may be shown as the constant dilution of concentrated protein, and the light transmittance inside the eggshell changes, which could be reflected by the change of the color information inside the egg. Therefore, color spaces R, G, B, H, I, and S were used to describe the change of color information of egg freshness. Furthermore, the expansion of the egg yolk area also follows the change of egg freshness. When egg freshness decreases, the egg yolk gradually increased and tends to be flat in terms of morphological characteristics. The freshness changes of eggs could be analyzed by morphological features in their images. In this work, presented by Lin et al. [20], the freshness of eggs was detected by computer vision. A total of 112 eggs with storage times ranging from 1 to 14 days was used in this work. As the color of eggs greatly influences the results, the color of the eggs is consistent and the shape is uniform. The R, G, B, H, S, and I color components of egg transmission images are shown in Fig. 4.16. The mean value and variance of HIS and RGB color components were extracted for a total of 12 variables to describe their color information.

With the extension of storage time, the area of egg yolk would gradually increase. The egg yolk image area could be counted through morphological features to describe the information of egg freshness change. Therefore, the variables from morphological features were also used as input vectors for modeling. The ANN models were employed based on the variables from color components and morphological features of egg images, and the results from modeling are shown as Fig. 4.17. The results proved that the freshness egg was related to the image features obtained by machine vision.

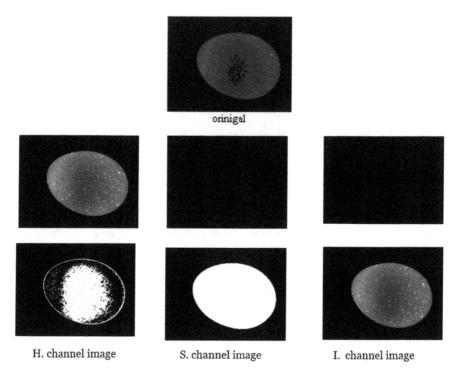

orinigal

H. channel image   S. channel image   I. channel image

Fig. 4.16 Color component diagram of egg image

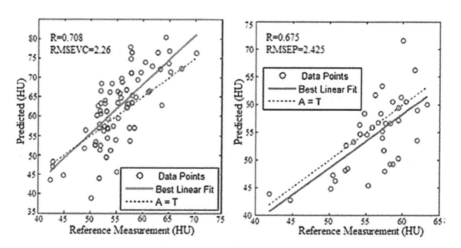

Fig. 4.17 Reference measured versus predicted of egg's Haugh Units by BP-ANN in calibration set (**a**) and prediction set (**b**)

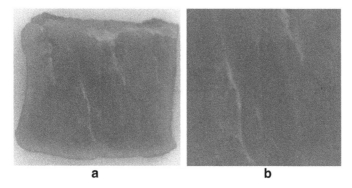

Fig. 4.18 Original (**a**) and pre-processed (**b**) images of a pork

3. **Application in Meat Freshness Detection**

Freshness changes of pork would be reflected in characters of color and texture, which could be analyzed by their images. The pork used in the experiment was the chilled tenderloin meat from the "Changbai" pig slaughtered on the same day, which was purchased from the meat products counter in Zhenjiang China. The meat was stored in the refrigerator for 1 day, 3 days, 5 days, 7 days, 9 days, and 11 days, respectively. Each meat sample was taken for computer vision data collection and TVB-N content determination.

Pork images were collected and then stored in BMP format. Before extraction of image characteristics of pork, the original image was preprocessing [21, 22]. Figure 4.18 showed the pork images before and after preprocessed. The mean values and their standard deviation were extracted from RGB components of pork images.

Then, the images were converted from the RGB to HIS model images (Fig. 4.19), and six color characteristic variables including mean values and their respective standard deviation δ_H, δ_I, and δ_S were extracted from the HIS models. A total of 12 characteristic variables reflecting the color change of meat sample were included. To reflect the characteristics of the pork texture changes of metamorphic process information, convert RGB image to gray image. Six textural feature parameters were extracted from each greyscale images using the image statistical moments, which were mean gray level (m), standard deviation (δ), smoothness (R), third-order moment (μ_3), consistency (U), and entropy (e), respectively. Based on the above color and texture features, 18 feature variables were extracted from the image of each meat sample. Since most of the 18 variables were of different dimensions, make these variables comparable, these feature variables were standardized.

The 18 characteristic variables described the color and texture state of the meat sample image from different aspects. Pearson correlation analysis was conducted between the 18 characteristic variables and TVB content. It was found that most of the characteristic variables were significantly correlated with each other, and the 11 characteristic variables such as H, I, S, δ_H, δ_I, δ_S, m, δ, R, U, and e were

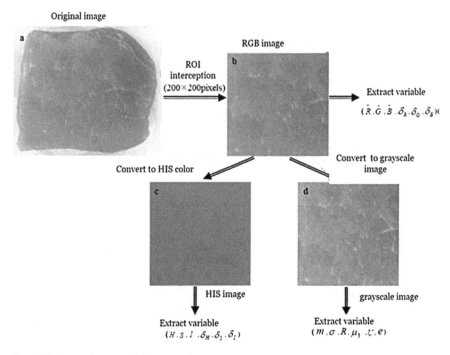

Fig. 4.19 Image feature variable extraction

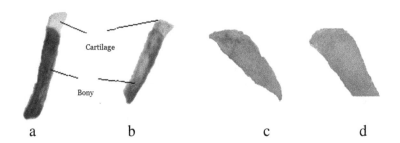

Fig. 4.20 (**a–d**) Schematic diagram of the degree of osteogenesis of thoracic vertebrae

significantly correlated with the content of pork TVB-N. Therefore, computer vision combined with the image processing method was a high potential for pork freshness discrimination.

4. **Application in Beef Physiological Maturity Detection**

Physiological maturity, namely the physiological age of cattle, is an important index of carcass quality grade, closely related to carcass quality, beef tenderness, health, and other indicators. The physiological maturity of cattle carcasses may be divided into five grades (A–E). The terminal cartilage of grade A and B of the thoracic spine process is white and transparent (Fig. 4.20a, b). The cartilage above grade C has gradually become osteoid (Fig. 4.20c, d), and the white and

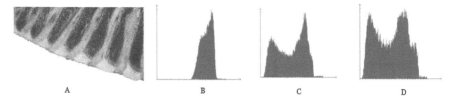

Fig. 4.21 Physiological maturity is the histogram distribution of grade A beef image (small image) and its *R*, *G*, and *B* components. (**a**) The original image. (**b**) *R* component histogram distribution. (**c**) *G* component histogram distribution. (**d**) *B* component histogram distribution

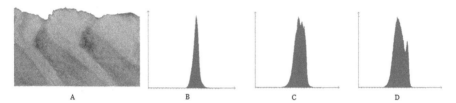

Fig. 4.22 Physiological maturity is the histogram distribution of grade C beef image (small image) and its *R*, *G*, and *B* components. (**a**) The original image. (**b**) *R* component histogram distribution. (**c**) *G* component histogram distribution. (**d**) *B* component histogram distribution

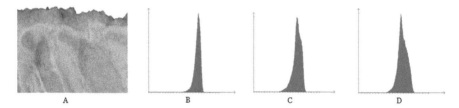

Fig. 4.23 Physiological maturity is the histogram distribution of grade E beef image (small image) and its *R*, *G*, and *B* components. (**a**) The original image. (**b**) *R* component histogram distribution. (**c**) *G* component histogram distribution. (**d**) *B* component histogram distribution

transparent cartilage gradually disappears. During the change of physiological maturity from A to E, young bone's color is mainly dark red. With the increase of age, the bone's color gradually fades and turns yellow and white. The classification of physiological maturity could be achieved by analyzing the images of beef [23].

First, image segmentation was employed to separate beef samples with the background. Ohta system can achieve good segmentation results among different colors and has a good distinguishing ability in image color feature recognition, edge detection, and other aspects. Therefore, the Ohta color system was used to segment the images.

After the Ohta color system segments the bone image of the bovine thorax, small images containing cartilage and a small part of bone were attempted to be segmented. The chondral ossification was determined by histogram distribution of small images, and cartilage and bone can be segmented. The moment feature of the RGB component histogram is useful to estimate the histogram distribution feature.

Table 4.3 Training results of test samples

Degree of ossification	Number of samples	Number of correct identification	Correct recognition rate (%)
A	12	11	91.7
B	12	10	83.3
C	12	11	91.7
D	12	12	100
E	12	11	91.7

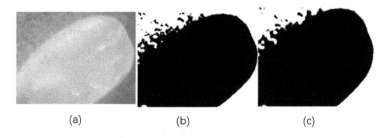

(a)   (b)   (c)

Fig. 4.24 Picking out of rice kernel image from gray image. (**a**) The gray level image. (**b**) Image after threshold segmentation. (**c**) Image after filtering operation

Figures 4.21, 4.22, and 4.23 were typical histogram distribution examples of A, C, and E levels of physiological maturity. According to these histogram distributions, the lower the physiological maturity, the more complex the histogram's contour, due to the obvious differences between the color of cartilage and bone, and between the color of soft bone and other tissues. The higher the physiological maturity, the more uniform the color of thoracic vertebrae, and the simpler the contour of the histogram.

With 60 beef samples used for testing, the features of beef images were used as input vectors, and support vector machine was used to identify beef physiological maturity. The identification results were shown in Table 4.3. The identification rate of physiological carcass maturity reached 91.7%.

4.4.3 Applications in Cereals

1. **Identification of Rice Variety**

   At present, all rice varieties grains are graded manually, which is a huge workload, and the testing results are highly subjective and inconsistent. Inspectors mainly observe the detection of rice grain shape with naked eyes. Although this detection method is simple and low cost, it is more subjective and less efficient.

   The rice used in the experiment was "Meihe" brand rice produced in "Jilin" Province. The samples are divided into two kinds: whole grain rice and broken rice. Sun Yu et al. [24] presented rice shape identification by using computer vision. Extracted rice grain shape characteristic parameters by image processing method, analyzed the extracted grain shape characteristic values by PCA, and then recognized rice grains by using neural network. Figure 4.24 showed the gray image, images after segmentation and filtering operation. The BP neural network was used to construct the classifier to recognize the grain shape of rice. The classifier had good fault tolerance and adaptability and was suitable for dealing with complex classification characteristics. A three-layer BP network structure was constructed. Two hundred grains of rice were selected for training and learning of the neural network, including 100 grains of whole rice and 100 grains of broken rice. In addition, another 600 grains of rice (300 grains of whole rice and 300 grains of broken rice) were taken as test samples and input into the system for identification. The results showed that the recognition accuracy of the whole rice reached 98.67%, which could meet the requirements of rapid and accurate detection of rice grain shape.

2. **Detection of Mildew Degree Rice**

   To realize nondestructive detection of mildew in rice storage, the rice types with different mildew degrees were determined and divided into control group (no mildew), slight mildew group, and severe mildew group. The computer vision system was used for image acquisition and image processing of three rice sample groups to extract grayscale, color, and texture features, and 68 image features were obtained.

   First, the region of interest (ROI) was used to select the target area. Then, the image of the ROI obtained was transformed into RGB components, and then the brightness information and color information were also extracted. The gray-level co-occurrence matrix was used to describe and extract the texture and information of the moldy image of the region of interest, and SVM was used to build an identification model. The results showed that the SVM model constructed by using all parameters could distinguish the control group and the mildew group well. The overall accuracy of the modeling set and the verification set were 99.7% and 98.4%, respectively.

4.4.4 Application in Aquatic Products

Pretreatment is an important step to ensure the quality and commodity value of freshwater fish processing, and fish body classification is one of the main procedures of pretreatment. At present, the classification of freshwater fish mainly relies on artificial methods, which have high labor intensity, low efficiency, and low accuracy. Zhang et al. [25] used machine vision technology to obtain the images of the crucian carp, and image processing was used for fish grading based on their projected areas. The experimental results showed that the fish body mass was highly correlated with the projected area, and its determination coefficient R^2 was 0.9878. The quality prediction model was verified, and the mean of relative error was 3.89%, and the mean of absolute error was 6.81 g. The experimental results showed that the machine vision technology can reference the quality grading method of freshwater fish.

There are many varieties of freshwater fish bred, with different shapes and sizes of fish. Before the processing of freshwater fish, such as "three removal," the species classification and size grading of freshwater fish are all completed by manual operation, with high labor intensity and low production efficiency [26, 27]. Machine vision technology was used by Wang et al. to obtain fresh water fish's body image for identifying carp, crucian carp, grass carp and bream. Morphological feature parameters and color feature parameters were used for freshwater fish species recognition according to images of fish body features, 92.50% of samples could be correctly identified.

4.4.5 Others

Computer vision technology has been widely used in many fields, in addition to the detailed examples described above, likewise, in other fields, as shown in Table 4.4.

Table 4.4 Computer vision technology applications in food detection

Name	Direction	Method
Ying [28]	Pork freshness	When the classification threshold is 0.88, that is to say, if the classification threshold is greater than this, it is fresh pork, and if the classification threshold is less than this, it is rotten pork
Alçiçek and Balaban [29]	Detection of translucent aquatic products	The "two image" method overcomes segmentation difficulty by defining the object from the silhouette obtained from a backlighted image (image I), and performs color analysis using a front lighted image (image II) using segmentation from image I
Nguyen et al. [30]	Detection of red and bicolored apples on tree with an RGB-D camera	An algorithm based on color and shape features is proposed to detect and locate red and two-color apples
Abdullah et al. [31]	Fresh-cut starfruits	The color spectra of matured and unmatured fruits were characterized using all color features ranging from hue 10 to hue 74, and, using principal hues generated by Wilks-lambda analysis
Wan et al. [32]	A methodology for fresh tomato maturity	A method for detecting the maturity levels (green, orange, and red) of fresh market tomatoes (Roma and Pear varieties) by combining the feature color value with the backpropagation neural network (BPNN) classification technique

4.5   Conclusion and Future Trends

With the development of industrial intelligence, more attention was paid to grading and classification by machine vision. The traditional manual detection method has the disadvantages of low production efficiency, large manual detection error, and no data storage after detection. However, the intelligent machine vision detection technology has the advantages of real-time, noncontact, high precision, and high efficiency, which has a very wide application prospect. In recent years, computer vision technology has been widely used in automatic detection and screening system. However, most of the research is based on static, collecting static images from indoor or field, and then processing the images, which is an ideal laboratory environment. However, the actual agricultural scene is usually very complex, so it is necessary to segment and extract the fast-moving image, and process the dynamic image in real-time [33]. Computer vision technology has good capacity for detection, classification, and grading of agricultural products, aquatic products, livestock products and grains. At present, it has realized the extraction and recognition of the dynamic characteristics of agricultural products, according to international standards, online detection. The detection speed and accuracy continue to improve, the compatibility of the system continues to upgrade, which can be applied to the detection of different types of agricultural products. In short, improving the recognition accuracy of image processing in the future development trend.

References

1. Bao X, Zhang R, Zhong L (2004) Research on apple recognition method based on artificial neural network and image processing. Trans Chin Soc Agric Eng 20(3):109–112
2. Hu C, Li P (2004) Research on identification of nitrogen deficiency and magnesium deficiency in cucumber based on image processing. J Jiangsu Univ (Nat Sci Ed) 25(1):9–12
3. Jia D, Jia X (2008) Image color eigenvalue extraction and matching based on C#. J Hebei Softw Vocat Techn Coll 10(2):52–55
4. Zhou Z, Zhao SAZ (2004) Defect extraction of X-ray images based on sub-region adaptive median filtering. Acta Aeronaut 24(5):420–424
5. Zang J (2014) Application of image processing technology in the design of organic gas sensor. Electron Comp Mater 33(1):77–78
6. Luo X, Wang D, Bai X (2011) Embedded gas detection system and its image analysis algorithm. Comp Appl 31(8):2270–2274
7. Yao M (2006) Digital image processing. China Machine Press, Beijing
8. Li J, Cheng J, Zhou X (2003) Adaptive median filtering under multiple windows in digital image processing. Comput Eng 29(17):154–156
9. Hamed HAL (2003) Feature vector based analysis of hyperspectral crop reflectance data for discrimination and quantification of fungal disease severity in wheat. Biosyst Eng 86 (2):125–134
10. Chen Q (2008) Research on fast nondestructive detection of tea quality based on near infrared spectroscopy and machine vision technology. Jiangsu University
11. Zou X, Zhao J, Li Y (2010) In-line detection of apple defects using three color cameras system. Comput Electron Agric 70(1):129–134

12. Guo Z (2015) Research on nondestructive testing method and device of apple quality based on near infrared spectroscopy and imaging. China Agricultural University
13. Xie W, Wei S, Wang F (2020) Research on carrot surface defect recognition method based on machine vision. Trans Chin Soc Agric Machin (51):451–456
14. Lv Q, Zhang J (2009) Design of automatic fruit classification system based on neural network. J Anhui Agric Sci 37(35):17392–17394. 17439
15. Li P (2007) Research progress of fruit grading based on computer vision technology. Agric Mechan Res (11):10–15
16. Zhao M, Hou W (2007) Research progress of automatic fruit grading technology based on machine vision in China. Pack Food Machin 25(5):5–8
17. Su B, Hua X (2018) Research on strawberry image processing based on machine vision. J Cap Norm Univ (Nat Sci Ed) (4):42–45
18. Xin H (2019) Research on the application of computer visual perception in the quality measurement of Amonkfruit. Res Agric Mechan (9):190–193
19. Lin J, Puri VM, Anantheswaran RC (1995) Measurement of eggshell thermal mechanical properties. Trans ASAE 38(6):1769–1776
20. Lin H (2010) Research on non-destructive testing of egg quality based on tapping vibration, machine vision and near infrared spectroscopy. Jiangsu University
21. Li H (2015) Nondestructive detection of total bacteria in pork and rapid identification of dominant rotting bacteria. Jiangsu University
22. Lin Y, Fang F (2011) Review of pork freshness detection methods. Meat Res 25(05):62–65
23. Shi Z, Tong Y, Chen D et al (2009) Electronic nose detection technology for beef freshness. J Agric Machin 40(11):184–188
24. Wan P, Sun Y, Sun Y (2008) Rice shape recognition method based on computer vision. J Jilin Univ (Eng Ed) 38(2):490–492
25. Zhang Z, Niu Z, Zhao S (2011) Quality grading of freshwater fish based on machine vision technology. Acta Agric Sin 27(2):351–354
26. Li L, Zong L, Wang J et al (2010) Research status and direction of pretreatment technology and equipment for bulk freshwater fish processing. Fisher Modern 37(5):43–46. 71
27. Wan P, Pan H, Long C (2012) Design of freshwater fish species online identification device based on machine vision technology. 28(06):164–167
28. Ying J, Wang P, Lu Y (2019) Application of computer vision technology in agriculture. Sichuan Agric Agric Machin (1):25
29. Alçiçek Z, Balaban M (2012) Development and application of the two image method for accurate object recognition and color analysis. J Food Eng (111):46–51
30. Nguyen TT, Koenraad V, Niels W et al (2016) Detection of red and bicoloured apples on tree with an RGB-D camera. Biosyst Eng (146):33–44
31. Abdullah MZ, Mohamad-Saleh J, Fathinul-Syahir AS et al (2005) Discrimination and classification of fresh-cut starfruits (Averrhoa carambola L.) using automated machine vision system. J Food Eng 4(76):506–523
32. Wan P, Toudeshki A, Tan H et al (2018) A methodology for fresh tomato maturity detection using computer vision. Comput Electron Agric (146):43–50
33. Xiao K, Duan X, Gao G et al (2012) Nondestructive detection of pork freshness based on image features. J Hebei Agric Univ 35(4):111–113. 122

Chapter 5
Spectral Imaging Technology in Food

What Is Spectral Imaging Detection Technology?

Spectral imaging is a technology that uses multispectral channels and different sensitivities of spectral reflection (absorption) rates of target objects in various bands for image acquisition, display, processing, analysis, and interpretation. It combines image analysis with spectral analysis. The image information describes the size, shape, and color, appearance characteristics, and spectral information to reflect its internal structure characteristics, composition content, and other information. Therefore, spectral imaging technology can visualize the internal and external quality characteristics of food and agricultural products and has unique advantages in the quality and safety detection of food and farming products. This chapter mainly describes different spectral imaging technologies and their applications in food quality and safety detection, as shown in Fig. 5.1.

Abbreviation	Full name
ACO	Ant colony optimization
ANN	Artificial neural network
BP	Back propagation
FAO	Food and Agricultural Organization
HSI	Hyperspectral imaging
HMI	Hyperspectral microscope imaging
IPLS	Interval partial least squares
KNN	K-nearest neighbor
PCA	Principal component analysis
PCs	Principal components
PLS	Partial least squares
ROI	Region of interest
RMSEP	Root mean squared error on prediction
RPD	Relative percent different
SCAE	Stack convolution auto encoder
SVDD	Support vector data description

(continued)

Abbreviation	Full name
SPA	Successive projections algorithm
SIPLS	Synergy interval partial least squares

5.1   Introduction

5.1.1   Brief Introduction of Spectral Imaging Technology

Spectral imaging is a new technology developed in the 1980s, which organically combined two-dimensional imaging technology and spectral technology. Although the spectral technique is gradually applied to food safety detection, this detection method based only on optics still has some limits. For example, near-infrared spectroscopy can characterize the internal quality information of food well. Still, it is often inadequate to describe food external characteristics (such as water power, pH value) [1]. On the contrary, imaging technology can detect food external features (such as flesh color, texture) with good results, but it cannot obtain internal quality information of food. The focus of securing information is different for the above two techniques, resulting in obtained information for both of them being incomplete. These limitations will inevitably affect the accuracy and stability of the detection results. Therefore, combining various detection technologies to improve the high throughput, sensitivity, and reliability of the detection technology is a research trend for achieving rapid, nondestructive food safety detection.

Hyperspectral imaging (HSI) technology has the characteristics of multi-bands, a high spectral resolution, and map and spectrum integration. In 1983, the Jet Propulsion Laboratory developed the first aerial imaging spectrometer (AIS-1), and it showed a great potential for image acquisition and analysis. Spectrum, called "optical spectrum" in full, is composed of the sequential monochromatic light formed after natural light is split through a prism, grating, and other dispersive systems. The spectrum can be roughly divided into ultraviolet, visible, and infrared bands. Among them, below 400 nm is ultraviolet, 400–780 nm is visible light, 780–3000 nm is near-infrared light, and above 3000 nm is far-infrared light. The spectral imaging technology with spectral analysis and imaging analysis has its unique advantages in food quality and safety detection.

5.1.2   Generation of Spectral Imaging Technology

Spectral imaging technology takes advantage of the sensitivity differences of spectral reflectance (absorption) rate of the target object in different wavebands to collect, display, process, analyze, and explain images. The generation of spectral imaging in

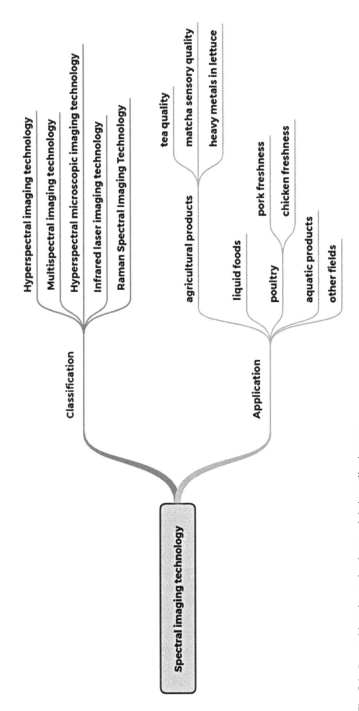

Fig. 5.1 Spectral imaging technology and their application areas

the composition of hardware facilities mainly depends on the specific light source, filter unit, and spectral image acquisition unit. The appropriate wavelength range of the excitation light source is an essential factor affecting the production of effective spectral images. Wavelengths beyond visible ranges are beneficial to enhancing the image characteristics of different parts of the target object, thus facilitating the quality detection of the target object. The spectral image acquisition unit is mainly divided into three types according to their scanning methods: point scanning, line scanning, and area scanning. The spectral imaging speed of point scanning type is slow, while the speed of area scanning type is fast, but hardware facilities' requirements are high. On the other hand, the generation of spectral imaging also relies on the assorted software control system, which is mainly used to control the exposure time of light source, the moving speed of mobile imaging platform, and the starting and ending physical positions of spectral imaging. Too high exposure time leads to the loss of image information, while too low exposure time causes a poor spectral response effect. Similarly, the moving speed of the platform is closely related to the formed image information, and inappropriate, moving speed distorts the image, reducing the advantages of spectral imaging technology.

The generated spectral image is a three-dimensional data block which is composed of a series of two-dimensional images at continuous wavelengths obtained by a spectroscopic system within a specific spectral range. At each wavelength, spectral data provides two-dimensional image information. Besides, the grayscale of the same pixel at different wavelengths provides spectral information. Image information can reflect appearance characteristics such as size, shape, and color, and spectral information can represent characteristic information like internal structure and component content. It can be concluded that spectral imaging technology can visually analyze food's internal and external quality characteristics.

5.1.3 Principle of Spectral Imaging Technology

Based on the fundamental principle of spectral imaging, spectral imaging technology includes many different systems. With the development of science and technology, the resolution accuracy of imaging spectral instruments is getting higher. According to the difference of spectral resolution, spectral images are divided into multispectral images, hyperspectral images, and ultra-hyperspectral images. It is generally believed that an image with a spectral resolution in the range of $10^{-1}\lambda$ is called a multispectral image. An image in the range of $10^{-2}\lambda$ is called a hyper-spectra image with a resolution in the order of $10^{-3}\lambda$ called an ultra-spectral image. Spectral imaging technologies with different resolutions can be selected according to additional detection accuracy and requirements. For the quality and safety detection of food and agricultural products, HSI technology is generally used, and it has great development potential in the food field.

The three-dimension data block obtained by spectral imaging technology contains an abundant component and image feature information, widely studied and applied in the rapid detection of food quality and safety. Therefore, scholars further developed and upgraded this technology, such as microscopic hyperspectral imaging, infrared laser imaging, and Raman spectral imaging. Microscopic hyperspectral imaging technology combines microscopic imaging with HSI. Thus, it can deepen HSI technology research into the microscopic level, providing a new means for studying or observing microscopic objects, e.g., microorganisms, tissues, and cells. Specifically, a translational fine-tuning lens is designed between the objective lens of the microscope and the C-Mount interface under the instantaneous view field to realize broom-like imaging. Infrared laser imaging technology is mainly based on the change process of physical optical path to realize imaging [2]. When parallel infrared laser beams irradiate the samples, optical paths of diffraction, interference, and refraction are projected onto the diffuse reflection screen. After which the near-infrared camera captured them. Therefore, the physical attributes of an object to be detected, such as shape, texture, and biochemical composition, effect the image information during the imaging process. Raman imaging technology combines spectral imaging technology with Raman scattering, which is a unique effect in the physical phenomenon of light scattering. The interaction between molecules in substances and incident photons causes light propagation direction and frequency to deviate from the laser, and this phenomenon is called Raman scattering. Each substance may have unique Raman characteristic peaks. Raman scattering can more fully obtain the information of the detection object at the atomic and molecular levels, such as molecular rotation and vibration phenomenon. Therefore, imaging technology coupled with Raman scattering could achieve sensitive qualitative and quantitative analysis.

5.2 Instrumentation

5.2.1 Hyperspectral Imaging Technology

1. **Brief Introduction of Hyperspectral Imaging Technology**
   HSI is an image data technology that emerged in the 1980s. It was initially used in remote sensing and military fields and then gradually expanded in many civilian fields, such as crops and food, biomedicine, geophysics, material science, vegetation species, and environmental disaster reduction. In the application of food and agricultural products, HSI technology also shows excellent potential. In recent years, many researchers have applied it to the detection of agricultural production, achieving good research results. The HSI technique's spectral range is from the ultraviolet region to the near-infrared region (200–2500 nm). Its working principle is that the imaging spectrometer can obtain many short and continuous spectral image data in the mid-infrared, near-infrared, visible, and ultraviolet regions with high resolution, providing several short-band spectral information for each pixel to produce a complete and continuous spectral curve. HSI is an interdisciplinary technique based on many narrow-band image data and

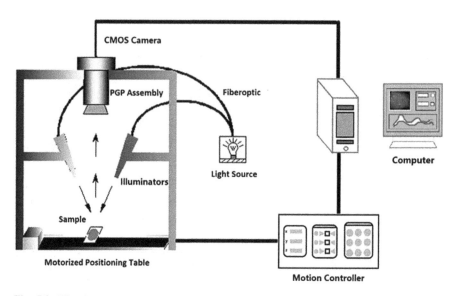

Fig. 5.2 Filter-based hyperspectral imaging system (**a**) and spectral ritual-based hyperspectral imaging system (**b**)

combined optics, optoelectronics, electronics, computer science, information processing, and so on.

2. **Hyperspectral Imaging Detection System**

The HSI system is generally composed of hardware and software. The hardware includes a spectral imager, light source, sample carrier, dark box, and computer. And the software consists of data acquisition and data processing software. At present, there are two kinds of HSI systems based on the filter and spectral ceremony. Two HSI systems are illustrated in Fig. 5.2 [3]. HSI is a combination of two-dimensional image and spectral technology. The hyperspectral image is a three-dimensional data block composed of two-dimensional images under a series of wavelengths with spectral information of each pixel. It can reflect the physical structure and chemical composition of the sample. There is also image information at various wavelengths, reflecting external features such as sample shape, color, and texture [4, 5], as shown in Fig. 5.3. The hyperspectral data block is composed of hundreds of adjacent bands of each pixel in the sample space. The spectral information at each point can represent the composition of the point, and the image information at each wavelength represents the spatial characteristics of the object surface.

Hyperspectral images exhibit considerable information and outstanding advantages in composition analysis, and thus it is very suitable for evaluating the quality and safety of food and agricultural products [6]. Its main benefits can be summarized as follows:

(a) It is a non-destructive testing technology without chemical reagents, almost no sample pretreatment, so save time, effort, and cost.

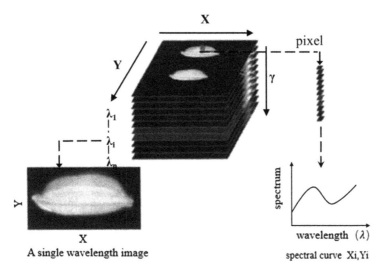

A single wavelength image

spectral curve Xi,Yi

Fig. 5.3 Hyperspectral data block

(b) It is similar to spectral technology, both qualitative and quantitative analyses can be carried out. The full spectrum curve of each pixel in the field of vision collected by hyperspectral imaging is an obvious advantage. It can not only describe the content of different substances in the sample but also describe its distribution.

(c) Operators could realize simple and rapid detection of samples by establishing and verifying the model. In the process of grading operation, the content and distribution of several different substances in the same sample can be determined at the same time, and the additional sample objects can be labeled and priced.

(d) The region of interest, the characteristic region, or the spectral curve at a pixel can be easily selected as spectral features and saved in the spectral database.

(e) Hyperspectral imaging has a good effect in distinguishing different objects with similar color, spectral curve overlap, and morphological characteristics, since it has rich spectral and spatial information.

However, like other technologies, still some problems in HSI technology need to be solved in future research. First, HSI is an indirect method like spectrum, which requires accurate correction data and robust model algorithms, which does not have a low detection limit like chemical analysis methods; second, hyperspectral data contains a lot of redundant data, which poses a significant challenge to data concentration, and the speed of hyperspectral hardware needs to be improved to meet the rapid acquisition and analysis of hyperspectral data blocks. Data acquisition and analysis are time-consuming, so direct online application is not recommended. The multispectral imaging system that obtains images at several optical wavelengths meets the speed requirement of quality detection. The optimized multispectral imaging system has a lower dimension and shorter data acquisition time.

5.2.2 Multispectral Imaging Technology

1. **Brief Introduction of Multispectral Imaging Technology**
   Hyperspectral image data has a large amount of information, and hundreds or even thousands of pictures can be collected for a detection target. Hyperspectral image technology has significant advantages in feature recognition, but it also brings great challenges. Massive data processing greatly reduces the detection speed. In other words, the massive data processing greatly reduces the detection speed, and it is challenging to meet the requirements of rapid online detection of modern food processing. Multispectral imaging is the abbreviation of multi-channel spectral imaging technology. If each pixel of channels is proportional to the spectral reflection value of the corresponding object point on the channel, the image is called the multispectral image. The three-dimensional data block composed of images with multiple discrete wavelengths can be used to observe and analyze the sample's composition, distribution, and surface texture. With the combination of multispectral imaging and chemometrics, a prediction model is established based on spectral and spatial information, which makes the prediction performance more stable than that of near-infrared spectroscopy [7]. Studies have shown that multispectral imaging is particularly suitable for noninvasive and rapid analysis of a series of quality-related components, providing spectral responses of spatial variables for components. Multispectral imaging technology is a new nondestructive detection technology that integrates traditional imaging and spectral imaging; spectral information and spatial information can be simultaneously obtained without destroying the sample [8]. Compared with the single-wavelength imaging system, the multispectral imaging system can get more comprehensive sample information. Therefore, multispectral imaging technology has been widely used in geological exploration, environmental monitoring, life state observation, and other fields. Compared with the traditional methods to detect food quality, multispectral imaging is a fast, environmentally friendly, and nondestructive technique. Compared with traditional hyperspectral technology, multispectral technology, with the merits of more straightforward data and low-cost, is applied to real-time online detection of food quality.

2. **Multispectral Imaging System**
   The multispectral system is generally composed of hardware and software, the hardware includes spectral imager, wavelength selection system, light source, sample carrier, dark box, and so on, and the software provides data acquisition and data processing software [9]. Like the hyperspectral system, the multispectral system is an imaging system involving a number of wavelengths. Still, the difference is the wavelength selected by the multispectral system is smaller than that of the hyperspectral system. At present, two kinds of multispectral systems are the most predominantly employed. One is a filter-based multispectral system (shown in the following Fig. 5.4) [10], selecting the full-band emitting

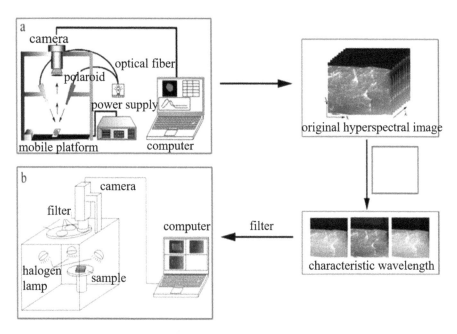

Fig. 5.4 The process for establishing multispectral imaging detection system. (**a, b**) Full wavelength spectral system

light source on request of the detection object (visible light or visible near-infrared band) [11], and then install filters of different wavelengths in front of the camera, realizing the need of different band selection by rotating the filter through the motor. The other is to emit different single-band at the light source, and the multispectral camera directly detects the reflection spectrum irradiated by a single wavelength on the object. Multispectral images are three-dimensional blocks of data composed of images with multiple discrete wavelengths, which can observe and analyze the sample's composition, distribution, and surface texture. Compared with the single-wavelength imaging system, the multispectral imaging system can find more comprehensive sample information. Therefore, multispectral imaging technology has been widely used in geological exploration, environmental monitoring, life state observation, and other fields. Compared with traditional HSI, the advantages of multispectral imaging are:

(a) Fewer spectral bands and shorter time to acquisition
(b) Simple data structure, easy to transmit, save, and process
(c) Low composition and operation cost

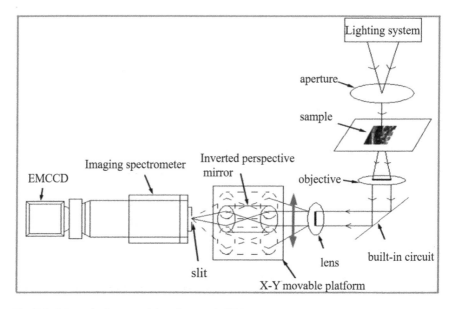

Fig. 5.5 Schematic diagram of the microscopic HSI system

5.2.3 *Hyperspectral Microscopic Imaging Technology*

1. **Brief Introduction of Microscopic Hyperspectral Imaging**

   Microscopic hyperspectral imaging involves HSI and microscopic imaging, which can extend the research content of HSI technology to the microscopic field and provide a new method for studying and observing small objects. It can be used to observe the morphological changes of tissues, cells, and microorganisms. Recently, it has been widely used in medical research. For example, food spoilage is usually a process in which microorganisms invade and consume the nutrients in pork to release metabolites. Therefore, during the process of spoilage, there are some changes in the distribution, the number of microorganisms, and internal tissues (such as muscle cells, connective tissue, intermuscular fat, etc.) differences in their absorption spectra. For example, when muscle cells are invaded by microorganisms, there are changes in the composition, structure, and function of the cell membrane, which inevitably leads to changes in the content and configuration of cell components such as protein, nucleic acid, sugar, lipid, and water molecules. Compared with normal cells, there are significant differences in spectral absorption frequency, peak intensity, and other spectral parameters. These differences must be captured when detecting pork freshness and tracking the deterioration process. They can only be shown in the micro-scale images of specific bands. Microscopic hyperspectral image integrates spectrum and microscopic image, representing the changes of internal information and microstructure in the process of quality change.

2. **Hyperspectral Microscopic Imaging System**

The design principle of the HMI system is shown in Fig. 5.5 [10]. The sample is placed on the carrier stage of the inverted microscope and illuminated by the Kohler lighting system. The stripe in the instantaneous field of view designs a translation fine-tuning lens between the eyepiece of the microscope and the C-Mount interface, which leads to the corresponding translation of the imaging position of the sample on the carrier stage by moving it. Combined with the slit, the traditional broom imaging can be realized, and finally, three-dimensional blocks of data can be obtained.

The HMI system comprises the following parts: inverted microscope for data acquisition; imaging spectrometer; electron multiplier CCD camera, quartz mercury lamp lighting system; X-Y mobile platform; data acquisition and preprocessing software; computer for data storage and display. In the micro-spectral imaging system, the micro-spectral images in the visible/near-infrared spectral range can be obtained, each hyperspectral fiber data contains hundreds of visible/near-infrared spectral images. The two ends of the imaging spectrometer are connected with a movable optical lens (incident slit end) and an EMCCD camera. The other end of the optical lens is associated with the inverted microscope through the C-Mount interface. The computer is connected with the stepper motor through the serial port to control the movement of the X-Y mobile platform, which can synchronize the scanned image of the camera with the movement of the platform. The specific physical drawing is shown in Fig. 5.6 [10].

5.2.4 Infrared Laser Imaging Technology

1. **Introduction of Infrared Laser Imaging Technology**

Laser imaging is a technique which mainly based on the physical optical path change process. When a parallel laser beam irradiates an inhomogeneous medium, a series of optical phenomena such as diffraction, interference, and refraction will occur, as shown in Fig. 5.7 [12]. When the laser illuminates the detected object, one part of the laser directly passes through. The other part of the light will have a series of optical phenomena such as diffraction, interference, refraction. Both form the unique image information related to the quality and safety of the detected object by projecting onto the diffuse reflection screen.

Meanwhile, the obtained image information can also be regarded as the propagation of plane wave or Gaussian beam. On the one hand, the near-infrared laser has strong penetration ability, and no damage to biological tissue and biological tissue hardly absorbs near-infrared light. Thus it reduces the interference of biological tissue autofluorescence and improves the signal-to-noise ratio [13]. On the other hand, the texture image is not clear if the excitation wavelength and the optical change is inappropriate. The physical properties such as refractive index, shape, and biochemical composition of the object are the main factors, which could affect the intensity distribution and polarization law of the texture

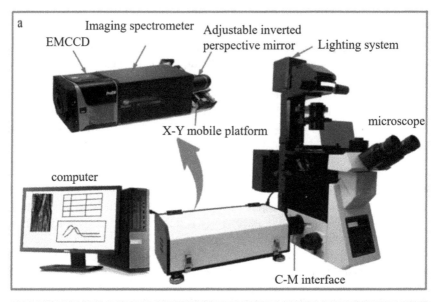

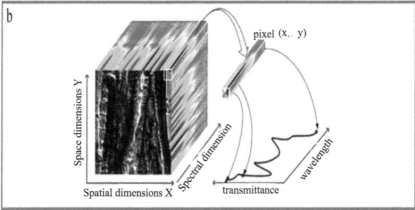

Fig. 5.6 Physical image of the microscopic hyperspectral imaging system (**a**); Fiber hyperspectral Image 3D Data Block (**b**)

image. Therefore, the image information of the detected objects such as size, intensity, and shape of the outer ring is different.

2. **Infrared Laser Imaging System**

Figure 5.8 is the schematic diagram of the near-infrared laser imaging system [12]. The system chooses the laser emitter of 980 nm as the excitation light source to collect the image of the object, which including the following four parts:

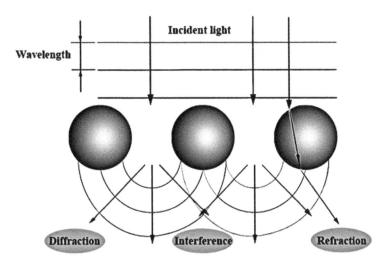

Fig. 5.7 Principle of light scattering

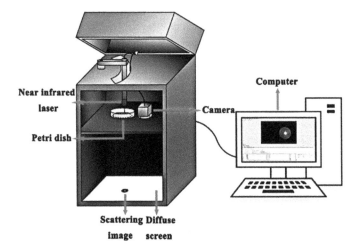

Fig. 5.8 Near-infrared laser imaging system

(a) Near-infrared laser emitter (excitation wavelength 980 nm, power 1 mW, excitation light is a positive circular spot with diameter 5 mm, Changchun Xin ye Optoelectronic Technology Co, Ltd.).

(b) Near-infrared camera (wavelength response range is 900–1700 nm, the resolution is 320 × 256 pixel).

(c) Diffuse reflection screen.

(d) Computer.

The distance between the laser emitter and the plate and the distance from the camera lens to the diffuse screen remain constant throughout the experiment, determined since the specific experimental conditions. The parallel beam emitted by the laser emitter irradiates the colony on the plate vertically. One part of the light directly passes through the detection object to the diffuse reflection screen. The other part of the light is scattered on the detection object and projected onto the diffuse reflection screen. The texture images formed by transmitted light and scattered light on this screen are collected and transmitted to the computer by a near-infrared camera.

5.2.5 *Raman Spectral Imaging Technology*

1. **Overview of Raman Spectral Imaging Technology**

Raman imaging is a new generation of fast, high-precision, surface scanning laser Raman technology [14], which perfectly combines confocal microscope technique with a laser Raman spectral technique. As the third-generation Raman technology, the characteristics of it are high speed and high resolution. The combination of microscopic technology and Raman spectrometer is a revolutionary breakthrough in Raman experimental technology. The microscopic objective lens has been used to focus the laser beam on the sample and collected the Raman scattered light in the microscopic Raman system. On the one hand, it also reduces the laser power needed for measurement, thus significantly broadening the application range of Raman spectroscopy. Furthermore, microscopic technology promotes the spatial resolution of Raman measurement to submicron and micron scale, which introduces a new experimental method for Raman spectroscopy, known as Raman spectral imaging.

Raman spectral imaging technology, combining the Raman spectroscopy and machine vision, which can obtain the Raman spectra of the sampling points in the image, and the Raman spectra can obtain the information of the vibration and rotation of material molecules. Thus, it could represent the chemical bond and functional groups of substances, which can be used for qualitative analysis, quantitative analysis, and determination of molecular structure. The sample information obtained by Raman spectral imaging can be analyzed and processed by the spectrum, image, or a combination of both, and make full use of the information provided by the object. Raman spectroscopy is an emerging analytical technique, which can achieve real-time image synthesis and composition distribution of target chemical substances, providing fingerprint information of chemical and biological molecular structures.

The basic principle of Raman imaging technology: benefit by the progress of laser, Monochromator and weak light signal detection technique, Raman spectrum can currently not only distinguish all kinds of chemical composition information in trace mixtures with the help of characteristic Raman frequency. Moreover, the spatial distribution information of various components can be given, and its spatial resolution is close to the diffraction limit of light, which is Raman imaging technology [15]. The main chemical components and the spatial distribution of each component in the sample could be revealed by Raman images, which can also give the size and number of particles (aggregates) in the sample and show the stress distribution on the semiconductor material and the molecular orientation on the micron scale.

2. **Raman Imaging Technology DuoScan and SWIFT**

DuoScan is a new imaging technology, which uses two high-speed rotating mirrors to control the laser beam to scan the sample surface in a user-selected mode to create a Raman image. The working methods of the DuoScan includes:

(a) Average mode: The measuring area (1300 μm, square, circular, or arbitrary shape) is selected by the user since the condition of the sample, and the laser spot is scanned quickly and continuously on the sample by the high-frequency rotation of two mirrors, which is equivalent to a large area of the light place, thus the average spectrum of all components in this region is obtained, which is suitable for Raman imaging of photosensitive biological samples.

(b) Point-by-point mode: A part of the model is selected in advance, and then fine point-by-point scanning is carried out in this area to construct a Raman image, which is suitable for submicron chemical imaging.

(c) Large-scale scanning mode: With the help of the movement of the automatic platform, the large light spot is used to scan the larger sample, record the spectrum on the whole sample surface and construct the Raman image. This model is especially suitable for component distribution analysis or locating small-size objects such as pollutants and nanomaterials on the substrate.

The DuoScan mode can be used for fast and uniform scanning for all sample areas. Because the scanning step can be as low as 50 nm, Raman imaging can be extended from the submicron scale to the macroscopic scale. Optical elements such as lenses should not be used in this mode. The spectral range operations are 220 mm to 1600 nm, so it is suitable for any laser from ultraviolet to near-infrared. In all the modes of DuoScan, the Raman signal is kept on the confocal axis, so the confocal characteristics of measurement results maintain reasonable, and the generation of confocal images becomes faster, easier, and more flexible.

SWIFT means "incredibly fast scan." In general, the data acquisition time of each sample point cannot be less than 500 ms, but the new data acquisition method of SWIFT mode can reduce this time to less than 5 ms; therefore, the samples suitable for this mode can only be those signals that are strong and can acquire enough signals in a short time.

5.3 Spectral Imaging Processing and Data Analysis

5.3.1 Composition of Spectral Image Data

As mentioned earlier, spectral image data is three-dimensional (as shown in Figs. 5.9 and 5.10), known as image block (hyperspectral cubes) [16]. The two-dimensional

Fig. 5.9 Hyperspectral
image data

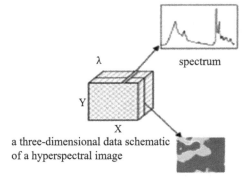

a three-dimensional data schematic
of a hyperspectral image

the image at a particular wavelength

Fig. 5.10 Schematic
diagram of image data
obtained by hyperspectral
image system based on
spectrometer

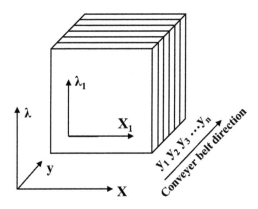

data blocks are the horizontal and vertical coordinate information of the image pixel
(represented by coordinates x and y). The third dimension represents the wavelength
information (represented by λ). For example, the data of spectral images are outlined
by image and spectral dimension.

1. Image dimension: In the image dimension, the hyperspectral image data is similar
   to the general image.
2. Spectral dimension: Corresponding to each pixel of the hyperspectral image,
   there is a continuous spectral curve.

For example, an image array detector with a resolution of 512×512 pixels
obtained sample image information at 100 wavelengths, and the image block is a
$512 \times 512 \times 100$ three-dimensional array.

The characteristics of hyperspectral images include multiple bands, high spectral
resolution, and a strong correlation between spectra, so that the data obtained by
hyperspectral images required to remove redundant information between adjacent bands.

5.3.2 Spectral Image Data Description Model

The spectral image technique can obtain the image and spectral information of the detected object with many narrow electromagnetic wavebands. The hyperspectral image technology with hundreds of imaging wavelengths is almost equivalent to continuous sampling on the spectral interval of the sample, and the obtained data is different from the traditional image [17]. Therefore, defining the hyperspectral data accurately is the premise and foundation of data processing and analysis. At present, hyperspectral image data can be outlined as follows.

1. **Image Description**

   The image description is one of the most intuitive ways to present hyperspectral image data information. It describes the spatial position relationship between detection objects. A fraction of the hyperspectral image data can be represented by the image description, since only the gray image of one wavelength or the color image composed of three or four wavelengths can be observed at a time, and the correlation between wavelengths is hardly reflected from the image. One of the most significant merits of this data display method is displaying the measured agricultural products in the form of images, providing spatial knowledge for hyperspectral image processing and analysis.

2. **Spectral Curve Description**

   The spectral curve has been used to describe the hyperspectral image data's information, depending on the relationship between the spectral response of the detected object and the wavelength. In the description, the change of the gray value of each pixel with different wavelengths reflects the spectral information of the detected object at the pixel, which can be abstracted as an approximately continuous spectral curve. The value of each point on the curve is the spectral response value of the detected object at the corresponding wavelength. Spectral curve description is mainly employed in data analysis methods based on spectral space.

3. **Eigenvector Description**

   Each pixel in the hyperspectral image corresponds to the reflection values of multiple imaging bands, which can be described as an N-dimensional vector in the multidimensional data space (N represents the number of imaging bands). It is equivalent to transforming the approximately continuous spectral curve into an N-dimensional vector in the multidimensional data space (as shown in Fig. 5.11). The advantage of this method is that objects with different distribution characteristics are distributed in various areas in the feature space, which is conducive to the quantitative description of the detected objects' spectral characteristics and their changes in the feature space.

Among those methods, the description method based on feature space has been applied well in spectral image processing and analysis. For example, the spatial distance judgment method is formed from the point of view of spatial geometry, statistical discriminant models established from multivariate statistics, and the data

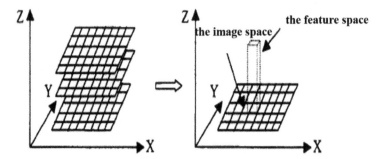

Fig. 5.11 Spectral image feature vector expression

analysis tools such as fuzzy theory, fractal geometry, and neural network are based on feature space. Feature space is an ideal method to describe hyperspectral image data. It represents the changes of the spectral characteristics of the detection objects and describes the distribution structure of the detection objects.

The merits of spectral image technique are embodied in the amount of information. Still, the characteristics of spectral image data are usually in several bands (dozens or even hundreds), high spectral resolution (nanometer level), and massive data which are inconvenience for application and analysis. Therefore, it is significant to choose a reasonable image data processing algorithm for the design of agricultural products detection system. Spectral image signal processing methods are generally divided into image feature data processing methods and spectral feature data methods.

1. **Image Feature Data Processing Method**

   The image features at different wavelengths are processed in the collected spectrum. At present, there are various algorithms for hyperspectral image processing. Several commonly used methods are outlined below.

(a) Principal component analysis (PCA)

   PCA is widely used in hyperspectral image data processing. Hyperspectral image data usually have some redundancy; PCA works to make the original redundant data into mutually independent data. Meanwhile, it could select less data to replace the original data depending on the actual requirement. Therefore, it is a mathematical dimensionality reduction method. The characteristic wavelengths for detecting the quality of agricultural products are employed by the technique. The hyperspectral images at these characteristic wavelengths are scaled to acquire the proportional image, and the features of the proportional image are extracted in the end. For example, extracting the gray co-occurrence matrix of spectral images at each wavelength and texture features based on statistical moments constitute the input of the pattern recognition model.

(b) Quadratic differential analysis

To highlight the quality characteristics of agricultural products and make use of the advantages of hyperspectral image information, quadratic differential analysis is currently most popular. The methods of quadratic difference analysis contain symmetrical second difference and nonuniform quadratic difference analysis.

The uniform quadratic difference (symmetrical second difference) analysis method is as follows:

$$
\begin{aligned}
S''(\lambda_n, g) &= S'(\lambda_n + g) - S'(\lambda_n) \\
&= [S(\lambda_n + g) - S(\lambda_n)] - [S(\lambda_n) - S(\lambda_n - g)] \\
&= S(\lambda_n + g) - 2S(\lambda_n) + S(\lambda_n - g)
\end{aligned}
$$

In the formula, the uniform quadratic difference gray value at the center wavelength λ_n is $S''(\lambda_n, g)$.

The wavelength interval is g (nm).

The uniform primary difference gray value at the center wavelength λ_n is $S'(\lambda_n, g)$.

The gray value at the center wavelength λ_n is $S(\lambda_n, g)$.

The quadratic difference algorithm can be applied to identify pollution and defects in agricultural products images, reducing the influence of uneven light caused by the uneven shape of agricultural products. The principle of selection is to ensure that the defective and contaminated part differs in the normal part. Generally, λ_n, the strongest wavelength of the absorption spectrum of the defection or contamination, with the fixing of wavelength interval g, cannot select the effective feature image, contributing to nonideal detection effect, while the nonuniform quadratic difference can cope with it.

The method of asymmetrical second difference analysis is as follows:

$$
S''(\lambda_n, g) = S(\lambda_n + g_1) - 2S(\lambda_n) + S(\lambda_n - g_2)]
$$

In the formula, the grayscale value of the asymmetrical second difference image at the center wavelength λ_n is $S''(\lambda_n, g_1, g_2)$.

g_1, g_2 represents the wavelength interval, $g_1 \neq g_2$.

(c) Band ratios

Band ratio algorithm can effectively reduce the uneven light reflection caused by uneven surface and enhance the spectrum difference between the selected bands simultaneously, providing useful information, which a single band. The process of this algorithm is to divide the image in one band from the image in another band, which generate a more precise image that contains the required target feature information. And its mathematical expression is shown in the formula.

$$BW_{(i,j,r)} = \frac{BW_{(i,j,m)}}{BW_{(i,j,n)}}$$

where

$BW_{(i, j, r)}$—The ratio of pixels in the corresponding position of the image (i, j)

$BW_{(i, j, m)}$, $BW_{(i, j, n)}$—The grayscale value of pixels in the same position in them and n bands (i, j)

2. **Spectral Characteristic Data Processing**

The software cuts the hyperspectral image properly, taking the center of the image as the center, intercepts the x (pixel) $\times$ y (pixel) rectangular region, which contains the image with the wavelength in the 450–900 nm range and obtains a $200 \times 200 \times 525$ three-dimensional block of data. Moreover, mean and extreme values can be gained from the selected data block, then one-dimensional spectral data block obtained can be employed for further analysis.

5.4   Application of Spectral Imaging Technique in Food Quality and Safety Detection

Spectral imaging is a new technique for obtaining spatial and spectral information from objects that combines traditional computer imaging and spectral technology [18]. Spectral imaging is a potentially appealing process analysis technique for food applications because of its robust, nondestructive and flexible nature. Therefore, it has been employed extensively for agricultural products, defect identification, component analysis, quality evaluation, and other food quality and safety detection.

5.4.1   Application in the Detection of Agricultural Products

In the growing and processing of agricultural products, many factors are causing product quality pollution and deterioration, which are mainly reflected in the pollution and particle size of heavy metals in vegetables (lettuce) and in different processing methods. The sensory quality discrimination model for tea was established using spectral imaging and chemometrics, and the image, spectrum, and olfactory characteristics of tea were collected. HSI technology has been applied for nondestructive detection of heavy metals successfully in vegetables and fruits. Combined with a variety of algorithms, effectively extract in-depth spectral features and establish a robust heavy metal content prediction model for early detection of toxic stress in crops to monitor its growth effectively for high-quality crops and increased cultivation yields [19].

	Grade 1	Grade 2	Grade 3	Grade 4
762.67 nm				
793.26 nm				
838.70 nm				

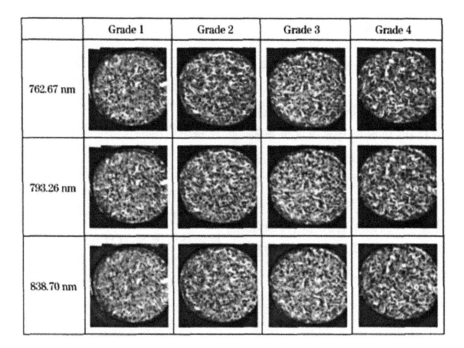

Fig. 5.12 Four grades of tea sample images at three feature bands

1. Detection of Tea Quality

In order address shortcomings in sensory evaluation method for tea quality and realize the rapid nondestructive testing of internal and external quality information, a quantitative analysis of tea sensory quality using hyperspectral images was proposed, and the discriminant model of tea grade was established. Chen et al. [20] suggested a HSI system based on spectrometer which was developed to perform HSI data acquisition. Three optimal band images were selected by PCA. Then, from each optimal band image, six texture features based on statistical moments (mean, standard deviation, smoothness, third moment, uniformity, and entropy) were extracted, giving each tea sample a total of 18 variables. Finally, PCA was carried out on 18 characteristic variables, and eight principal components (PCs) were extracted as inputs of BP neural network. In the experiment, the wavelengths corresponding to three large weight coefficients were chosen from hyperspectral image data in the 700–850 nm band for analysis. Based on PCA, these three wavelengths were chosen for characteristic images [20]. Figure 5.12 depicts images for different grades of tea leaves at these three characteristic wavelengths.

Figure 5.13 shows the training and prediction results of back propagation (BP) neural network model under different number of principal component factors. It can be seen that as the number of principal component factors increased, the recognition rate of the training set and prediction concentration

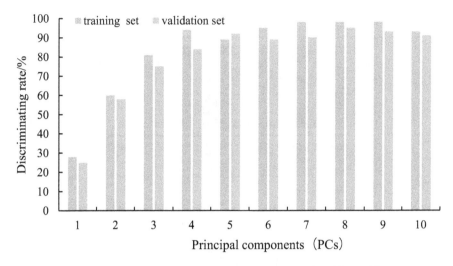

Fig. 5.13 Discriminating results in training set and prediction set with different PCs

increased. The model's recognition rate changed when the number of principal component factors reached 8. The first eight PCs could explain 99.75% of the original data information and were independent of each other, eliminated redundant information, and improved the stability of the model.

2. **Detection of Matcha Sensory Quality**

Different cultivation and production methods of matcha may lead to different sensory qualities. Ouyang et al. [21] used hyperspectral microscopy to study the particle size of matcha powder to express the changes of its microstructure. Figure 5.13 illustrates the process of selecting feature spectral images and ROI. Figure 5.14a, b shows the color image and spectrum of a representative matcha sample obtained from the man–machine interface system, respectively. The spectral image at 524 nm was selected as the feature image to express the particle size distribution, as shown in Fig. 5.14c, d shows the ROI of the corresponding sample, which was marked green, dark and white areas were excluded for further analysis due to non-conformities [21].

In order to explore the feasibility of evaluating matcha particle size distribution by man–machine interface technology, matcha particle size models based on D10, D20, D30, D40, D50, D60, D70, D80, and D90 were established. The key spectral features were taken as the original data of the fusion model, and 28 texture features were extracted from the feature image. Figure 5.15 shows the scatter diagram of the correlation between the predicted particle size and the reference value in the ANN model for each particle size distribution.

3. **Detection of Heavy Metals in Lettuce**

Heavy metal pollution is a serious issue for the environment. According to Xin et al. [22], with the growing demand for lettuce, it was necessary to improve lettuce cultivation management and pollution control. For the extraction of depth

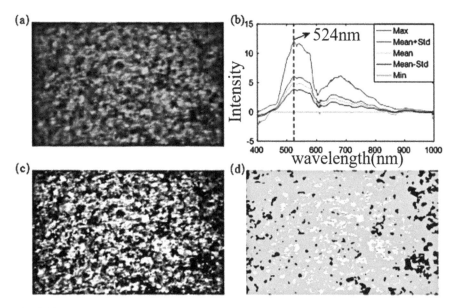

Fig. 5.14 The characteristic spectral image and ROI selection process: (**a**) the color image, (**b**) spectra, (**c**) image at 524 nm, and (**d**) the ROI marked with the green of one representative matcha sample

features of heavy metals in lettuce leaves, a depth learning method based on wavelet transform (WT) and stack convolution automatic encoder (SCAE) was proposed. The optimal wavelength was obtained by successive projection algorithm (SPA) and variable iterative space contraction method (VISSA), and the deep features were extracted by the combination of WT and SCAE [22]. The characteristic wavelengths of heavy metals extracted by SPA and VISSA are shown in Fig. 5.16.

A support vector machine prediction model of Cd and Pb content was established using the deep characteristics of deep learning, and the results are shown in Table 5.1. Table 5.1 shows that the support vector machine prediction model of Cd content based on the depth characteristics of WT provide good prediction results with R_P^2 of 0.9319, RMSEP of 0.04988 mg/kg, and RPD of 3.187. In addition, the support vector machine prediction model of lead content based on the depth characteristics of WT had also achieved good prediction results, with R_P^2 of 0.9418, RMSEP of 0.04123 mg/kg, and RPD of 3.214. The results showed that heavy metals Cd and Pb could be detected in lettuce using hyperspectral technology combined with deep learning algorithm.

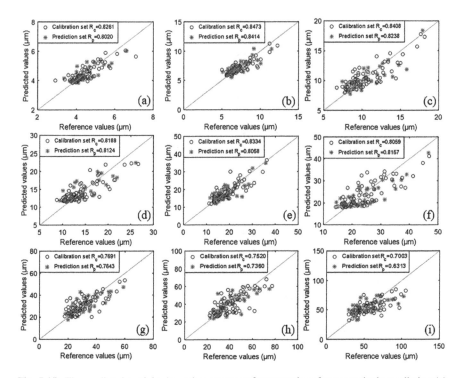

Fig. 5.15 The predicted particle size values versus reference values for respectively predicting (**a**) D10, (**b**) D20, (**c**) D30, (**d**) D40, (**e**) D50, (**f**) D60, (**g**) D70, (**h**) D80, and (**i**) D90 in the ANN models with data fusion

5.4.2 Application in the Detection of Liquid Foods

Solid-state fermentation is an important step in vinegar production that directly affects its quality and yield. In addition, the total acid, pH, and non-volatile acid are also key indexes to measure the quality of fermented vinegar grains. Based on their changes, the fermentation status of vinegar-fermented grains could be determined in time, and problems such as hardening or bad fermented grains can be reduced or avoided. Zhu et al. [23] exploited HSI technology to study the image and spectral information of vinegar grits. According to the different PCs of image information, three feature images were selected, and four texture feature variables were extracted, including contrast, correlation, angular second moment, and consistency. *K*-nearest neighbor (KNN) method was used to establish the identification model of vinegar grains. As shown in Fig. 5.17, when $K = 1$ and the number of PCs was 10, the optimal KNN model predicts the samples in the fermentation stage. The recognition rate of the prediction set was 90.04% [23].

The preprocessed hyperspectral image was divided into a training set and a prediction set. Then, PLS, IPLS, and SIPLS methods were used for interactive

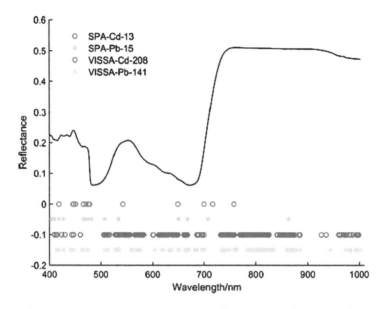

Fig. 5.16 Composite heavy metals characteristic wavelengths extracted by SPA and VISSA

Table 5.1 The results of the SVR model based on deep features extracted by WT-SCAE

HM	BL	Model scale	Calibration set		Prediction set		RPD
			R_C^2	RMSEC (mg/kg)	R_P^2	RMSEP (mg/kg)	
Cd	3	478–240-112	0.9603	0.03618	0.9176	0.05483	3.151
		478–240-112–95	0.9771	0.03023	0.9319	0.04988	3.187
		478–240-112–95-67	0.9702	0.03315	0.9234	0.05344	3.166
		478–240-112–95-67–35	0.9437	0.04094	0.905	0.05503	3.099
Pb	4	478–240-127	0.9627	0.03659	0.9281	0.05205	3.178
		478–240-127–90	0.9756	0.02617	0.9361	0.04612	3.203
		478–240-127–90-55	0.9863	0.02281	0.9418	0.04123	3.214
		478–240-127–90-55–40	0.9688	0.03547	0.9307	0.05176	3.183

Note: (1) Heavy metals are abbreviated as HM. (2) The best wavelet decomposition layer is abbreviated as BL. (3) The number of neurons in each layer of WT-SCAE is represented by the model scale

verification modeling, respectively. The results demonstrated the best effect for the optimized SIPLS model. The RMSEP of lumped acid, pH value, and nonvolatile acid was predicted to be 0.75, 0.05, and 0.3, respectively, which could realize the rapid prediction of important physical and chemical indexes. Therefore, the HSI technology could quickly predict the fermentation state of fermented vinegar grits, thereby optimizing the process operation and improving the vinegar quality.

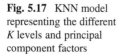

Fig. 5.17 KNN model representing the different K levels and principal component factors

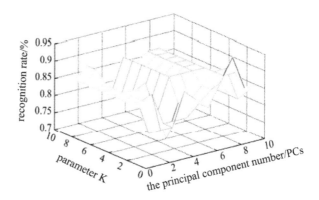

5.4.3 Application in the Detection of Poultry

China, according to the Food and Agricultural Organization (FAO), is a major producer and consumer of poultry [17]. The poultry meat is considered a healthy dietary choice as it is rich in protein (15–20%), mineral salts, and vitamins. The protein components in meat will gradually decompose to produce histamine, tyramine, putriamine, tryptamine, and other toxic small molecule components, making it vulnerable to deterioration. Fats could be degraded into aldehydes and aldehydes acids, and carbohydrates might be converted alcohols, carboxylic acids, ketones, aldehydes, etc. These substances along with other basic nitrogen compounds together make up total volatile elemental nitrogen (TVB-N) during storage [24].

1. **Detection of Pork Freshness**

    Xu et al. [25] proposed a method for rapid microscopic detection of pork freshness, using self-assembled HMI system to detect TVB-N in different storage periods, and regarded it as the classification standard of pork freshness [25].

    Pork samples were prepared in advance, each sample was vacuum packed in a sealed bag, with the air extracted using a vacuum pump. After 1, 3, and 5 days of cold storage at 4 °C, the main components such as protein, fat, and carbohydrates in pork would be decomposed into toxic small molecules with the extension of storage time. The main compounds were TVB-N compounds, mainly ammonia, dimethylformamide, and methyl trimethacrylate. Figure 5.18 shows the reference measurements of pork for different storage times, and it could be seen that the difference in TVB-N content among the three groups was very significant [25].

    Before classification modeling, PCA for the whole image matrix was carried out, and three PCs were extracted with a total variance of principal component 1 (PC_1) image description was as high as 96.35% making the data set's information easy to visualize while retaining the majority of the original data set's information. Therefore, according to these PC_1 images, the dominant bands were estimated by investigating all the weighting coefficients, five dominant bands with higher weight coefficients (i.e., 521.08, 589.69, 636.88, 687.58, and

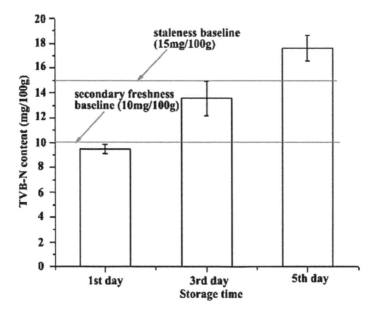

Fig. 5.18 TVB-N content in pork meat for the first, third, and fifth storage day

738.66 nm) were selected (see Fig. 5.19). The results showed that based on the extracted data, back propagation artificial neural network (BP-ANN) gave better results with 100% accuracy for freshness classification.

2. **Detection of Chicken Freshness**

Chicken is very popular among consumers because of its high protein content, low fat content, and easy digestion [26]. However, raw chicken is an extremely perishable food with a short shelf life of only 3–5 days, even if it is refrigerated at 4 °C [27]. The amount of TVB-N in chicken is served as a chemical index for determining the degree of spoilage. The higher the content of TVB-N, the higher the degree of microbial spoilage. Urmila et al. [27] used HSI system to evaluate the quality of chicken. PCA and ant colony optimization (ACO) were compared in data dimensionality reduction.

As shown in Fig. 5.20a, the HSI system developed by Jiangsu University's Agricultural Product Processing and Storage Laboratory was used to collect sample data. In this paper, five ROIs were selected from the hypercube, with the size of each sample being 50 × 50 pixels, and five ROIs were obtained from each sample. The average of the five ROI spectra represented the original spectra of the sample, and all 75 samples were processed. Figure 5.20b shows the average spectrum of a sample.

The five main texture feature images were based on the selected wavelengths of 544.01, 638.75, 705.25, 726.08, and 855.33 nm, as determined in Fig. 5.21a. The wavelength selection of the five texture feature images based on PCA is shown in Fig. 5.21b. PCA and algorithm were used to extract the principal

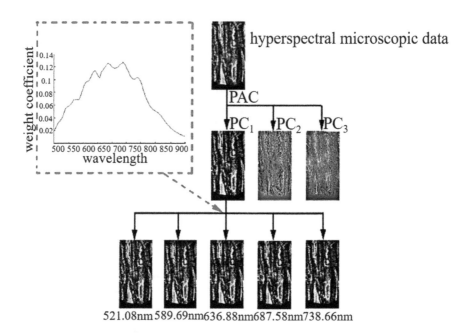

Fig. 5.19 Dominant wavelengths selected by PCA. PC1, principal component 1; PC2, principal component 2; PC3, principal component 3

wavelength image of chicken hypercube. Then, according to the statistical moment, six texture variables were extracted from each dominant wavelength image, with a total of 30 variables. Next the classical BP-ANN algorithm was used for modeling. The optimal ACO-BPANN model was obtained by the prediction set RMSEP = 6.3834 mg/100 g and $R = 0.7542$. The results indicated the applicability of the proposed HSI system for rapid and nondestructive quantification of TVB-N content in chicken.

5.4.4 Application in the Detection of Aquatic Products

Fish products are a staple of most people's diets all over the world. Fish bone is often present in fish products, which is regarded to have a serious hazard. To investigate the Raman spectral differences between fish bones and fish, a new method for fish bone detection based on Raman HIS was developed by Song et al. [28]. The fuzzy rough set model (FRSTCA) based on heat charge algorithm was used to select the optimal band information followed by the establishment of the support vector data description (SVDD) classification model [28].

At the unique Raman peaks of five-layer fishbone slices, single-band Raman spectra of five groups of raw and pretreated fish bones were obtained (i.e., fish fillets

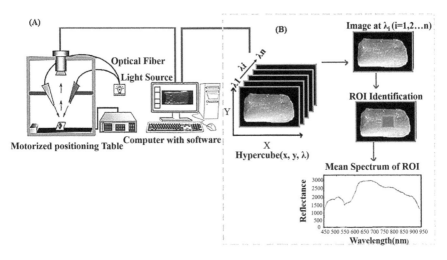

Fig. 5.20 (**a**) The developed HIS system. (**b**) ROI selection from a hypercube and spectral data extraction

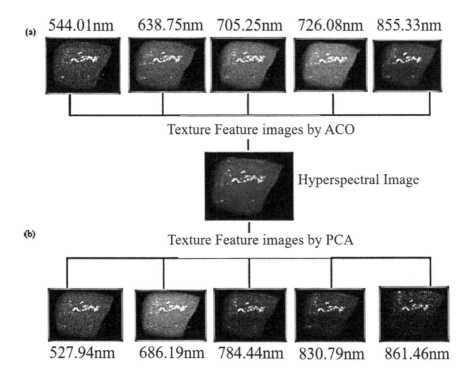

Fig. 5.21 (**a**) Five texture feature images extracted from the characteristic wavelengths selected by ACO. (**b**) Five texture feature images extracted from the characteristic wavelengths selected by PCA

| Fish bones on the surface of fillet | Fish bones under 1mm fish slice | Fish bones under 2mm fish slice | Fish bones under 2.5mm fish slice | Fish bones under 3mm fish slice |

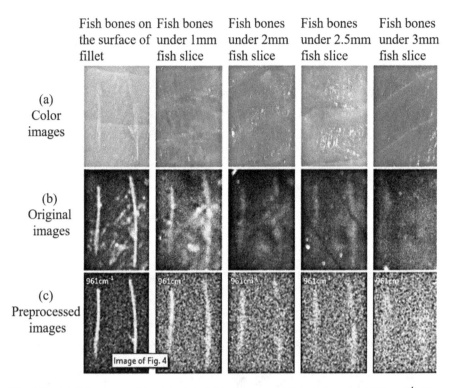

(a) Color images

(b) Original images

(c) Preprocessed images

Fig. 5.22 (a) Color image, (b) original, and (c) preprocessed Raman images at 961 cm^{-1}. (For interpretation of the references to color in this figure legend, the reader is referred to the Web version of this chapter.)

with thickness of 0, 1, 2, 2.5, and 3 mm) are shown in Fig. 5.22b, c, respectively. Figure 5.22a is a comparison of the color images corresponding to the layered samples. The Raman signal of fish bones was the strongest in the layered samples with fish bones on the surface of fish fillets, as shown in Fig. 5.22c.

Figure 5.23 shows the final test results of five cases. The image of the detection result is shown in Fig. 5.23c. The classification method distinguishes the texture of fish bones from that of fish fillets and detects fish bones that are not visible to the naked eye (Fig. 5.23a). The anatomical images obtained from the destructive experiment are shown in Fig. 5.23b. Experimental results distinguished the fish bones with a depth of 2.5 mm, and the detection performance was up to 90.5% [28].

5.4.5   Applications in Detection of Other Foods

The imaging technology can fully reflect the external characteristics of the sample, while the spectral technology can detect the physical structure and chemical

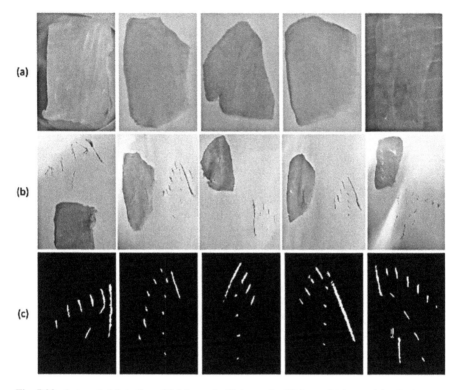

Fig. 5.23 Automated detection of fish bones in fillet sample of 2.5 mm thickness: (**a**) color images, (**b**) anatomical images, and (**c**) graphical result. (For interpretation of the references to color in this figure legend, the reader is referred to the Web version of this chapter.)

composition of the sample. In addition to the detailed examples described above, it is widely used in many areas, as shown in Table 5.2. The spectral imaging technology in food detection has a number of interesting avenues. It can extract image information from natural foods and detect the quality changes of processed foods in order to obtain comprehensive product quality data and improve the food quality testing level.

5.5 Future Trends and Conclusion

Currently, most spectral image detection systems used for agricultural product quality inspection use spectral regions between 400 and 1100 nm. The spectral images in these ranges include the special absorption information of water, protein, peroxide value, chlorophyll, and carotenoid. These absorption bands are also very effective for detecting surface defects or contamination of agricultural products and food. It will become more important to study the long-wavelength near-infrared

Table 5.2 Application of spectral imaging detection technology in regional food detection

Class	Product	Detection result	Reference
Natural food	Eggs	The correct identification rate of egg samples by LDA analysis was 95.7%	[29]
	Navel orange	The correlation coefficient R of navel orange sugar content prediction model was 0.831	[30]
	Ginkgo biloba	Flavonoid content correction model ($R = 0.9307$)	[31]
	Ophiocordyceps sinensis	Recognition accuracy of the support vector classification (SVC) models was 96.30%	[32]
	White maize	The overall accuracy of the full NIR spectrum (288 wavebands) model was 93.3%.	[33]
Processed food	Dry-cured ham	The accurate ones of the PLS predictive models for predicting (R2p) were 0.89, 0.85, 0.83, and 0.63	[34]
	Enriched pasta	Results showed R2V between 0.28 and 0.89, % LOF <6%, variance explained over 99%	[35]
	Milk powders	Detection of lower concentrations in milk powders (e.g., <0.02% or 200 ppm)	[36]
	Chocolate powder	Detection for the lowest concentration level (0.1% of peanut)	[37]
	Fried lotus root	The correlation coefficient and root mean square error of model prediction set were 0.819 and 0.682 mg/g	[38]

(1100–2500 nm) spectral images as nondestructive detection technology progresses from visual inspection and appearance to internal quality. Spectral imaging technology can simultaneously obtain spectral and image information from samples, which is useful for more detailed analysis. Although the computation time will be lengthened due to the large amount of data and redundant information, influencing the rapid detection goal. As a result, a series of information filtering is required to reduce the redundant data and decrease analysis time. First, the effective wavelength ranges could be obtained by screening characteristic wavebands, and then a filter-multispectral imaging system with multi-band characteristic wavelengths is designed and constructed, which fundamentally solves the shortcomings of huge spectral data and time-consuming computation. On the other hand, when analyzing spectral image data, some new chemometrics algorithms, such as independent component analysis and uneven quadratic difference, are introduced to improve spectral imaging detection's effectiveness and accuracy and enhance the robustness of the detection system. Besides, the online inspection equipment requires the coordination of the material delivery unit. For instance, during the delivery process, the sample can be turned over so that the multispectral camera can collect the spectral images of the sample from different angles. In this way, the online detection equipment reasonably evaluates the quality according to the comprehensive spectral image information. The goal of online, rapid, nondestructive and accurate detection of the complete quality of agricultural products can be further achieved using the above optimization methods.

Spectral imaging technology is an optoelectronic has attracted a large amount of attention in recent years and has become the frontier research field in nondestructive detection of food and agricultural products. It does not necessitate any complicated pretreatment prior to testing. It almost does not use any chemical reagents and instead relies on optoelectronic signals and chemometrics algorithms to obtain sample information for discrimination. It produces a huge amount of data, contains abundant information, and has the characteristics of "map integration." The development and application research of spectral imaging technology has a lot of potential in the processing and detection of food and agricultural products in the future, especially given the rapid development of data mining and artificial intelligence technology.

References

1. Xu X, Cheng F, Ying Y (2009) Application and research progress of near infrared spectroscopy in meat detection. Spectrosc Spectr Anal 29(07):1876–1880
2. Zhang Y, Yang H, Kong C (2004) Laser scanning confocal spectral imaging system. Opt Precis Eng 22(06):1446–1453
3. Xu Y (2015) Study on nondestructive detection of peanut seeds quality based on hyperspectral imaging technology. Jiangsu University
4. Liu S (2015) Study on nondestructive detection method of chilled pork quality based on hyperspectral imaging technology. Huazhong Agricultural University
5. Zhang C (2014) Classification of paddy rice species and pest in different processing state using hyperspectral imaging technique. Jiangsu University
6. Li Q, Xiao G, Xue Y, Zhang J (2008) Study of human blood cells based on microscopic hyperspectral imaging. Optoelectronic Engineering (5):98–101
7. Zhang Y (2015) Research on key technologies of image processing in multispectral imaging system. Chinese Academy of Sciences, Beijing
8. Liu J (2016) Rapid detection of the quality and safety of watered meat and frozen meat based on multispectral imaging technology. Hefei University of Technology
9. Wan H (2011) Study on multispectral imaging system. Zhejiang University
10. Huang Q (2016) Study of pork quality based on spectral imaging technology. Jiangsu University
11. Liu K, Cheng F, Lin H, Sun T, Xu K, Hu L, Ying Y, Xu H (2009) Detection of fat, protein and water content in chilled pork by visible/near infrared spectroscopy. Spectrosc Spectr Anal 29 (01):102–105
12. Pan W, Zhao J, Chen Q (2015) Classification of foodborne pathogens using near infrared (NIR) laser scatter imaging system with multivariate calibration. Sci Rep 5:9524
13. Kamruzzaman M, ElMasry G, Sun D, Allen P (2012) Prediction of some quality attributes of lamb meat using near-infrared hyperspectral imaging and multivariate analysis. Anal Chim Acta 714:57–67
14. Li X, Gao G, Niu L, Lin C, Qin Z, Liu J, Yao H (2012) Study on Raman imaging of living mouse ears. Anal Chem 40(10):1494–1499
15. Zhai C, Peng Y, Li Y, Zhao J (2017) Non-destructive testing of chemical additives in food based on Raman spectroscopy imaging. J Chem Coll Univ 38(03):369–375
16. Chen Q (2007) Study on tea's quality rapid non-destructive inspection based on NIR spectroscopy and machine vision. Jiangsu University
17. Chen Q, Zhang Y, Wan X, Cai J, Zhao J (2010) Study on pork tenderness detection based on hyperspectral imaging technology. J Opt 30(09):2602–2607

18. Zhao J, Chen Q, Cai J, Ouyang Q (2009) Automated tea quality classification by hyperspectral imaging. Appl Opt 48(19):3557
19. Gowen AA, O'Donnell CP, Cullen PJ (2017) Hyperspectral imaging-an emerging process analytical tool for food quality and safety control. Trends Food Sci Technol 18(12):590–598
20. Chen Q, Zhao J, Cai J, Vittayapadung S (2008) The quality grade of tea was evaluated by hyperspectral image technology. J Opt (4):669–674
21. Ouyang Q, Yang Y, Park B, Kang R, Wu J, Chen Q, Guo Z, Li H (2020) A novel hyperspectral microscope imaging technology for rapid evaluation of particle size distribution in matcha. J Food Eng 272:109782
22. Xin Z, Sun J, Tian Y, Lu B, Yingying H, Chen Q (2020) Hyperspectral technique combined with deep learning algorithm for detection of compound heavy metals in lettuce. Food Chem 321:126503
23. Zhu Y, Zou X, Shi J, Zhao J, Xu Y, Lin T (2014) Study on solid fermentation process of Zhenjiang vinegar based on hyperspectral image technology. Modern Food Sci Technol 30 (12):119–125
24. Li H, Felix K, Zhao J, Chen Q (2016) Quantifying total viable count in pork meat using combined hyperspectral imaging and artificial olfaction techniques. Food Anal Methods 9 (11):3015–3024
25. Xu Y, Chen Q, Liu Y, Sun X, Huang Q, Ouyang Q, Zhao J (2018) A novel hyperspectral microscopic imaging system for evaluating fresh degree of pork. Korean J Food Sci Anim Resour 38(2):362–375
26. Zhao J, Hui Z, Huang L, Zhang Y, Chen Q (2013) Detection of volatile base nitrogen in chicken by hyperspectral imaging. Adv Laser Optoelectron 50(07):158–164
27. Urmila K, Zhao J, Hu W, Chen Q (2016) Nondestructive quantifying total volatile basic nitrogen (TVB-N) content in chicken using hyperspectral imaging (HSI) technique combined with different data dimension reduction algorithms. Food Chem 197(Part B):1191–1199
28. Song S, Liu Z, Huang M, Zhu Q, Qin J, Moon SK (2020) Detection of fish bones in fillets by Raman hyperspectral imaging technology. J Food Eng 272:109808
29. Bi X, Zhao J, Lin H, Sun L, Guan B (2013) Portable near infrared spectrometer to determine the storage time of eggs. Food Sci 34(22):281–285
30. Guo E, Liu M, Zhao J, Chen Q (2008) Hyperspectral image nondestructive testing technique for sugar content of navel orange. J Agric Machin (5):91–93. 103
31. Shi J, Zou X, Zhang D, Chen Z, Zhao J (2014) Hyperspectral image detection of total flavonoids content distribution in different colors of ginkgo biloba leaves. J Agric Machin 45(11):242–245. 33
32. Duan H, Tong X, Cui R (2020) On-site identification of Ophiocordyceps sinensis using multispectral imaging and chemometrics. Int J Agric Biol Eng 13(6):166–170
33. Sendin K, Manley M, Marini F, Williams PJ (2021) Hierarchical classification pathway for white maize, defect and foreign material classification using spectral imaging. Microchem J 162:105824
34. ElMasry GM, Fulladosa E, Comaposada J, Al-Rejaie SS, Gou P (2021) Selection of representative hyperspectral data and image pretreatment for model development in heterogeneous samples: a case study in sliced dry-cured ham. Biosyst Eng 201:67–82
35. Badaró AT, Amigo JM, Blasco J, Aleixos N, Ferreira AR, Clerici MT, Pedrosa S, Barbin DF (2020) Near infrared hyperspectral imaging and spectral unmixing methods for evaluation of fiber distribution in enriched pasta. Food Chem 343:128517
36. Fu X, Moon SK, Chao K, Qin J, Jongguk L, Hoyoung L, Ana GV, Dolores PM, Ying Y (2014) Detection of melamine in milk powders based on NIR hyperspectral imaging and spectral similarity analyses. J Food Eng 124:97–104
37. Antoine L, Francesc PC, Delphine JRB, Luc E, Christophe C, Benoît J (2021) Detection of chocolate powder adulteration with peanut using near-infrared hyperspectral imaging and multivariate curve resolution. Food Control 119:107454
38. Zhu Y, Zou X, Shen T, Hu X, Zhao J, Shi J (2016) Rapid prediction of oil content and three-dimensional reconstruction of microstructure of fried lotus root slices. J Agric Eng 32 (05):302–306

Chapter 6
Colorimetric Sensor Technology in Food

What Is CSA Array?

To objectively evaluate the odor quality, researchers have focused on finding a new way to design an odor sensor with the same performance as the human nose. This CSA array (CSA) technology mimics the mammalian olfactory system by digital imaging to quantify odorant. This chapter discusses how chemo-responsive dyes recognize odorant molecular features by the imaging process and the development, design, and application of CSA as well, as shown in Fig. 6.1.

Abbreviation	Full name
AA	Acrylic acid
BODIPY	Boron-dipyrromethene
CSA	CSA array
DMAC	N,N-dimethylacetamide
DFT	Density functional theory
EDS	Energy dispersive X-ray spectroscopy
GC-MS	Gas chromatography-mass spectrometry
HOMO	Highest occupied molecular orbital
HAC	Hierarchical cluster analysis
KNN	k-nearest neighbor
LUMO	Lowest unoccupied molecular orbital
LDA	Linear discriminant analysis
LOOCV	Leave-one-out cross-validation
M	Metal ions
M-porphyrin	Metalloporphyrin
MnTPP	5,10,15,20-Tetraphenyl-21H,23H-porphine manganese
PSA	Poly(styrene-co-acrylic acid)
PEG-400	Polyethylene glycol 400
PEG-600	Polyethylene glycol 600
PSN	Porous silica nanospheres
PSN-NH$_2$	Amino-modified porous silica nanospheres

(continued)

Q. Chen et al., *Advanced Nondestructive Detection Technologies in Food*,
https://doi.org/10.1007/978-981-16-3360-7_6

Abbreviation	Full name
PSN-CH$_3$	Methyl-modified porous silica nanospheres
PSN-COOH	Carboxy-modified porous silica nanospheres
PCs	Principal components
PDA	Potato dextrose agar
PCA	Principal components analysis
PLS	Partial least square
ROI	Region of interest
RGB	Red-green-blue
RMSE	Root mean square error
RMSECV	Root mean square error of cross-validation
RMSEP	Root mean square error of prediction
SEM	Scanning electron microscope
SVMR	Support vector machine regression
TEM	Transmission electron microscope
TVB-N	Total volatile basic nitrogen
UV-vis	Ultraviolet and visible
VOCs	Volatile organic compounds
VCs	Volatile compounds
ZnTPP	5,10,15,20-Tetraphenyl-21H,23H-porphine zinc

6.1 Introduction

6.1.1 General Concept

Odor is an important indicator of food quality and safety. During the storage and processing of food, due to some changes such as its own freshness, infection of microorganisms, gases of different components are usually volatilized [1, 2]. Because of slow detection efficiency and higher cost, traditional detection methods including artificial sensory evaluation method, chromatographic chemical analysis method, etc. are difficult to apply toward monitoring of food quality and safety [3, 4]. Odor imaging detection technology is currently a new branch in the research field of artificial olfaction technology. This technology is based on the sensitivity of volatile gases to chemo-responsive dyes sensitive to volatile gases, captures them by chemo-responsive dyes for expressing information in the form of imaging. Thence it can realize the rapid, economical, nondestructive detection of the food quality and safety [5]. Regarding some problems of the common odor sensors, CSA is capable to have the solution and make up for the deficiencies of the existing technology [6, 7]. More prominently, CSA enables olfactory information to convert into visual information, making odors intuitively visible and easily analyzed. Compared with traditional electronic olfactory technology, the odor sensing that can be used in CSA is broader and more sustainable, and the detection results are presented more intuitively and

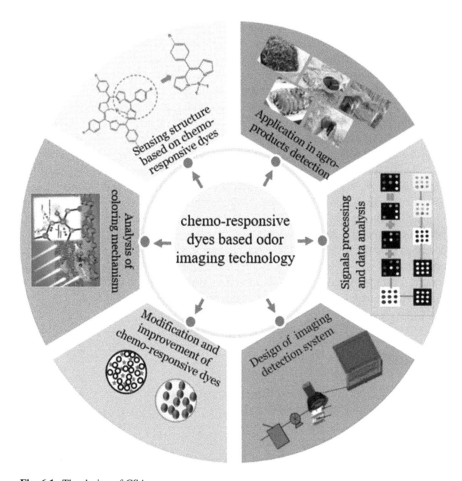

Fig. 6.1 The design of CSA

vividly [8–10]. Through the theoretical analysis, as well as the latest research introduction and application results, this chapter systematically reveals the basic principles, sensor production, sensitization processing, detection system as well as the processing and analysis of sensing signals. Overall, this chapter expounds the application prospects of CSA, in order to provide a certain reference value for its promotion in the field of food quality and safety.

CSA technology was originated in 2000 and was first proposed by Professor Kenneth S. Suslick of the University of Illinois at Urbana-Champaign [11, 12]. Metalloporphyrin was employed as a sensor by Suslick's group for the qualitative and quantitative detection of volatile organic substances, which lays the foundation for the application of odor imaging detection technology. Odor imaging detection technology is a novel olfactory system that imitates humans and mammals. It exploits the color change of chemical dyes before and after exposure of target gas for the visual qualitative and quantitative analyses. Traditional electronic olfaction technologies depend on weak forces such as physical adsorption or van der Waals forces, contrarily, odor imaging detection technology mainly relies on strong forces of covalent bonds. Also, the technology has a good anti-interference ability on water vapor in the environment, wherefore to well make up for the shortcomings of the existing biological and chemical sensor technology.

The CSA is mainly composed of some dyes with specific recognition capabilities. After dye molecules interacting with the detected object, the color of the dye molecules have significant changes. And the data and chemical index related to odor are both used to establish recognition pattern, then regression analysis is performed. With the advancement of material processing technology and the improvement of computer data processing level, odor imaging detection technology has been proved to have great application prospects in environmental monitoring, food and beverage quality monitoring, disease diagnosis and other fields. In recent years, odor imaging detection technology has been gradually applied for the odor detection in some highly volatile foods such as vinegar, white wine and mildew of grains [3, 13–35]. In order to overcome obstruction in the recognition ability of sensing materials in connection with odorant molecular features, it is necessary to study sensing technologies to recognize the odorant molecular features.

6.1.2 Sensing Materials

The basic principle of CSA is to utilize the color change induced by reaction between volatile compounds from samples and CSA array. The CSAs were composed of sensing materials, and chemo-responsive dyes are commonly used upon ligand binding for chemical vapor detection and differentiation. Chemo-responsive dyes are often selected according to their sensitivity to the specific volatile organic compounds (VOCs). The essence of odor imaging sensor to detect volatile substances is the interaction between molecules, and formation and fracture of chemical bond are involved in these effects. Therefore, the change of the data signal before and after the reaction between the sensor and the detected object is essentially determined by the physical and chemical properties of the chemically responsive dye molecules that make up the sensor.

Correspondingly, chemo-responsive dye become the key to CSA. The material used in CSAs demands to the following two basic conditions: (1) The dye should have at least one interaction center that involves interactions such as π–π molecular complexation, acid–base interactions, bond formation, van der Waals interaction. (2) A certain color change can occur after colorimetric material exposure to substance. The same amount of external groups or different amounts of the same group are proved to produce different color changes, so that volatile compounds (VCs) can be qualitatively or quantitatively analyzed according to the degree of color change [36, 37]. Normally, metalloporphyrins (M-porphyrin), boron-dipyrromethene (BODIPY), some natural pigments, and pH indicators could fulfill these requirements, and they are frequently used as sensing materials in experiments. There are

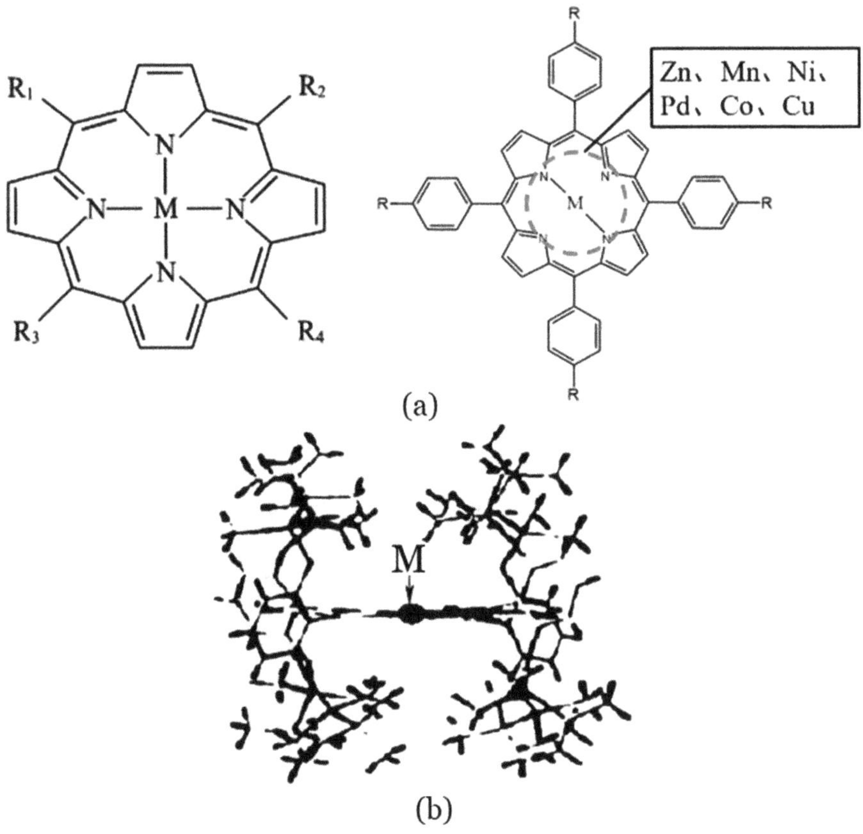

(a)

(b)

Fig. 6.2 (**a**) The chemical structure and (**b**) three-dimensional structure of M-porphyrin molecules

introductions for the characteristics of the functional groups and rendering principle of chemo-responsive dyes.

1. **M-Porphyrins**

For instance, M-porphyrin is a natural choice for the detection of volatile organic vapors [15, 16]. The array's responses are produced from partially selective and specific interactions after VOCs exposure to chemo-responsive dyes. M-porphyrins are composed of large π-conjugated molecules (porphyrin ring) and metal ions (M). They are a kind of macrocyclic polymer compounds containing metal ions, of which molecular structure is shown in Fig. 6.2. The metal ion is located in the central part of the porphyrin ring, and the ring system of porphyrin is basically on a plane and is simultaneously a highly conjugated system, thus porphyrins have stable internal structure and bright color performance [38]. The metal ion located in the center of the M-porphyrin is connected to the nitrogen atoms on the four pyrrole rings. On account of the broad axial ligand of the M-porphyrin, the metal ion yet have an approach to other groups above or below the porphyrin plane to take shape of metallic bond. More importantly, the color changes are produced with M-porphyrin as a chemical reaction acceptor when the external group enters the porphyrin ring to connect with the central metal ion or other group [34, 39]. The nitrogen-containing ligand is axially coordinated with the iron porphyrin, thereby the electron cloud density on the iron ion and pyrrole ring will both increase. It reflected as red shift occurred of the band peak in the spectrum. Hence, M-porphyrin has significant advantages as a receptor, which can be used to identify molecular size, shape, functional group, and chiral isomer [40, 41].

2. **BODIPY**

BODIPY was a general term for the derivatives of boron fluoride after being replaced by various substituents, first discovered in 1968. BODIPY as an intermediate product in the synthesis of porphyrins retains part of the chromophore (two pyrrole rings), undoubtedly, so it has strong color rendering properties. As shown in Fig. 6.3, while the substituents are grafted on the core of BODIPY, and they can be forced to separate from each other without affecting the properties of the chromophore. This prevents the $\pi-\pi$ stacking to form a conformationally restricted cyclic structure, enabling the boron fluoride compounds to have simple single-conjugated chemical structures [42]. This kind of compound has absorption mostly in the red to near-infrared region. Consequently, its spectral characteristics can be adjusted to obtain blue shift or red shift resulting from minor changes in structure. The substituted compounds with high sensitivity and specificity can be easily obtained in the application of ambient gas detection. The simple chemical structure makes it easy to be functionalized and nanomodified, recent research has excellent sensing effects achievements. On this basis, it has been widely used in the fields of fluorescent labeling, sensor, and laser dye production [24].

3. **Natural Pigments**

Natural pigments with lower toxicity are extracted from natural products, spinach (*Spinacia oleracea*), red radish (*Raphanus sativus* L.), winter jasmine (*Jasminum nudiflorum*), and black rice (*Oryza sativa* L. *indica*) for instance. Anthocyanins are often found in these natural products [43], the molecular structure of which is shown in Fig. 6.4. Anthocyanin can bind with bases like amines, which is ascribed to anthocyanin molecule containing carbonyl and hydroxyl groups [22]. Due to natural pigments with colors in the different range of visible light, it offers a promising prospect for the application of chemo-responsive dyes.

4. **pH Indicators**

pH indicators are generally weak acids or weak alkali. With the change of pH in the environment, the indicator gains protons so as to convert to acid type or loses protons to covert to base type [44, 45]. Due to the different structures of acidic and alkaline indicators, they appear in different colors. For example, methyl orange is an alkaline indicator with the function of two-color indicator. There are dissociation balance and color change as shown in Fig. 6.5. Obviously, when the H^+ concentration increases, the reaction proceeds to the right, and

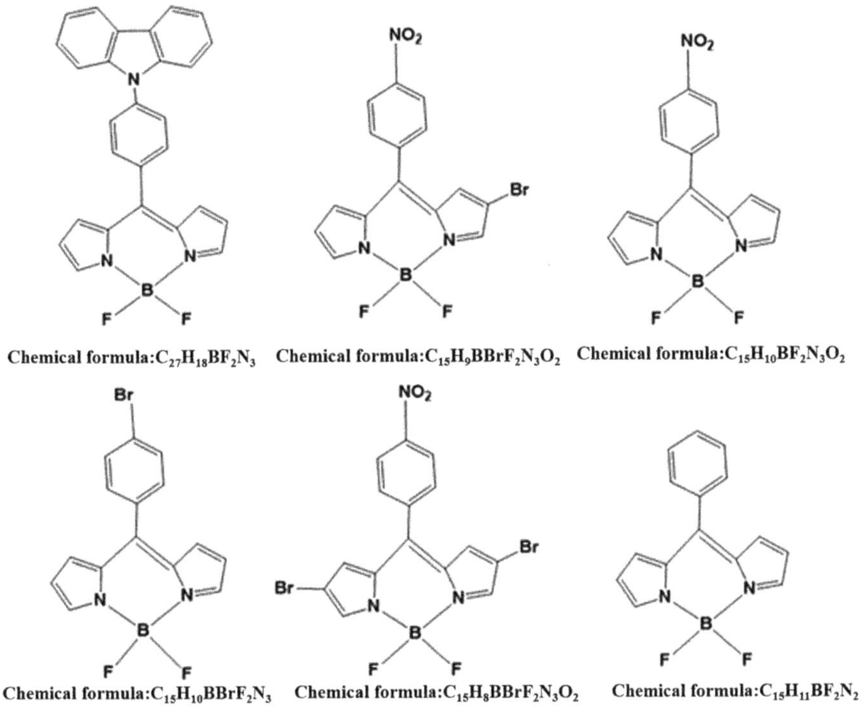

Chemical formula:$C_{27}H_{18}BF_2N_3$ Chemical formula:$C_{15}H_9BBrF_2N_3O_2$ Chemical formula:$C_{15}H_{10}BF_2N_3O_2$

Chemical formula:$C_{15}H_{10}BBrF_2N_3$ Chemical formula:$C_{15}H_8BBrF_2N_3O_2$ Chemical formula:$C_{15}H_{11}BF_2N_2$

Fig. 6.3 The structure of boron-dipyrromethene

methyl orange mainly exists as a quinoid (acid color type), showing red. The reaction proceeds to the left as the H^+ concentration decreases, in this case, methyl orange mainly exists in the azo (alkaline color type) and appears yellow [18]. The pH indicator not only senses the acid–alkali changes in the environment, but also arise the change in color with the change of polarity of the substances [46].

6.1.3 Coloring Mechanism

Chemo-responsive dyes are the core for building olfactory visualization sensors. M-porphyrins and BODIPY compounds are widely considered as the widely used materials with better performance. The M-porphyrins are in a π-shaped conjugated system composed of four pyrrole rings and have excellent color rendering properties. On the one hand, the diversified porphyrins provide a broad choice for chemo-responsive dyes and thereby are beneficial to the construction of specific CSAs, but

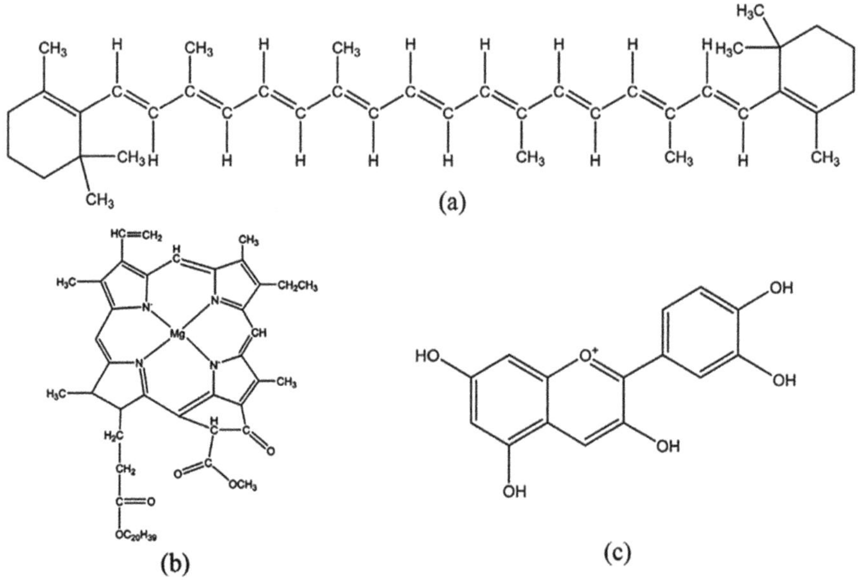

(a)

(b)

(c)

Fig. 6.4 The structure of natural pigments. (**a**) Molecular structure of anthocyanin anion; (**b**) the chemical structure of chlorophyll; (**c**) the chemical structure of B-carotene [22, 42]

Fig. 6.5 The discoloration principle of methyl orange in acid and alkali environment

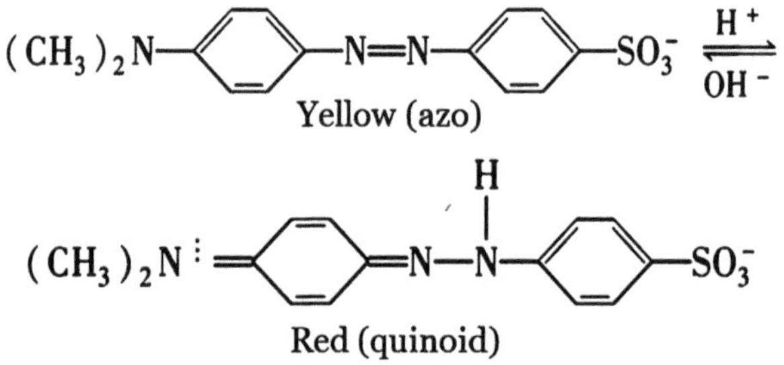

Yellow (azo)

Red (quinoid)

inevitably, they also bring certain difficulties in the selection of chemo-responsive dyes. Consequently, screening and synthesizing efficiently the low cost chemo-responsive dyes are of significant directions for the research of CSA.

1. **Dynamic Analysis of Ultraviolet and Visible Spectrum (UV-Vis) Spectroscopy**

    Chemo-responsive dyes mostly have large π-conjugated structures and rigid planar structures. They possess chromophore groups that can interact with a variety of substances through intermolecular interactions such as axial coordination, hydrogen bonding, and electrostatic interaction. In consequence, the configuration of the porphyrins changes and can be detected by absorption spectroscopy, fluorescence spectroscopy, etc. In addition, chemo-responsive dyes are united to the volatile gas through molecular coordination, which leads to a change in the energy level of electronic transition. That can be reflected in the

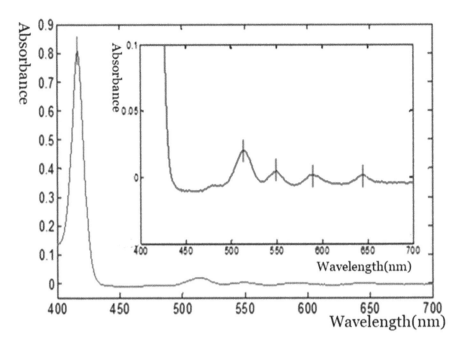

Fig. 6.6 UV-Vis spectrum of TPP

appearance as a color change, which is called the color reaction. The spatial structure of the material molecule and the activity of the central metal ion have a great influence on the result of the coordination reaction. Figure 6.6 shows the ultraviolet spectrum of the color reaction of porphyrin with volatile ethanol. It is noted that UV-Vis spectrum of porphyrin shows a band at 417 nm, the name of it is the Soret band (i.e., B-band) which is caused by the electron transition from the highly occupied orbital $\alpha_{1\mu}(\pi)$ to the lowest empty orbital $e_g(\pi^*)$. Furthermore, TPP exhibits six Q bands at 277, 302, 372, 426, 570, and 626 nm due to the transition from $\alpha_{1\mu}(\pi)$ orbital to $e_g(\pi^*)$ orbital. Because of the effect of transition dipole cancellation, the absorption of the Q-band is relatively weak. On the contrary, the absorption of the Soret band is stronger for the transition of the Soret band is the enhancement of the two transition dipoles. And the absorbance of the Soret band of TPP is about 40 times that of the Q band [47].

In the coordination reaction, the coordination number and equilibrium constant are vital parameters for studying the interaction between molecules, which can be calculated using the Benesi–Hildebrand equation, as shown in Formula (6.1).

$$\lg k = \lg \frac{A - A1}{A2 - A} - n\lg[C] \tag{6.1}$$

In the formula, $A1$ is the blank absorbance of the porphyrin without ethanol; $A2$ is the absorbance of the solution when the porphyrin is finally completely protonated; A is the absorbance of the porphyrin solution at any point in the titration process, and C is the concentration of the added ethanol. According to the change of absorbance at a certain wavelength, the slope of the curve obtained by plotting $\lg((A - A1)/(A2 - A))$ vs. $\lg[C]$ can be used to obtain the coordination number n, and the equilibrium constant $\lg K$ can be obtained from the intercept [48, 49].

2. **Density Functional Theory (DFT) Analysis**

The color changes after the colorimetric material exposure to volatile gas are involved with the molecular interactions, and the interaction has some relationship with the electron cloud distribution between nuclei. Therefore, the reaction mechanism of the colorimetric material and the volatile characteristic gas is theoretically calculated through DFT. Hohenberg–Kohn theorem and Kohn–Sham method are considered as the theoretical basis of DFT.

(a) *Hohenberg–Kohn* Theorem

Hohenberg and *Kohn* proposed two *Hohenberg–Kohn* theorems and published them in the authoritative journal *Physical Review* [50]. The theorem states that the electron density at the relatively lowest energy ρ is the ground state density of the system, and the corresponding energy is also the ground state energy of the system [51, 52]. Therefore, the common functional $E[\rho]$ will be minimized to get the energy and electron density of the ground state. It clarifies the feasibility of the DFT method, but no suggestion to obtain the function. Therefore, it is necessary to perform theoretical calculations by other methods on the basis of this theorem.

(b) *Kohn–Sham* Method

For a multi-electron system, its energy $E(\rho)$ is mainly composed of three parts, the equation is shown in the Formula (6.2).

$$E[\rho] = T_e[\rho] + V_{e-e}[\rho] + V_{ext}[\rho] \tag{6.2}$$

In the formula, $T_e[\rho]$ is the kinetic energy of electrons, and $V_{ext}[\rho]$ is the interaction energy between nuclei and electrons, and $V_{e-e}[\rho]$ is the interaction energy between electrons and electrons. Among these three energies, only $V_{ext}[\rho]$ is known, and $V_{e-e}[\rho]$ and $T_e[\rho]$ can only be approximated by functionals.

Kohn and Sham describe the electron density mainly through a series of independent molecular orbitals, as shown in Formula (6.3)

$$\rho(r) = \sum_i^{occ} |\Phi_i(r)|^2 \tag{6.3}$$

The occ in Formula (6.3) is the summation of the occupied orbital. Therefore, the effect of classical Coulomb and kinetic energy can be described as Formulas (6.4) and (6.5)

$$T_s[\rho] = -\frac{1}{2} \sum_1^{occ} < \Phi_1 \mid \nabla^2 \mid \Phi_1 > \tag{6.4}$$

$$V_H[\rho] = \frac{1}{2} \int \frac{\rho(r_1)(r_2)}{\mid r_1 - r_2 \mid} dr_1 r_2 \tag{6.5}$$

Therefore, the functional of the energy system can be transformed into Eq. (6.6):

$$E[\rho] = T_s[\rho] + V_H[\rho] + V_{ext}[\rho] + E_{XC}[\rho] \tag{6.6}$$

$E_{XC}[\rho]$ in the formula is a commutative-correlation (XC) functional, and its expression is shown in Formula (6.7).

$$E_{XC}[\rho] = (T[\rho] - T_s[\rho]) + (V_{e-e}[\rho] - V_H[\rho]) \tag{6.7}$$

Taking variational operation on the corresponding orbit, it can be obtained by using Formula (6.8).

$$\left[-\frac{1}{2}\nabla^2 + \int \frac{\rho(r')}{\mid r - r' \mid} dr' + V_{ext}(r) + V_{XC}(r) \right] \Phi(R) = \varepsilon_i \Phi_i(r) \tag{6.8}$$

In Formula (6.8), $V_{XC}(r) = \frac{\partial E_{XC}[\rho]}{\partial \rho(r)}$.

Here it developed the binding energy, orbital energy level gap between HOMO and LOMO, dipole moment, atomic charge, central metal ion deviation from the porphyrin molecular plane displacement and other parameters, to investigate the mechanism of the combination of chemo-responsive dyes and volatile gas molecules. The popularization of computers has led to considerable development in quantum chemical calculations. Researchers have successively developed calculation software such as Gaussian, Multiwfn, Molekel, VASP, Sybyl, Molpro, and HyperChem. Among them, Gaussian is currently one of the most widely used quantum chemical calculation software.

This program is suitable for the simulation analysis of molecules in the gas phase and solution and can realize the optimization of molecular configuration and conduct the energy, charge, molecular orbital, potential energy surface, substitution effect in the process of molecular interaction. The density functional theory based on the hybridization function B3LYP can be employed to optimize the structure of color-sensitive materials and hexanal and calculate the changes in energy, molecular orbital, charge, and other properties under the intermediate basis set level of 6–31 $G(d)$. Accordingly, Gaussian view realizes visualization of molecular structure [53–55].

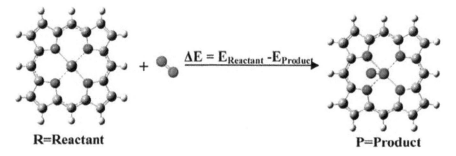

R=Reactant   **P=Product**

Fig. 6.7 Calculation of binding energy before and after the reaction of porphyrin and ethanol

1. Binding Energy

    The energy released during the reaction between porphyrin and ethanol can be used to analyze the binding ability of porphyrin compounds and ethanol. That is, the binding energy is the difference between the total energy of before the reaction with ethanol and the energy of the product in the system after the reaction. It is shown in Fig. 6.7 that the conversion formula between the energy unit Hartri (a.u.) and the commonly used energy unit is kcal/mol: 1 a. u. = 627.5095 kcal/mol.

2. Orbital Energy Level Gap

    As the frontier molecular orbital theory proposed by Fukui Kenichi stated, orbital energy level determines the electron transfer and other important chemical properties of molecules. In the view, there are the highest occupied molecular orbital (HOMO) and the lowest unoccupied molecular orbital (LUMO) in the molecule. The electron distributed on HOMO has the highest energy, while no electron is distributed on LUMO and the electron energy is lowest [56, 57]. In the chemical reaction, the electrons on the HOMO in the high-energy state are suffered least bound with the highest activity. And it is liable to transfer to the LUMO in the low-energy state, thus realizing the transfer of electrons, as presented in Fig. 6.8. To sum up, HOMO and LUMO orbitals are the key to the reaction between molecules in a system. There are studies showing that the less the energy required to excite electrons, the more favorable it is for electrons to transition from HOMO to LUMO. It indicated that the smaller energy gap is more positive to initiate chemical or physical reaction.

3. Distance and Charge Changes Between Atoms

    The essence of a chemical reaction is the sharing of electrons between the reactants, which can be characterized by the change in charge before and after the chemical reaction. The strength of the intermolecular interaction can be reflected in the distance between them to a certain extent [53]. Chemical calculation software packages (such as Mercury) are used to calculate the distances between the –F of BODIPY and the –C2, –C3, and –C4 of hexanal, so that the interaction strength of BODIPY and hexanal molecules can be explored. The breaking and formation of chemical bonds in the process of chemical reaction is essentially the

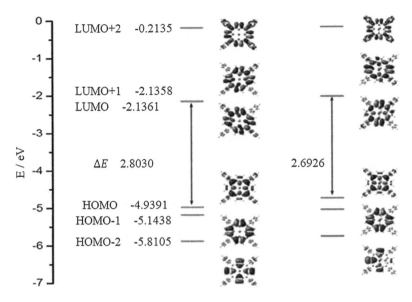

Fig. 6.8 Energy level changes before and after the coordination reaction between ZnTPP and ethanol

transfer of electrons between reactants. Reasonably, the transfer can be characterized by the charge change before and after the compound reaction and the
process displayed in Fig. 6.9.

4. Molecular Plane Angle

   In the combination of the chemo-responsive dyes and the volatile gas molecule, the change of the substituent on the heterocycle of the materials will lead to
change in the distance and the angle between the two molecular planes, as shown
in Fig. 6.10. This change is often closely related to the combination between the
material and the molecule to be tested. Therefore, the sensitivity of the colorimetric material to the tested molecule can be revealed in the distance and angle
between the two.

5. Dipole Moment

   The dipole moment refers to the result of distance between center of positive
charge (r) and negative charge (q) multiplied by the electric quantity given as
($\mu = r \times q$, D (debye)). The dipole moment can be employed to judge the polarity
of the molecule. If the dipole moment is equal to zero, it is a non-polar molecule;
otherwise, the molecule is a polar molecule with the dipole moment not equal to
0. The greater the dipole moment, the stronger the polarity of the molecule. The
comparison between porphyrin and various metalloporphyrins before and after
the action of ethanol is shown in Fig. 6.11.

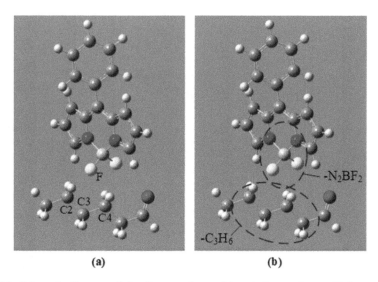

Fig. 6.9 Schematic diagram of the distance change (**a**) and charge change (**b**) between the colorimetric material and the volatile gas

6.2   Instrumentation

Conventional chemo-responsive dyes have a good detection effect on constant gas, but poor detection performance to trace gases. It primarily comes from that with small particle size of chemo-responsive dyes that the effective contact area with the tested gas is insufficient. Porphyrins and BODIPY compounds are heterocyclic compounds with a π–π conjugated structure formed by connecting pyrrole molecules, and the H atom of the NH situated in the center of the molecule can be replaced by metal ions to form M-porphyrin. With a good color sensitivity, easy modification, and stable properties of the π-conjugated system, the colorimetric material is usually under the treatment of nano-dispersion to increase the binding force of the colorimetric medium substrate and gas molecule ultimately to improve its sensitivity.

6.2.1   Nano-modification

Nanomaterials are a kind of typical mesoscopic systems, and macroscopic objects reduced to the nanoscale cause obvious changes in their optical, mechanical, and chemical properties [58–60]. Due to the large surface area of nanomaterials, nanoparticles have become a good choice for immobilizing and supporting biochemical molecules. Currently, porphyrins and BODIPY are used to fabricate CSAs. In terms of the application, they can be further modified to improve the sensitivity

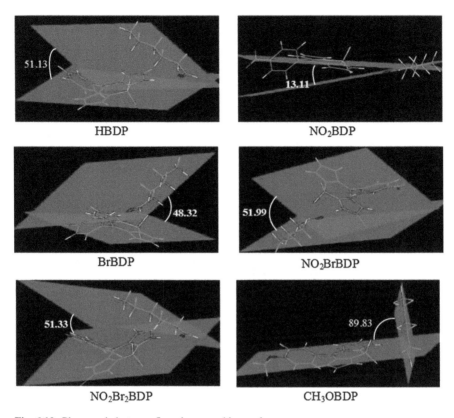

Fig. 6.10 Plane angle between fluoroboron and hexanal

and stability of the sensor. Based on the specific surface area effect, small size effect, quantum effect, and interface effect of polymer nanospheres [61–65], it can be polymerized with chemo-responsive dyes to form nanomaterials with higher sensitivity and chemical activity, which can interact stably with the detected volatile organic compounds to enhance the color rendering effect. The common nanomodification methods of chemo-responsive dyes are as follows.

1. **Nano Self-Assembly of CSA**

   Thanks to the intermolecular forces being non-covalent bonds, the nanoparticle self-assembly of chemo-responsive dyes can better ensure that the electronic structure of nano-scale materials cannot be destroyed. And the method is simple and easy to control the shape of nano-porphyrins, so it has been widely used. Self-assembly refers to a technology in which molecules spontaneously combine through a certain assembly method and then these molecules can have a certain ordered structure and a certain regular geometric appearance. It is mainly based on hydrogen bonds, coordination bonds, π–π stacking, and the role of non-covalent bonds such as van der Waals forces [66–69]. Currently, the commonly used self-assembly methods are systematized into solid-phase self-

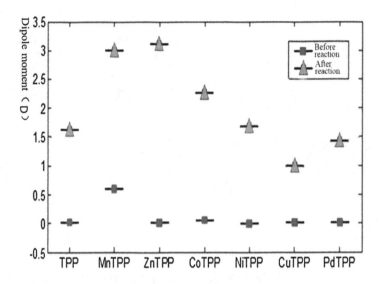

Fig. 6.11 Changes in the dipole moment of different porphyrins and ethanol

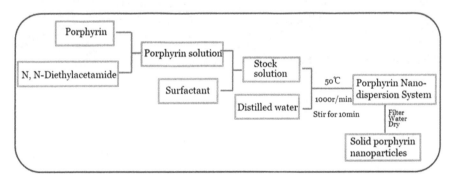

Fig. 6.12 The flowchart of nano-self-assembly porphyrin

assembly method and liquid-phase self-assembly method. The solid-phase self-assembly method is mainly a physical vapor deposition method, and the liquid-phase self-assembly method mainly includes reprecipitation method, surfactant-assisted method, and ion self-assembly method. Among them, the surfactant-assisted method is an improvement of the reprecipitation method, which is simple to operate and can be widely used in chemo-responsive dyes of nanoparticles self-assembly [70]. Figure 6.12 shows the flowchart of nano-self-assembly porphyrin.

2. **Nano-poly-modification of CSA**

Poly(styrene-*co*-acrylic acid) (PSA) is a kind of self-assembly polymer nanosphere with surface aggregation function. The PSA is prepared through soap-free emulsion copolymerization, styrene (St) and acrylic acid (AA) are used as monomers, and ammonium persulfate (APS) is used as non-buffer initiator in the medium.

Styrene and acrylic acid are used as raw materials, and chemo-responsive dyes are added, then to increase the temperature of the mixed system by heating. Due to the increase in temperature, the solubility of the chemo-responsive dyes can be increased [71]. On the other hand, the activity of the acrylic chains on the surface of the nanospheres is enhanced, resulting in more free volume. It can promote the interaction between the chemo-responsive dyes and the nanospheres in the solution. If they are in full contact with the surface of the nanospheres, the hydrogen bond will be formed between the material molecules and the carboxyl groups of acrylic acid, to increase the amount of material adsorbed on the outer hydration layer. Then a concentration gradient is created between the surface and the inside of the nanosphere. Due to the strong hydrophobic interaction between the material and the styrene segment, the chemo-responsive dyes diffuse from the surface of the nanosphere to the inside, reducing the concentration in the water phase [72]. To maintain balance, more material in the suspended particles will dissolve until all the molecules are absorbed by the nanospheres. The synthesis diagram is shown in Fig. 6.13. This research studied the influence of particle size, mass ratio and the selection of emulsifiers to nano chemo-responsive dyes, so as to enhance the capability for VOC detection by using nano-CSAs [26, 73].

The synthesized PSA nanosphere has a certain particle size, and small particle size is not conducive to the absorption of chemo-responsive dyes. If the particle size is too large, more chemo-responsive dyes will be needed and cause a waste of resources, in addition to weaken the nano-effect of the microspheres and have a certain impact on the nanosized chemo-responsive dyes. Therefore, it is of necessity to optimize the particle size of the polystyrene–acrylic acid nanospheres [26, 29]. According to the synthesis process of PSA, the copolymerization rate of styrene and acrylic acid is proportional to the concentration of acrylic acid.

3. **Nano-porous-modification of CSA**

Mesoporous silica nanosphere materials are porous, due to the characteristics of large specific surface and pore volume of porous silica nanospheres (PSN), as well as tunable pore size [74–76]. Herein, we described an approach to fabricate a CSA based on PSN for VC detection. The screened chemo-responsive dyes were mixed with PSN and dispersed in N,N-dimethylacetamide (DMAC) and ethanol, and then stirred at 50 °C for 2 h. The specific process is shown in Fig. 6.14.

As shown from the figure, experimentally synthesized PSN with ordered pores, the appearance is white powder [77]. Figure 6.15 shows the scanning electron microscope (SEM), TEM, and energy dispersive X-ray spectroscopy (EDS) image. The white powder is composed of spherical nanoparticles, the average diameter of which is estimated to be about 250 nm. The corresponding EDS (the small image in Fig. 6.15a) finds out the presence of Si and O elements, indicating that the spherical nanoparticles are composed of silica. At higher magnification (Fig. 6.15b), it can be

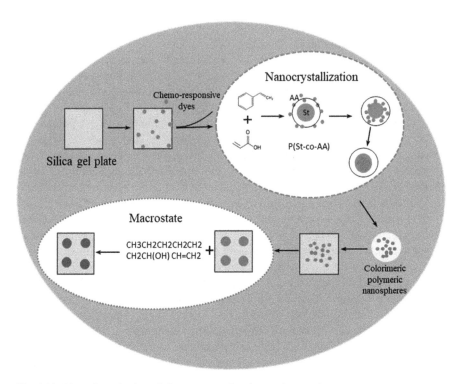

Fig. 6.13 The polymerization of chemo-responsive dyes and nanospheres

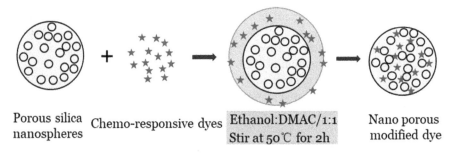

Fig. 6.14 Preparation process of nano porous modified dye

seen that the nanospheres are porous. Figure 6.15c–e shows the TEM images of different surface modified PSN. Referring to the particle size, pore size, and surface morphology, compared with PSN, the PSN-CH$_3$, PSN-COOH, and PSN-NH$_2$ have no significant changes. However, the shell of PSN-CH$_3$ is smoother than that of other samples (Fig. 6.15d). Since the surface chemistry of the particles has an important influence on the Zeta potential, the particles grafted with different coupling agents have a variety of surface chemistry.

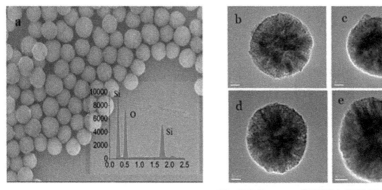

The SEM image and EDS of PSNs(a)   The TEM images of PSN(b), PSN-NH2(c), PSN-CH3(d), and PSN-COOH(e)

Fig. 6.15 The scanning SEM, TEM, and EDS image of PSN [26]

6.2.2 CSA System

Through the CSA technology, odor information is converted into visual information to make odors "visible." The CSA system is composed of supporting equipment and software. The schematic diagram of the portable system is presented in Fig. 6.16. Compared with traditional electronic nose technology, olfactory imaging technology is more intuitive and vivid. The sample is placed in the movable container to collect gas. Once the gas collection has been completed, the vacuum pump is turned on, which bring the volatile gas of the detected object into the reaction chamber through the pipeline. As the gas is volatilized by the sample exposure adequately to the colorimetric material in the reaction chamber, the color of the material will have a change. After the exposure, a final sensor image (after image) can be obtained by image capture compartment, and the pattern recognition is performed through the computer to analyze the tested sample [78–80].

1. **Supporting Equipment**

    The supporting equipment of CSA mainly includes gas collection chamber, vacuum pump, camera, light source, reaction chamber, CSA array, and computer (as shown in Fig. 6.17). We select devices that are closely related to system performance as the research object and optimize the equipment to improve the ability of odor imaging detection.

    (a) Choice of Image Acquisition

    Scanners and cameras, as the main methods of image acquisition at present, have their own advantages and disadvantages. Specifically, scanner has lower price and its own light source, moreover, the lamp tube moves with the stepping motor to provide uniform illumination. However, compared with camera, scanner is bulk, slow in scanning speed, and generally not supported to carry out secondary development. Given that, scanner is not conducive to

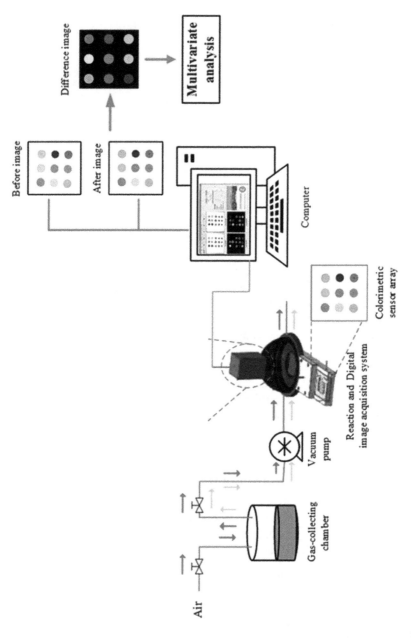

Fig. 6.16 Schematic diagram of the portable system [23]

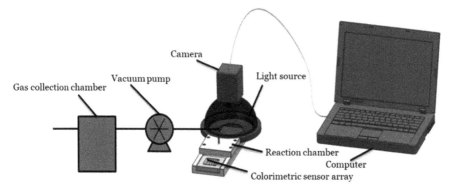

Fig. 6.17   The supporting equipment of odor imaging technology

(a)   **(b)**

Fig. 6.18   The choice of image acquisition method. (**a**) CCD, (**b**) 3CCD

developing into a portable system. The camera has the advantages of being portable, capable of collecting images online, and being able to be used for secondary development. Therefore, camera is selected as the image acquisition method to capture images.

Figure 6.18 is the physical image of the CCD and 3CCD camera. The prism in the 3CCD camera can divide the light source into the tricolor of red, green, and blue. There are three independent CCD image sensors in the 3CCD camera, which can process separately the tricolor to improve the accuracy of the acquisition of colors and the image quality. In addition, the light collection area of the three CCD image sensors of the 3CCD camera is relatively large, making the 3CCD camera have a higher signal-to-noise ratio, better sensitivity and a wide dynamic range than a CCD camera.

(b) Choice of Light Source

The uniformity of the light source will directly affect the quality of the image, thereby affecting the performance of the device. Uniformity, brightness, efficiency, service life, spectral characteristics, etc. are the major factors in the selection of the camera light source. Compared with other light sources, LEDs not only have high brightness and long service life but also can be designed into a complex structure to achieve different angles of light source

(a)   (b)

Fig. 6.19 The choice of light source. (**a**) Strip light source. (**b**) Diffuse reflection integrating sphere light source

illumination. Diffuse reflection plate is added in front of the light source, that is the constitution of diffuse reflection light source. The light emitted by the diffuse reflection light source can be reflected for multiple times through the reflection plate, realizing diffuse reflection illumination of the space area. Furthermore, it can make the light more uniform, completely eliminate shadows, and weaken the effect of specular reflection from the object's surface on the image. Generally, there are strip diffuse reflection LED light sources and diffuse reflection integrating sphere light source being used, as shown in Fig. 6.19. In general, concerning the latter, a circle of LED light sources on the bottom plane can be evenly reflected on the image, so that the brightness of the image is uniform

(c) Design and Optimization of Reaction Chamber

The stability, repeatability, signal response time, and degree of the sensor signal will be greatly affected by the structure of the reaction chamber of the CSA system. During the reaction process, it is necessary to ensure that the gas flow can be uniformly contacted with each chemo-responsive dyes. Therefore, the optimization of the reaction chamber is of great significance to improve the detection capability of odor imaging sensors. The fluid distribution simulation software (such as *Comsol*, etc.) are devoted to transform the physics solution problem into the partial differential equations solution problem to specialize the physics simulation. And then a two-dimensional model of the reaction chamber is established, the distribution of flowing gas under the action of the baffle is simulated. Meanwhile, the degree of baffle bending and the position of the baffle embedded in the reaction chamber are constantly changed, and the law of gas flow field changes is summarized. On the premise of placing CSAs to the place that the gas distribution should be as uniform as possible, the shape and position of the reaction chamber are continuously optimized.

It can be seen from Fig. 6.20a–e when the baffle is too close to the inlet, the airflow velocity on both sides of the sensor array becomes smaller. After continuous adjustment of the curvature and position of the baffle, the optimal results of the curvature and position of the baffle is shown in Fig. 6.20f. It is not conducive to improve the reaction between CSA and VOCs because the

Velocity fields in the reactor chambers under different conditions

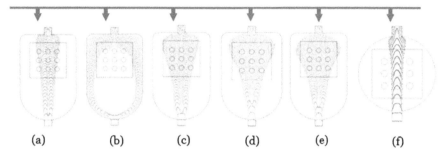

| (a) | (b) | (c) | (d) | (e) | (f) |

Particle distribution in the reactor chambers under different conditions

Fig. 6.20 The design and optimization of reaction chamber. (**a–f**) are simulation diagrams of different reaction chambers. (**a**) Large curvature; (**b**) Small curvature; (**c**) Close to the air inlet; (**d**) Far from the air inlet; (**e**) With baffle; (**f**) Without baffle

airflow was extremely concentrated in the middle of the chamber. The application of particle tracking which shows the particle distributions also supports the results on simulation of the velocity fields. The simulation of particle tracking matching to different kinds with baffles is shown in Fig. 6.20. All of the reaction chambers with a small or large curvature baffle and the one without baffle cannot meet the requirements, because their particle cannot cover the CSA and the distribution is not as uniform as the reaction chamber with an optimized baffle.

2. **Fabrication of Sensor Array**

The fabrication of the odor imaging sensor array is the core technology of the CSA system. For a sensor, its basic requirements include: high sensitivity, anti-interference stability (insensitive to noise), linearity, high reliability, repeatability, safety, interchangeability, high precision, high response rate, wide measuring range, wide working temperature range, etc. In addition, there should be special requirements for specific types of sensor [74]. To explore further, the process of obtaining and manufacturing odor imaging sensor materials is the core technology of the sensor. Because the dielectric substrate is used as a thin film to form the corresponding material, the selection of the substrate also has an impact on the performance of the sensor. At present, the substrates used in the production of odor imaging sensors mainly include hydrophobic materials such as reverse phase of silica gel plates and polytetrafluoroethylene. A chemical developer can be made into a single visual sensor. Sensors made of different color developers have different sensitivity characteristics. Multiple sensors made of different color

Fig. 6.21 The CSA

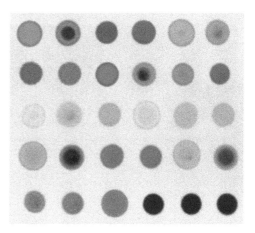

developers together are arranged to form a visual gas sensor array. The combination of the sensors has improved the detection accuracy and greatly expanded the application range of CSA [81]. Therefore, CSA generally appears in the form of arrays, as shown in Fig. 6.21. The manufacturing method of the CSA is as follows:

(a) Selection of gas chemo-responsive dyes: Hydrophobic porphyrins, BODIPY, and pH indicators that can have color changes to gases are selected as gas chemo-responsive dyes.
(b) Selection of solvent: According to the materials selected in (a), the corresponding solvent is selected, which can be used to dissolve the gas chemo-responsive dyes at the concentration of 0.1–0.5 mol/L.
(c) Selection and production of sensor substrate materials:

- White polytetrafluoroethylene material (3 mm $\geq$ thickness $\geq$ 1 mm) are chosen and processed into rectangles or squares.
- Each square is neatly engraved into a square with a side length of 0.2–0.5 mm (or a circle with a diameter of 0.2–0.5 mm) and a depth of 0.01–0.02 mm, which are treated as marks.
- Each mark is covered with thick hydrophobic white stable agent with a layer of 0.01–0.02 mm.

(d) Production of array: 0.1–10 μL of colorimetric material solution is taken through a micro-sampling device then fixed on the stabilizer at the mark of the substrate plate, and then dry the solvent to obtain CSA.

3. **Design of CSA System Software**

Through the image acquisition software of the portable CSA system, the images of the sensor array before and after exposure to gas can be timely collected and analyzed, and the gray value of the colorimetric material can be extracted in real time. A 3CCD camera is devoted to the image acquisition device, which is exploited under Windows system using *Microsoft Visual Studio 2008 + QT4.6*

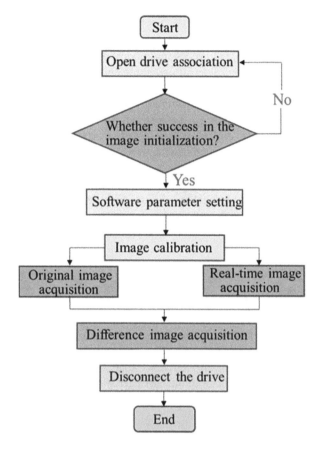

Fig. 6.22 The flowchart of the image acquisition software

and calling *Halcon* 8.0 for digital image analysis. In summary, the system is composed of image acquisition and display, relevant parameter setting, image storage, and analysis. Figure 6.22 shows a flowchart of the image acquisition software. At length, 3CCD camera parameters, camera acquisition speed, image correction, etc. are settled in order to obtain better image effects. The 3CCD camera is in the service of collecting and analyzing the image of the CSA in real time. With the threshold segmentation algorithm and the minimum bounding rectangle method, the background image of the image acquisition of the sensor is extracted. The original image, real-time detection image, and real-time difference image of the effective area can be displayed on the system interface [81]. Finally, according to the path set by the user, the sensor image obtained by the 3CCD camera and the difference image obtained by the computer are in storage [82].

6.3 Signals Processing and Data Analysis

The image signal processing module is the core constitute of the imaging system, which is responsible for extracting and analyzing the color-difference information. And characteristic difference matrix is generally used to represent the difference information before and after the reaction of each color area of the visual sensor array. On the premise of the requirement, the design and application of image processing algorithm are ought to occupy small calculation amount and shorter calculation time. The current image processing process is composed of the positioning of the graphic center point, the selection of the characteristic region, and the establishment of the color space model. The specific implementation steps include filter noise reduction, binarization, morphological processing, figure centering, characteristic region selection, etc., in order to segment the color regions of the sensor array and extract the corresponding color information, etc., which can be as shown in Fig. 6.23.

6.3.1 Characterization of Images of CSA

In the detection of gas with CSA, the characteristic values of target image is generally extracted from the color regions under the RGB, HSV, and Lab color space. Providing different color spaces are applied in detecting different objects, it would be beneficial to the characteristic extraction for pattern recognition analysis. Among the three spaces, the RGB color model (red-green-blue three primary color model) is the most commonly used [83]. In the HSV model, three-dimensional values of hue, saturation, and brightness indicate color. The Lab color space uses three elements to instruct colors, namely the coordinate L is the representative to the brightness of the color, but the a and b channels represent the chromaticity

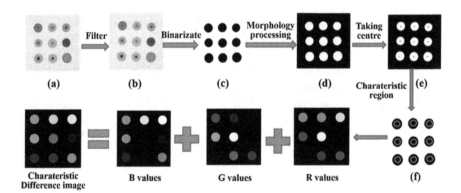

Fig. 6.23 The image processing flowchart of CSA. (**a**) Original image; (**b**) filtered image; (**c**) binarized image; (**d**) morphological processing image; (**e**) taking the central point; (**f**) taking the characteristic region

[84]. Namely, the "*a*" channel represents the degree of color changes between green and red, but the "*b*" channel reflects the different degrees of color between blue and yellow. In summary, the characteristic variable matrix of the sensor color area is calculated based on the RGB, HSV, and Lab color models, as shown in Fig. 6.24. Besides, for the convenience of data visualization and comparison, the values of each channel obtained in all spaces are normalized to the interval of [0, 255].

6.3.2 Selection of Region of Interest (ROI)

The color-difference variables are obtained by subtracting the two images before and after the reaction between the visualization sensor and the gas. The precise position of each sensor before and after the reaction is first determined, so that the color region in the image can be subtracted in one-to-one correspondence. The region of interest (ROI) can be selected as the mark of unique position of each sensor, which further affect the accuracy of extracting color characteristic difference. In view of this, the method of first moment, minimum bounding rectangle, and ellipse fitting were employed to find the coordinates of the ROI, as shown in Fig. 6.24. With the first moment method of obtaining the center of gravity of the image, the center of gravity of the color-developing region of the visualization sensor can be regarded as its heart [85]. Supposing that the image can be represented by the function $f(x, y)$, because the points on the image are two-dimensionally discrete, the $p + q$ moment of the image $f(x, y)$ can be defined as the Formula 1 of Fig. 6.24.

The basic idea of the least squares ellipse fitting algorithm takes advantage of the mathematical meaning of ellipse [86], which can be expressed as Formula 2 of Fig. 6.24. The optimal parameter can be solved by Formula 3, that is, the optimal solution of the linear equation system is also the optimal parameter of the ellipse, and the center is calculated. Using minimum bounding rectangle method, the contour of the target image needs to be scanned first to obtain the outer regular rectangle [87]. Then the minimum bounding rectangle is obtained by rotating and translating the main or auxiliary axis of the regular rectangle, and the center point coordinates o (x, y) of the minimum bounding rectangle are calculated according to the four vertex coordinates $p1–p4$, as shown in Fig. 6.24.

6.3.3 Image Processing After ROI Selection

In the actual production of CSA, the chemo-responsive dyes need to be formulated into a certain concentration solution. When the solution is spotted on the solid substrate with a capillary, the diffusion of the solution is a process, which results in different color layers surrounding the center of spotting. In order to accurately extract the color difference before and after the reaction between sensors and measured objects, it is necessary to perform image processing after ROI selection.

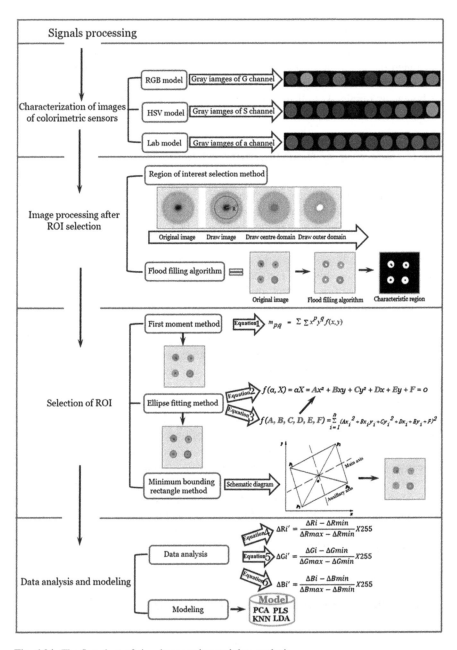

Fig. 6.24 The flowchart of signal processing and data analysis

Generally, after selecting the ROI of the color-developing area of the sensor, the center point of the sensor color-developing area is the center of a circle, of which the radius is a certain pixel length (it needs to be less than the smallest inscribed circle of the color area). The circular area is artificially set as a ROI, namely a characteristic area, as shown in Fig. 6.24. The flood filling method is a kind of region filling methods. With the former method, it can fill areas with very similar colors or brightness so as to achieve the region segmentation. The purpose of applying the flood filling method is to separate the central color layer and the peripheral color layer of the sensor, and the characteristic extraction regions (color layers) of the color change of sensor are shown in Fig. 6.24. After obtaining different regions of interest, the gray values of each channel of the ROI before and after the VCs exposure to chemo-responsive dyes are extracted and subtracted to characterize the change of channels. And the response to VCs could be characterized with a $3N$-dimensional vector (N dyes $\times$ 3 color component difference (RGB)).

6.3.4 Data Analysis and Modeling

After the imaging processing, the gray values of the three components of R, G, and B of ROI image before and after the CSA exposure to detected VCs are extracted. The feature matrix of gray values is enlarged by normalization, and the normalization formula is given in Formula 4–6. Here, taking the R component as an example, ΔRi and $\Delta Ri'$ are the characteristic variables before and after the normalization of the R component of the ith chemo-responsive dye, and ΔR_{max} and ΔR_{min} are the maximum and minimum original characteristic variables of the R component of all chemo-responsive dyes, respectively. The normalized matrix is used to generate grayscale difference images of the three components of R, G, and B; finally, the grayscale images are superimposed to obtain the difference feature image before and after CSAs exposure to the gas.

To identify different unknown odor, it is necessary to use the pattern recognition method to train and predict the known samples and construct a discriminant model. In general, some chemometric methods including principal component analysis (PCA), linear discriminant analysis (LDA), k-nearest neighbor (KNN), partial least squares (PLS), etc. are used to minimize the prediction variable complexity [88, 89]. PCA is a statistical method that has received considerable attention [24]. By orthogonal transform, a large number of variables are converted into a set of few variables, which is called principal components (PCs) [11, 31]. As a pattern recognition method, LDA is carried out based on minimizing the within-class distance and maximizing the between-class distance [90]. KNN is a kind of method, which stores all the sample data of the calibration set in the computer and the distance between each unknown sample. Each training sample is calculated one by one to find the nearest k based on the assumption that the same kind of samples in the simulation space closing to each other [91]. PLS is a supervised method used to build the linear and multivariate calibration model, which has the advantages of collinearity and interaction problems in the detection indexes data [22].

6.4 Application in Food Quality and Safety Detection

CSA is a novel e-nose for the detection and classification of food materials based on their VCs; it has gained multitudinous attention due to its intuitive detection results, excellent detection performance, and wide-spectrum detection range. Actually, CSA is playing an increasingly important role in rapid nondestructive testing of food quality and safety. Scholars at home and abroad have conducted a lot of researches on the application of this field. CSA has been proposed to evaluate the states of some commodities or agricultural products (agro-products), such as detection of cereals [27, 28], wine and vinegar flavored foods [13, 14, 23, 24], poultry meat [92, 93], aquatic products, fruits and vegetables, tea etc., indicating the technology as the considerable research development in food quality and safety control.

6.4.1 Application in Detection of Agro-Products

During the storage and transportation of agro-products, it will have a decline in the quality with the increasing storage time, which is reflected in the increase of lipids, sulfides, and furan compounds in the grains. The characteristic volatile gases of the agro-products with mold infection are captured to detect and analyze the condition of the grains, eventually providing guarantee for the quality and safety of the grains. VOCs can have a color-developing effect with chemo-responsive dyes, thereby realizing the "visualization" of odor information. Based on this, volatile odors would be present through visual images. The intuitive detection results and excellent detection performance of the CSA have gotten great attention and research in the field of grain quality detection.

1. **Detection of Grain Freshness**

     While grain is constantly aging, the important qualities such as color and smell will have a decrease. Lin et al. [24] used odor imaging method to detect and identify VOCs of rice with different storage. Figure 6.25 shows a characteristic diagram of rice. There appears different degrees of color change and forms a characteristic map corresponding to each rice storage. Each colorimetric unit of the image is converted into 3 colorful red-green-blue difference images (RGB), and a total of 18 variables are obtained (6 color sensitive materials × 3 color components).

     Figure 6.26a is a three-dimensional scatter plot with principal component scores of five types of samples as the input. The cumulative contribution rate reaches 95.41% (>85%). From Fig. 6.26b, we can conclude that the KNN model achieves the recognition rate of the calibration set and prediction set for 98%. As can be seen in Fig. 6.26c, the recognition rates of the calibration set and prediction set are the highest, respectively, 99% and 98%. There is only one sample in the prediction set with a storage duration of 0 month and was misjudged as 2 months.

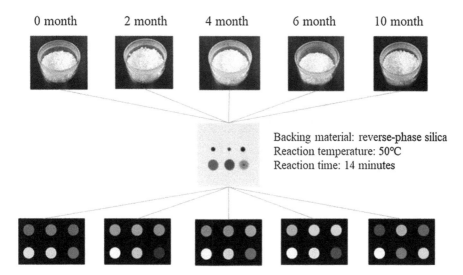

Fig. 6.25 The interactive sensor to distinguish the storage period of rice

It can be obtained that rice with different storage periods can be qualitatively distinguished by CSA.

2. **Detection of Grain Mildew**

Grain is susceptible to mold infection and invasion, leading to its some physical characteristics of the internal structure being change by the metabolism of mold. Lin et al. [26] have found that grains are infected by different molds during storage, accompanied by different types and content ratios of VOCs. It takes the early and rapid diagnosis of wheat mildew infection as an example. A total of 320 wheat samples with a mass of 8 g were studied, and these samples include 32 fresh wheat, while the remaining wheat samples were inoculated with *Aspergillus glaucus, Aspergillus candidus*, and *Aspergillus flavus*, which were grown on potato dextrose agar (PDA) medium for different days. In the case of wheat infected with three different *Aspergillus*, the average RGB difference response obtained from each group of samples is plotted as a characteristic image and shown in Fig. 6.27. The PCA algorithm is used to extract the image information of the CSA for further classification and discrimination.

Figure 6.28a is a three-dimensional principal component clustering diagram obtained by PCA for different mildew degrees of wheat infected with three different *Aspergillus glaucus*, and the cumulative contribution rate reached 93.33%. Figure 6.28b, c belongs to diagram of samples infected with *Aspergillus candidus* and *Aspergillus flavus*. In general, the PCA algorithm have to distinguish the degree of wheat mildew with three *Aspergillus* infections, especially performing no overlap between the mild mildew samples (degree II) and moderate mildew (degree III), which indicates that CSA can have an effective performance in the early detection wheat mildew.

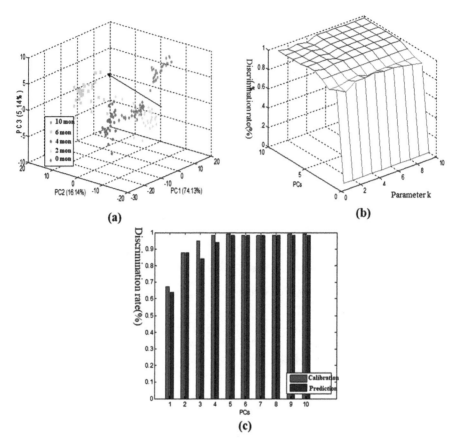

Fig. 6.26 Different model results. (**a**) Diagram of three-dimensional principal component scores from VCs of rice with different storage time; (**b**) identification rates from KNN model; (**c**) identification rates from LDA model

3. **Detection of Tea Variety Identification**

Due to the difference of processing techniques, tea can be divided into many different types, of which green tea, oolong tea, and black tea are the three most common types. Provided three color components of R, G, and B as the characteristic value by each color reagent, there are a total of 27 characteristic variables (9 color reagent × 3 color components) according to the study of Chen et al. [16]. Figure 6.29a shows the initial image, final image, and characteristic image of green tea, oolong tea, and black tea. Utilizing PCA and LDA pattern recognition methods, classification results were compared and analyzed. As can be clearly seen in Fig. 6.29b, compared with the boundary between black tea and green tea, the boundary between oolong tea and green tea is more obvious and can be better distinguished. As shown in the figure, a good classification effect has been achieved, and the interactive verification of different fermented tea reached a 100% correct rate.

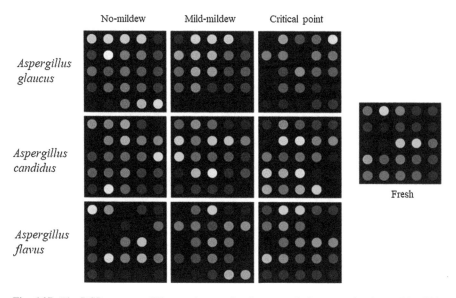

Fig. 6.27   The RGB response difference image of each group of wheat samples detected by CSA array

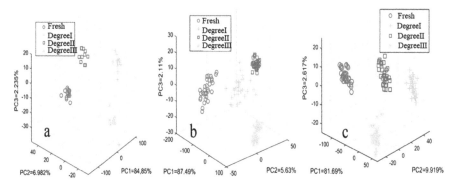

Fig. 6.28   The three dimensional principal component clustering of wheat samples infected with *Aspergillus glaucus, Aspergillus candidus*, and *Aspergillus flavus*

4. **Detection of Fruit Damage**

Fruits, such as banana, apple, peach, produce characteristic odor during storage and deterioration. Therefore, CSA can be applied to discriminate the quality classification. Peaches are susceptible to damage during picking and transportation, resulting in decreased flavor. The appearance characteristics of damaged peaches (especially mechanical damage) have no obvious difference from the characteristics of normal fruits in the early storage period. According to the principle that M-porphyrins have cross-sensitivity to aromatic substances in peach odor, a corresponding sensor array is designed to test three types of peaches

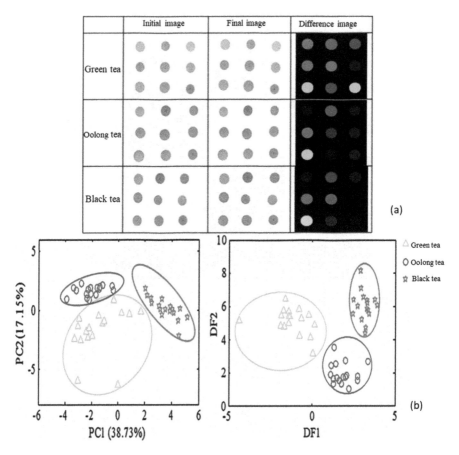

Fig. 6.29 (a) The initial image, final image, and characteristic image of green tea, oolong tea, and black tea; (b) the classification results of green tea, oolong tea, and black tea

with different damages in Huang group. The peaches are divided into three types: the first type of fruit (15 pieces) no treatment; the second type of fruit (20 pieces) do free fall from different heights to simulate mechanical damage, and after being placed at 30 °C for 24 h; For the third type of fruits (15 pieces), the peach peel is pierced with a needle, and a small amount of gray mold is inoculated on the damaged part of the peel with an inoculating needle. More than 20 kinds of metals sensitive to these gases are screened out from hundreds of porphyrins compounds to form a 5 × 5 sensor array.

After the data matrix is standardized, the principal component analysis is performed on the image. One fruit of the first category and one fruit of the second category were misjudged. Because the first and second types of fruit are edible fruits, and the third type of fruit is inedible due to fungal infection. In practical applications, the correct identification of the third type of fruit is more meaningful and the correct rate is 100% as shown in Table 6.1.

Table 6.1 Recognition results of different grades of fruits

Peach grades	Correct number of predictions	Error number of prediction	Correct rate(%)
First grade	5	1	85.7
Second grade	9	1	90
Third grade	7	0	100

6.4.2   Application in Detection of Liquid Foods

Odor is an important evaluation basis for authenticity identification and quality classification of liquid foods such as vinegar and wine. Chemo-responsive dyes are employed to interact with the volatile gas of the detected object and the color of the dye molecule changes significantly. After computer processing, a specific digital signal is formed, and finally the corresponding pattern recognition method is used to perform qualitative and quantitative analysis [13–17].

1. **Quality Control of Vinegar Fermentation**

   To monitor the acetic fermentation, it is necessary to consider the changes in the content of characteristic VOCs such as alcohols, esters, and acids to grasp the main condition and also pay attention to the detection and characterization of other volatile gases to obtain the overall information. Guan et al. [13] screened out nine kinds of porphyrins and three kinds of pH indicators to fabricate a sensor array with four rows and three columns (4 × 3). As shown in Fig. 6.30a, the characteristic image of the CSA before 14 min has been changing, indicating that the reaction has not yet reached equilibrium. When the reaction time reaches 14 min, the color of the difference graph tends to be stable. Based on gas chromatography-mass spectrometry (GC-MS) to detect the alcoholic strength of vinegar during culture fermentation, Fig. 6.30b shows the change of ethanol content during acetic fermentation. Figure 6.30c shows the correlation between the measured value of alcohol and the predicted value of the BP-ANN model. It can be seen from the figure that the odor information of different fermentation days collected by the odor imaging method is highly correlated with the ethanol content measured by GC-MS.

2. **Quality Control of Rice Wine**

   In the process of rice wine fermentation, it is absolutely necessary to monitor some vital physical and chemical indicators, such as the dynamic changes of alcohol and aromatic substances [94, 95]. CSA is not susceptible to changes in humidity easily, thus suitable for analyzing and detecting the characteristic odor of liquid food. According to the characteristics of the rice wine odor, nine porphyrins and six pH indicators are preferably screened as gas-sensitive materials to form a sensor array with good selectivity and sensitivity. After 16 min of rice wine sample exposure to CSA, the characteristic images of samples with different storage time is obtained, as shown in Fig. 6.31a. Different degrees of color changes occurred after each chemo-responsive dyes reacting with rice wine of different storage time. *RGB* color components of each color reagent were

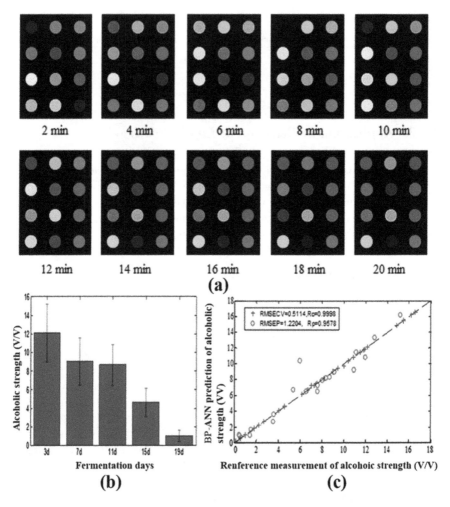

Fig. 6.30 (a) Difference images of CSA array exposure to vinegar substrate to different time arranged from 2 to 20 min; (b) variation of alcoholic strength through acetic acid fermentation; (c) correlation between the measured value and the BP-ANN model prediction of ethanol

extracted as the characteristic value, then a rice wine sample would get 45 characteristic variables (15 color reagents × 3 color components). Then, these variables are processed by LDA to distinguish rice wine of different storage time. Figure 6.31b is the scores of LD1 and LD2. And the two discriminant functions embrace 93.28% of the whole information. As displayed in the figure, the four types of rice wine with different storage time can be completely distinguished, and there is no overlap between different categories.

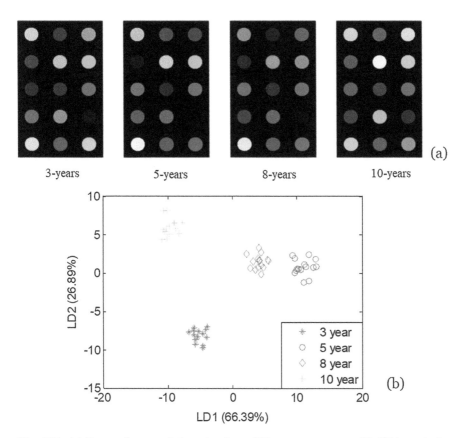

Fig. 6.31 (**a**) Feature images of rice wine from different storage ages; (**b**) LDA result for differentiating rice wine of different storage ages by olfactory imaging sensor

6.4.3 Application in Detection of Poultry

The demand for poultry is one of the fastest-growing foods in many places of the world. With the continuous increase of the demand for chicken, more and more attention from consumers and researchers has been gradually attracted in the freshness of chicken, especially focusing on the content of total volatile basic nitrogen (TVB-N), an important reference index for chicken freshness. New electronic noses based on low-cost CSA have also been explored and applied to rapid detection in this field. The national standard GB/T 5009.44–18 was adopted as the determination method. In terms of odor imaging method, the 4×3 sensor array is made and a scan is performed to obtain the images before and after the reaction between the CSA and chicken samples. Figure 6.32a shows the RMSEP of the AdaBoost-BPANN model under different principal component numbers and thresholds. It can be seen that the lowest RMSEP can be obtained when the principal component number is 6 and φ is 0.13. In the prediction set, when Rp equaled to 0.888 and RMSEP equaled to

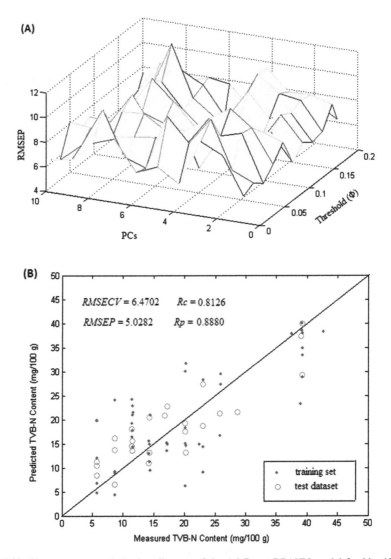

Fig. 6.32 The parameter optimization diagram of the AdaBoost-BPANN model for identifying chicken samples (**a**) and the recognition diagram (**b**)

5.0282 mg/100 g, the best BP-ANN model is obtained. The scatter diagram between the measured value of TVB-N content and the predicted result is shown in Fig. 6.32b. Therefore, CSA can quantitatively detect the freshness of chicken.

6.4.4   Application in Detection of Aquatic Products

Aquatic products such as freshwater fish have a relatively high content of unsaturated fatty acids and are easily oxidized. At the same time, the protein in the fish body is easily decomposed to produce bad odors, resulting in flavor changes and even endangering consumers' health. During the spoilage process of fish, the volatile components mainly include hydrocarbons, alcohols, aldehydes, ketones, esters, phenols, and sulfur and nitrogen compounds. A visual sensor array is made of porphyrin compounds and a pH indicator, which can detect changes in fish freshness. Huang et al. [21] show the sensor array measurements of fish were classified by radial basis function (RBF) algorithm into three groups, corresponding to day 1 (group 1), day 2, 3, 4, and 5 (group 2), and day 6 and 7 (group 3), which is shown in Table 6.2. The total classification accuracy was 87.5%. It can suggest that the system is capable of inspecting quality of fish and perhaps other food containing high protein.

6.4.5   Applications in Detection of Other Fields

CSA, as a rapid detection method, has a qualitative and quantitative analysis of various food volatile gases, so as to provide guarantee for food quality and safety monitoring. It has been widely used in many fields, in addition to the detailed examples described above, likewise, in other fields, as shown in Table 6.3. CSA has embodied some effectiveness of applying in the food detection. It is not only capable of acquiring imaging information from natural food, but it is also possible to explore the odor changes of processed foods to monitor their quality, so as to meet the challenges food sample diversity and volatile component complexity.

Table 6.2 Radial basis function (RBF) neural network classification of fish freshness

Test day	Group 1	Group 2	Group 3	Total	Classification accuracy (%)
1	33	7	0	40	82.5
2	8	32	0	40	80
3	5	35	0	40	87.5
4	3	37	0	40	82.5
5	3	36	1	40	90
6	0	3	37	40	92.5
7	0	1	39	40	97.5
Total					87.5

Table 6.3 The odor imaging detection technology applications in the area of food detection

Class	Product	Detection result	References
Natural food	Rice	Correct identification of 85% in the storage time of rice samples	[96]
	Wheat	Root mean square errors were 3.05 and 1.65 g L^{-1}	[25]
	Chicken	Quantification detection of total TVB content with Rp of 0.898	[30]
	Pork	SVMR prediction model with the determination coefficient of 0.9055	[97]
	Milk	Detection of streptomycin from simply-treated milk at 100 nmol/L	[98]
	Honey	Detection of streptomycin from honey at 125 nmol/L	[98]
	Herbal medicine	Discrimination (with 100% accuracy) of 46 herbal distillates	[99]
Processed food	Vinegar	Monitoring vinegar acetic fermentation days with LDA and HAC models	[70]
	Chinese liquor	Identification rate of geographic of 100% by PCA, LDA, and HAC models	[94]
	Beer	Error rate of categorization identification <3%	[34]
	Pork sausage	Detection accuracy rate of storage time of 93.81 by PLS model	[33]
	Tea	Classification of category 100% by LOOCV	[15]
	Fruit pickle	Analysis accuracy of toxic materials in adulterated of 0.99	[100]
	Coffee beans	In quintuplicate runs of 10 commercial coffees and controls, no confusions or errors in classification by HCA in 55 trials	[80]

6.5   Conclusion and Future Trends

The work presents a comprehensive insight into CSA, including the sensing principle, imaging principle, and application for food quality and safety. The implementation of nanotechnology in the sensor field for rapid sensing gives impetus to the development of CSA. The response signal of the technology is timely and easy to be characterized by color imaging method. CSA has proved itself as a leading detection technique with no or less sample preparation, nondestructive and rapid nature. In order to break through the lack of intelligence and dexterity of odor imaging equipment, it has emerged to replace the camera as an imaging device for collection. To make odor imaging technique more analytically useful, the trend in future will be oriented toward CSA combined with spectroscopy methods with higher developments for the application in food industry.

The foregoing contents suggest that CSA paves the way for detecting and differentiating complex components. In CSA, chemo-responsive dyes are generally utilized to perform as the probes to interact with given analytes, this interaction with odorants could be quantified by digital imaging. The applications of CSA using chemo-responsive dyes as probes in the evaluation of food safety and quality have many advantages; however, there are still some unovercomed limitations. For example, the equipment of CSA is not petite and intelligent to track real-time food quality. To solve the associated problem, related studies have employed cellphone camera to capture imaging data as collecting-imaging tool of CSA. Based on this, a small, less expensive hand-held device for use of CSA was developed to ensure food quality and safety [37]. The CSA composed of chemo-responsive dyes possess a certain timeliness, that is, it is applied in CSA for immediate fabrication and could not be prepared in advance, which affects the portability of the technology application. Fluorescent immunochromato graphic assay has the integration of high sensitivity of fluorescence detection and high specificity of traditional immunochromato graphic assay. Combining CSA, a small, portable, high-performance fluorescent immunochromatographic test paper would be developed for on-site testing in the food industry.

In addition, foods with quality problems would have some complicated changes in their odor components, and however CSA only collect few the odor data by the tri-color imaging method. Currently, applications of NIR spectroscopy and hyperspectral imaging have been explored for food quality detection and monitoring. Therefore, it is possible to combine spectroscopy methods and CSA to capture more information. There is a need to capture the odor information in sample by odor imaging method, and the subsequent use of spectroscopy methods for the detection of the response signal via data processing presents a way forward. This might solve the problem of fewer variables available for screening in odor imaging methods. The model established by CSA is more stable and accurate in the near future and can be used for quantitative analysis of food detection.

References

1. Pico J, Bernal J, Gómez M (2015) Wheat bread aroma compounds in crumb and crust: a review. Food Res Int 75:200–215
2. Moreira N, Lopes P, Ferreira H et al (2016) Influence of packaging and aging on the red wine volatile composition and sensory attributes. Food Packag Shelf Life 8:14–23
3. Feng T, Zhuang H, Ye R et al (2011) Analysis of volatile compounds of Mesona Blumes gum/rice extrudates via GC–MS and electronic nose. Sens Actuators B Chem 160(1):964–973
4. Sanaeifar A, Mohtasebi SS, Ghasemi-Varnamkhasti M, Ahmadi H (2016) Application of MOS based electronic nose for the prediction of banana quality properties. Measurement 82:105–114
5. Sung J, Kim BK, Kim BS, Kim Y (2014) Mass spectrometry-based electric nose system for assessing rice quality during storage at different temperatures. J Stored Prod Res 59:204–208
6. Radi CS, Litananda WS, Rivai M, Purnomo MH (2016) Electronic nose based on partition column integrated with gas sensor for fruit identification and classification. Comput Electron Agric 121:429–435
7. Deshmukh S, Bandyopadhyay R, Bhattacharyya N et al (2015) Application of electronic nose for industrial odors and gaseous emissions measurement and monitoring—an overview. Talanta 144:329–340

8. Liu Q, Zhao N, Zhou D et al (2018) Discrimination and growth tracking of fungi contamination in peaches using electronic nose. Food Chem 262:226–234

9. Majchrzak T, Wojnowski W, Dymerski T et al (2017) Electronic noses in classification and quality control of edible oils: a review. Food Chem 246:192–201

10. Ezhilan M, Nesakumar N, Babu KJ et al (2018) An electronic nose for royal delicious apple quality assessment—a tri-layer approach. Food Res Int 109:44–51

11. Rakow NA, Suslick KS (2000) A CSA array for odor visualization. Nature 406:710–713

12. Neal AR, Avijit S, Michael C et al (2005) Molecular recognition and discrimination of amines with a colorimetric array. Angew Chem Int Ed 29(44):4528–4532

13. Guan B, Zhao J, Lin H et al (2014) Characterization of volatile organic compounds of vinegars with novel electronic nose system combined with multivariate analysis. Food Anal Methods 7 (5):1073–1082

14. Guan B, Zhao J, Cai M (2014) Analysis of volatile organic compounds from Chinese vinegar substrate during solid-state fermentation using a CSA array. Anal Methods 6(23):9383–9391

15. Chen Q, Liu A, Zhao J (2013) Classification of tea category using a portable electronic nose based on an odor imaging sensor array. J Pharm Biomed Anal 84:77–83

16. Chen Q, Liu A, Zhao J et al (2013) Monitoring vinegar acetic fermentation using a CSA array. Sens Actuators B Chem 183:608–616

17. Chen Q, Sun C, Ouyang Q et al (2014) Classification of vinegar with different marked ages using olfactory sensors and gustatory sensors. Anal Methods 6(24):9783–9790

18. Chen Q, Li H, Ouyang Q et al (2014) Identification of spoilage bacteria using a simple CSA array. Sens Actuators B Chem 205:1–8

19. Gu H, Huang X, Chen Q et al (2019) Prediction of the property of CSA array based on density functional theory. Sens Mater 31(10):3067–3073

20. Huo D, Wu Y, Yang M et al (2014) Discrimination of Chinese green tea according to varieties and grade levels using artificial nose and tongue based on CSA arrays. Food Chem 145:639–645

21. Huang X, Xin J, Zhao J (2011) A novel technique for rapid evaluation of fish freshness using CSA array. J Food Eng 105(4):632–637

22. Huang X, Zou X, Shi J et al (2014) Determination of pork spoilage by colorimetric gas sensor array based on natural pigments. Food Chem 145:549–554

23. Lin H, Man Z, Guan B et al (2017) In situ quantification of volatile ethanol in complex components based on CSA array. Anal Methods 40:5873–5879

24. Lin H, Man Z et al (2018) A novel CSA array based on boron-dipyrromethene dyes for monitoring the storage time of rice. Food Chem 268:300–306

25. Lin H, Kang W, Kutsanedzie FYH et al (2019) A novel nanoscaled chemo dye–based sensor for the identification of volatile organic compounds during the mildewing process of stored wheat. Food Anal Methods 12(12):2895–2907

26. Lin H, Duan Y, Yan S (2019) Quantitative analysis of volatile organic compound using novel chemoselective response dye based on Vis-NIRS coupled Si-PLS. Microchem J 145:1119–1128

27. Lin H, Yan S, Song B (2019) Discrimination of aged rice using CSA array combined with volatile organic compounds. J Food Process Eng 42:13037

28. Lin H, Wang Z (2020) Identification of rice storage time based on CSA array combined hyperspectral imaging technology. J Stored Res Product 85:101523

29. Lin H, Kang W, Jin H et al (2020) Discrimination of Chinese Baijiu grades based on CSA arrays. Food Sci Biotechnol 39:1039–1043

30. Khulal U, Zhao J, Hu W et al (2016) Comparison of different chemometric methods in quantifying total volatile basic-nitrogen (TVB-N) content in chicken meat using a fabricated CSA array. RSC Adv 6(6):4663–4672

31. Qin H, Huo D, Zhang L et al (2012) Colorimetric artificial nose for identification of Chinese liquor with different geographic origins. Food Res Int 45(1):45–51

32. Qian S, Lin H (2015) CSA Array for detection and identification of organophosphorus and carbamate pesticides. Anal Chem 87(10):5395–5400
33. Salinas Y, Ros-Lis JV, Vivancos JL et al (2014) A novel CSA array for monitoring fresh pork sausages spoilage. Food Control 35(1):166–176
34. Zhang C, Bailey DP, Suslick KS (2006) CSA arrays for the analysis of beers: a feasibility study. J Agric Food Chem 54(14):4925–4931
35. Zhang C, Suslick KS (2007) CSA Array for Soft Drink Analysis. J Agric Food Chem 55:237–242
36. Askim JR, Mahmoudi M, Suslick KS (2013) Optical sensor arrays for chemical sensing: the optoelectronic nose. Chem Soc Rev 42:8649–8682
37. Askim JR, Suslick KS (2015) Hand-held reader for CSA arrays. Anal Chem 87:7810–7816
38. Prieto I, Pedrosa JM, Martín-Romero MT et al (2000) Characterization and structure of molecular aggregates of a tetracationic porphyrin in LB films with a lipid anchor. J Phys Chem B 104(43):9966–9972
39. Dinten O, Spichiger UE, Chaniotakis N et al (1991) Lifetime of neutral-carrier-based liquid membranes in aqueous samples and blood and the lipophilicity of membrane components. Anal Chem 63:596–603
40. White WI (1978) Aggregation of porphyrins and metalloporphyrins. In: Dolphin D (ed) The porphyrins, vol 5. Academic, New York, pp 303–339
41. Verrelli G, Francioso L, Siciliano P et al (2007) Identification of wine defects by means of a miniaturized electronic tongue. The International Society for Optical Engineering, Bellingham, p 65891I
42. Swamy PCA, Mukherjee S, Thilagar P (2014) Multichannel-emissive V-shaped boryl-BODIPY dyads: synthesis, structure, and remarkably diverse response toward fluoride. Inorg Chem 53(10):4813–4823
43. Huang X, Zou X, Shi J et al (2018) CSA arrays based on chemo-responsive dyes for food odor visualization. Trends Food Sci Technol 81:90–107
44. Gabor G, Chadha S, Walt DR (1995) Sensitivity enhancement of fluorescent pH indicators using pH-dependent energy transfer. Anal Chim Acta 313(1–2):131–137
45. Stone RD, Vaughn JL (1986) The effect of pH indicators chlorphenol red and bromocresol purple on the in vitro production of lepidopteran nuclear polyhedrosis viruses. J Invertebr Pathol 48(3):381–383
46. Motellier S, Toulhoat P (1993) Modified acid-base behaviour of resin-bound pH indicators. Anal Chim Acta 271(2):323–329
47. Li X, Xu W, Wang X et al (2000) Ultraviolet-visible and surface-enhanced Raman scattering spectroscopy studies on self-assembled films of ruthenium phthalocyanine on organic monolayer-modified silver substrates. Thin Solid Films 457(2):372–380
48. Kuntz JID, Gasparro FP, Johnston MD Jr et al (1968) Molecular interactions and the Benesi-Hildebrand equation. J Am Chem Soc 90(18):4778–4781
49. Wong C, Horrocks WDW (1975) New metalloporphyrins. Thorium and yttrium complexes of tetraphenylporphin. Tetrahedron Lett 16(31):2637–2640
50. Brandenburg J, Ggimme S (2013) Dispersion corrected Hartree-Fock and density functional theory for organic crystal structure prediction. Top Curr Chem 345:1–23
51. Parr RG (1983) Density functional theory. Annu Rev Phys Chem 34(1):631–656
52. Marques M, Gross E (2004) Time-dependent density functional theory. Annu Rev Phys Chem 55:427–455
53. Gu H, Huang X, Yao L et al (2014) Study on the binding ability of cobalt-porphyrin with small volatile organic compounds based on density functional theory. Anal Methods-UK 6(10):3360
54. Gu H, Sun Y, Liu S et al (2018) A feasibility study of the rapid evaluation of oil oxidation using synchronous fluorescence spectroscopy. Food Anal Methods 11(12):3464–3470
55. Gu H, Huang X, Chen Q et al (2020) Rapid assessment of total polar material in used frying oils using manganese tetraphenylporphyrin fluorescent sensor with enhanced sensitivity. Food Anal Methods 13(11):2080–2086

56. Houk KN (1975) Frontier molecular orbital theory of cycloaddition reaction. Acc Chem Res 8 (11):361–369
57. Walsh PJ, Gordon KC, Officer DL et al (2006) A DFT study of the optical properties of substituted Zn(II)TPP complexes. J Mol Struct Theochem 759(1–3):17–24
58. Zhou H, Baldini L, Hong J et al (2006) Pattern recognition of proteins based on an array of functionalized porphyrins. J Am Chem Soc 128(7):2421–2425
59. Zhang X, Guo C, Li Z et al (2002) An optical fiber chemical sensor for mercury ions based on a porphyrin dimer. Anal Chem 74(4):821–825
60. Hu Y, Xue Z, He H et al (2013) Photoelectrochemical sensing for hydroquinone based on porphyrin-functionalized Au nanoparticles on graphene. Biosens Bioelectron 47:45–49
61. Nakada K, Fujita M, Dresselhaus G et al (1996) Edge state in graphene ribbons: nanometer size effect and edge shape dependence. Phys Rev B 54(24):17954
62. Roach P, Farrar D, Perry CC (2006) Surface tailoring for controlled protein adsorption: effect of topography at the nanometer scale and chemistry. J Am Chem Soc 128(12):3939–3945
63. Awschalom DD, Vincenzo DP, Smyth JF (1992) Macroscopic quantum effects in nanometer-scale magnets. Science 258:414–414
64. Scholl JA, García-Etxarri A, Koh AL et al (2013) Observation of quantum tunneling between two plasmonic nanoparticles. Nano Lett 13(2):564–569
65. Sousa AA, Hassan SA, Knittel LL et al (2016) Biointeractions of ultrasmall glutathione-coated gold nanoparticles: effect of small size variations. Nanoscale 8(12):6577–6588
66. Beer PD, Cormode DP, Davis JJ (2004) Zinc metalloporphyrin-functionalised nanoparticle anion sensors. Chem Commun 4:414–415
67. Palmer LC, Stupp SI (2008) Molecular self-assembly into one-dimensional nanostructures. Acc Chem Res 41(12):1674–1684
68. Li D, Huang J, Kaner RB (2008) Polyaniline nanofibers: a unique polymer nanostructure for versatile applications. Acc Chem Res 42(1):135–145
69. Shimizu T, Masuda M, Minamikawa H (2005) Supramolecular nanotube architectures based on amphiphilic molecules. Chem Rev 105(4):1401–1444
70. Guan B, Zhao J, Chen Q et al (2020) Preparation of zinc porphyrin nanoparticles and application in monitoring the ethanol content during the solid-state fermentation of Zhenjiang Aromatic vinegar. Microchemical J 153:104353
71. Yu D, An J, Bae JY et al (2005) Negatively charged ultrafine black particles of P(MMA-\r, c\r, o\r, -EGDMA) by dispersion polymerization for electrophoretic displays. Macromolecules 38 (17):7485–7491
72. Ruenraroengsak P, Tetley TD (2015) Differential bioreactivity of neutral, cationic and anionic polystyrene nanoparticles with cells from the human alveolar compartment: robust response of alveolar type 1 epithelial cells. Particle Fibre Toxicol 12(1):19
73. Jimenez AS, Jaramillo F, Hemraz UD et al (2017) Effect of surface organic coatings of cellulose nanocrystals on the viability of mammalian cell lines. Nanotechnol Sci Appl 10 (11):123–136
74. Wang Y, Yang Q, Zhao M et al (2018) Silica-nanochannel-based interferometric sensor for selective detection of polar and aromatic volatile organic compounds. Anal Chem 90 (18):10780–10785
75. Dumanoğulları FM, Tutel Y, Küçüköz B et al (2019) Investigation of ultrafast energy transfer mechanism in BODIPY–porphyrin dyad system. J Photochem Photobiol A Chem 373:116–121
76. Von Baeckmann C, Guillet-Nicolas R, Renfer D et al (2018) A toolbox for the synthesis of multifunctionalized mesoporous silica nanoparticles for biomedical applications. ACS Omega 3(12):17496–17510
77. Xu S, Hartvickson S, Zhao J (2011) Increasing surface area of silica nanoparticles with a rough surface. ACS Appl Mater Interfaces 3(6):1865–1872
78. Su Z, Shaw WL, Miao YR et al (2017) Shock wave chemistry in a metal-organic framework. J Am Chem Soc 139(13):4619–4622

79. Zhang C, Suslick KS (2005) A CSA array for organics in water. J Am Chem Soc 127 (33):11548–11549
80. Suslick BA, Feng L, Suslick KS (2010) Discrimination of complex mixtures by a CSA array: coffee aromas. Anal Chem 82(5):2067–2073
81. Jiang H, Xu W, Chen Q (2019) Evaluating aroma quality of black tea by an olfactory visualization system: selection of feature sensor using particle swarm optimization. Food Res Int 126:108605
82. Felix YHK, Lin H, Yan S et al (2018) Near infrared chemo-responsive dye intermediaries spectra-based in-situ quantification of volatile organic compounds. Sens Actuators B Chem 254:597–602
83. Lana MM, Tijskens LMM, van Kooten O (2005) Effects of storage temperature and fruit ripening on firmness of fresh cut tomatoes. Postharvest Biol Technol 35:87–95
84. Leon K, Mery D, Pedreschi F, Leon J (2006) Color measurement in La*b* units from RGB digital images. Food Res Int 39:1084–1091
85. Phelan JP (2011) First moment. Mindfulness 2(1):68–70
86. Wang Y, Xu C, Chen L (2008) Relations between non-abelian chern-simons fielsd and fractional spins. Mod Phys Lett B 22(1):45–50
87. Jo B, Jung S (2018) Quadrant-based minimum bounding rectangle-tree indexing method for similarity queries over big spatial data in HBase. Sensors 18(9):3032
88. Domínguez-Aragón A, Olmedo-Martínez JA, Zaragoza-Contreras EA (2018) CSA based on a poly(ortho-phenylenediamine-co-aniline) copolymer for the monitoring of tilapia (Orechromis niloticus) freshness. Sens Actuators B Chem 259:170–176
89. Kim SY, Ko JA, Kang BS et al (2018) Prediction of key aroma development in coffees roasted to different degrees by CSA array. Food Chem 240:808–816
90. Berrueta LA, Alonso-Salces RM, Héberger K (2007) Supervised pattern recognition in food analysis. J Chromatogr A 1158:196–214
91. Škrbi B, Đurišić-Mladenović N (2007) Principal component analysis for soil contamination with organochlorine compounds. Chemosphere 68(11):2144–2152
92. Alimelli A, Pennazza G, Santonico M et al (2007) Fish freshness detection by a computer screen photoassisted based gas sensor array. Anal Chim Acta 582(2):320–328
93. Huang X, Zou X, Zhao J et al (2015) Monitoring the biogenic amines in Chinese traditional salted pork in jelly (Yao-meat) by CSA array based on nine natural pigments. Int J Food Sci Technol 50(1):203–209
94. Ouyang Q, Zhao J, Chen Q et al (2013) Classification of rice wine according to different marked ages using a novel artificial olfactory technique based on CSA array. Food Chem 138 (2–3):1320–1324
95. Ouyang Q, Zhao J, Chen Q et al (2013) Classification of rice wine according to different marked ages using a portable multi-electrode electronic tongue coupled with multivariate analysis. Food Res Int 51(2):633–640
96. Guan B, Zhao J, Jin H, Lin H (2016) Determination of rice storage time with CSA array. Food Anal Methods 10(4):1054–1062
97. Li H, Kutsanedzie F, Zhao J et al (2016) Quantifying total viable count in pork meat using combined hyperspectral imaging and artificial olfaction techniques. Food AnalMethods 9:3015–3024
98. Liu Z, Zhang Y, Xie Y et al (2017) An aptamer-based CSA for streptomycin and its application in food inspection. Chem Res Chin Univ 33(5):714–720
99. Bahram, H., ·Javad, T.,·Mohammad, MB. et al. Development of CSA array for discrimination of herbal medicine. J Iran Chem Soc, 14, (2017), 595–604
100. Bordbar MM, Tashkhourian J, Hemmateenejad B (2018) Qualitative and quantitative analysis of toxic materials in adulterated fruit pickle samples by a CSA array. Sens Actuators B Chem 257:783–791

Chapter 7
Acoustic and Vibrating Signal Analysis Technologies in Food

What Is Acoustic and Vibrating Signal Analysis Technologies?

Agro-product's vibration mechanical characteristics refer to the characteristics of vibration signals produced by agro-products under external excitation. The vibration signal analysis technology analyzes the characteristics of scattering, transmission, absorption, attenuation coefficient and propagation speed, acoustic impedance, and natural frequency of agro-products' vibration signals. They reflect the basic principle of the interaction between vibration signals and qualities of agro-products (as shown in Fig. 7.1).

Abbreviation	Full name
Agro-product	Agricultural product
FFT	Fast Fourier transform
FIR	Finite impulse response
GA-PLS	Genetic algorithm-partial least squares
GA-SiPLS	Genetic algorithm-synergy interval partial least squares
LMS	Least mean square
PLS	Partial least square

7.1 Instruction

7.1.1 Mechanical Vibration Waves of Objects

The mechanical vibration of objects transmitted in the medium with particles and elasticity is called volatility, and human auditory organs feel the volatility is called a sonic wave. It is known from general physics that acoustic energy is a form of mechanical energy. The generation should have two conditions: first is the signal excitation source, and the next is the elastic medium. When the vibration is caused

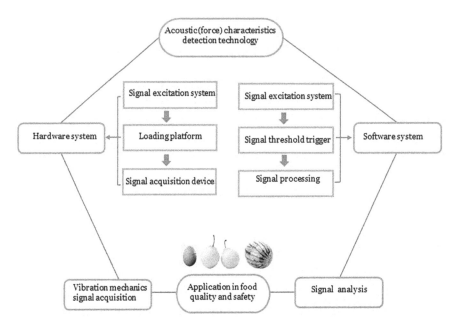

Fig. 7.1 The design of acoustic and vibrating signal analysis technologies

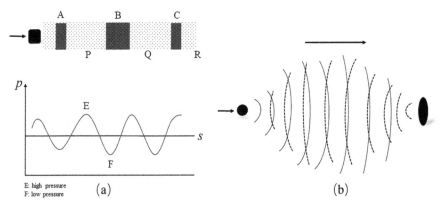

Fig. 7.2 (**a**) The density of medium causes the propagation of sonic wave and (**b**) the instantaneous situation after sonic wave is emitted

by external signal excitation, the surrounding medium particles will vibrate and produce displacement. The media density will be formed in the fluid medium space, and the sonic wave propagation will be formed (Fig. 7.2a). When the vibrator surface vibrates to the right, the medium particles are compressed, as shown in A, B, and C in the figure; when the vibrator surface vibrates to the left, negative pressure is formed and sparsity occurs, as shown in P, Q, and R in the figure. Note that the average value of the sound pressure variation along the propagation direction in the figure is not zero. It is the atmospheric pressure in air and the static pressure in water. When the sonic wave is emitted in an infinite homogeneous medium, its instantaneous state is shown in Fig. 7.2b.

The circular arc in the figure represents the wavefront, connected by the points with the same vibration phase in the sound field. The figure shows a special case, representing the instantaneous peak position of vibration, where the particle displacement ($\xi 0$) is the largest. The distance between the two peaks is λ, which is called the wavelength. Sound rays indicate sonic wave propagation. It can also be seen from the figure that far away from the sound source, the curvature of the wavefront is very small. Within a limited range, the wave front is nearly flat, the sound ray is approximately parallel, and the wave properties are similar to those of the plane wave. However, we should pay attention to the following two issues.

The transmission of vibration waves is not the propagation of medium particle itself but is the propagation of particle vibration. Therefore, it is necessary to distinguish particle's vibration velocity from the propagation velocity (sound velocity) of the sonic wave in the medium.

Figure 7.2 uses the curve to represent the wave's concept, which is completely different from electromagnetic wave and string vibration. The latter particle's displacement is perpendicular to its propagation direction, called a transverse wave. The removal of an acoustic particle in a fluid medium is parallel to its propagation direction called a longitudinal wave. Therefore, the displacement ($\xi 0$) drawn in Fig. 7.2a, b is perpendicular to the x-axis only for the convenience of expression, and the concepts of the two cannot be confused.

7.1.2 Vibration of Particle and Elastomer

To produce a sonic wave in an elastic medium, vibration shall be produced first. In a particular situation, a disturbance is stimulated by the external environment in a local area of a flexible medium, making the medium particle A move away from the equilibrium position. This particle A pushes the adjacent medium particle B to compress this part of the adjacent medium (Fig. 7.3). Due to the elastic action of the medium, when this part of the adjacent medium is compressed. It will produce a force against compression, which acts on particle A and makes it return to its original equilibrium position. On the other hand, because particle A has mass and inertia, particle A will appear "overshoot" when passing through the equilibrium position, which will compress the adjacent medium on the other side. Particle A forces compression will also be generated in the adjacent medium so particle A will return to the equilibrium position. It can be seen that due to the elastic and inertial effects of the medium, particle A was initially disturbed, vibrates back and forth near the equilibrium position. For the same reason, particles B pushed by A, and even other particles C and D vibrate near the equilibrium position, only lagging for some time in turn. The propagation of mechanical vibration of medium particles from near to far is called a sonic wave. A visible sonic wave is a kind of mechanical wave. The

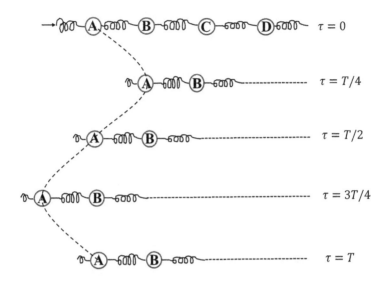

Fig. 7.3 Schematic diagram of particle vibration in elastic medium

propagation process of this kind of particle vibration in the elastic medium is very similar to the mass-spring-mass-spring chain system formed multiple oscillators coupling. In this system, one oscillator motion will affect the process of other oscillators following the motion [1]. Figure 7.3 shows the mass of oscillator A at four different time positions. The mass of other oscillators also vibrates in the equilibrium position's vicinity but lags for some time in turn.

7.2   Instrumentation

7.2.1   Acoustic Vibration Technology

Generally, agro-products with different structural strengths will produce other sound signals under external excitation. Under the external excitation, the sound generated by the stronger detection object is very crisp; the sound generated by the smaller detection object is relatively dull. Due to the different surface structure or internal agro-products organization, the acoustic signals generated by free vibration under external excitation show different characteristics [2]. The quality of agro-products with different qualities can be judged by the differences in the characteristics of reflection, scattering, transmission, absorption, attenuation coefficient, propagation speed, acoustic impedance, and natural frequency. By analyzing the difference of sound signals produced by external impact, structural stiffness could be analyzed.

7.2.2 *Vibration Response Signal Acquisition and Analysis Platform*

The detection system based on vibration mechanical signal analysis realizes eggshell detection with different quality by collecting and analyzing the response signal generated by agro-products knocking [3]. The structure and principal diagram of the detection system is shown in Fig. 7.4. Its main components include a signal excitation device, support device, signal collector, signal amplifier, sound card, and computer data acquisition system. The microphone collects the vibration response signal generated by the excitation detection object, and then it is sent into the computer after amplification, filtering, and A/D conversion. The program compiled by Labview language collects and processes the vibration response signal of agro-products.

1. **Hardware System**

   The vibration signal analysis platforms' hardware system is composed of three parts: the acoustic module, support module, and signal acquisition and processing module. The acoustic module pulses the inspected agro-product from the external environment so that the agro-product generates free vibration after being impacted. The function of the support module is to support the detected object so that the object could achieve the greatest possible vibration without damaging the surface structure of the object under the external impact. From the vibration, the collection and processing of the signal realizes the collection of the response signal and the corresponding filtering processing, then converts the analog signal

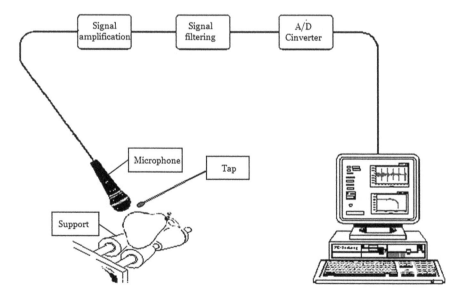

Fig. 7.4 Structural diagram of acoustic vibration detection system for agro-products

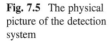
Fig. 7.5 The physical picture of the detection system

into a digital signal for the next step of processing. The physical picture of the detection system is shown in Fig. 7.5.

(a) Signal excitation system

   The acoustic module that is mechanically excited by agro-products' external impact is usually composed of an adjustable linear power supply, electromagnet, single-chip microcomputer, and cracking rod. Cracking is driven by electromagnets and combined with the linear power supply to achieve control. The electromagnet adopts the two-way corner permanent magnet electromagnet (SXCT38-15/18) of Nanjing Kailing Automation Factory. The performance of this electromagnet covers the main functions of a rotating electromagnet and a two-way self-holding electromagnet. Due to high-performance permanent magnets and grain-oriented high-silicon steel strips as the main material of the magnetic circuit, and the coil winding adopts the winding process, these aspects keep the main characteristic parameters of the electromagnet "Ampere Turn" and "Magnetic Power" at the best state. When the coil is energized, the shaft core could immediately produce rotational displacement. After the power is turned off, under the permanent magnet's closed magnet field, the shaft core still maintains its position. Rotation and reset are driven by forward and reverse pulse power. The two-way corner electromagnet could hold the shaft core in two positions at the beginning and end of the formation. The two different positions have the same output torque (with bistable characteristics). Therefore, compared with the rotating electromagnet of the same size, the two-way angle electromagnet has a large rotation angle (the action-angle can reach 100°), rotate in both directions (both left and right), fast response time, energy-saving, and temperature control. Notable features such as the small rise and high power meet the demand for automatic and quick cracking of inspected agro-products.

   When the cracking system executes the cracking operation, the system needs to signal the electromagnet to make the electromagnet execute the command. At the same time, the rate of cracking is adjusted according to the needs of the system. The action execution instruction and rate are relatively simple, and the system uses a single-chip microcomputer to control these instructions.

(b) Loading platform

The inspected agro-product support module comprises support rollers, steel pipes, steel bases and brackets, stepping motors with speed regulators, and other components. The supporting rollers are composed of four pillar-shaped hollow rollers made of nylon material. Their shape is suitable for normal agro-products placement; the hollow rollers are sleeved with the steel pipe. The position on the steel pipe is adjusted relative to find the best support position for the agro-products. The speed adjustment range of the motor with a governor is 20–60 rpm/min. The stepper motor drives the rotation of the agro-products by driving roller's rotation so that the longitude of the agro-products to be measured could be fully cracked.

(c) Signal collection

The tapping response signal is collected by the sound card, amplified by the amplifier, filtered and A/D converted by the sound card, and then input to the computer. The system's vibration response signal is collected by the microphone (parameter index: moving coil microphone, unidirectional, sampling frequency 20 Hz to 20 kHz). Since the microphone output is a weak voltage signal, this signal is generally about tens to hundreds of millivolts, amplified before further processing. This system uses an analog signal amplifier, which directly amplifies the collected analog signal and transmits it into the sound card. Then, use the sound card (16-bit A/D sampling accuracy, 44.1K sampling frequency) to convert the analog signal from the microphone into a digital signal processed by the computer.

2. **Software System**

The software system includes three functional modules: signal acquisition, signal threshold triggering, and signal processing. When the vibration response signal of agro-products is transmitted to the host through the sound card, the software system reads the signal. This provides the time–amplitude two-dimensional time domain diagram of the strength of the vibration response signal and displays the real-time changes of the time domain signal on the man–machine interface of the software. The software system sets a threshold trigger in the process of time domain and frequency domain signal conversion. When the system recognizes that the collected signal is egg cracking vibration response signal, rather than environmental noise, it performs fast. Fourier transform (FFT) on the time domain signal transforms the time domain signal into the frequency domain signal and uses the power spectrum as the frequency domain signal's statistical parameter. The spectrum of the response signal could also be observed in real-time on the system's man–machine interface. After obtaining response signal spectrum, the system uses a filter to filter and denoise the signal and store the data in the hard disk for subsequent processing.

7.2.3 The Acquisition and Processing Flow of Vibration Signal

The cracking vibration response signal acquisition and processing flow of each inspected agro-product is as follows: the response signal generated by the external cracking the detected object and converted the response signal into an analog signal by a moving coil microphone, then the analog signal is amplified by an amplifier, and then the signal enters the sound card for A/D conversion, the software system program obtains the analog time domain signal. The frequency domain signal of the detected object is obtained by threshold trigger, and then filtered and de-noised these two signals by virtual hardware, and the required data could be obtained and stored in the hard disk for further analysis. The specific flowchart is shown in the Fig. 7.6.

7.3 Signal Processing and Data Analysis

7.3.1 Signal Processing

The cracking vibration response signal can measure agro-products with external structure qualities. During the detection, there is an unavoidable interference response signal by the external environment. Due to the influence of noise, the detected objects' signal characteristics are relatively vague, which affect the detection of quality. Therefore, the de-noising is accomplished by filtering methods to improve the resolution and sensitivity of the signals.

Adaptive filtering is widely used at present, and its system can change or adjust its structure and parameters. This system improves its signal processing performance by contacting the external environment called the adaptive system. This system can automatically adapt to the changing environment and requirements of signal transmission without knowing the structure and prior knowledge or accurately design the signal processing system's structure and parameters. An adaptive system is a time-varying nonlinear system, whose characteristics are mainly determined by the system's self-adjustment to different signal environments. In the adaptive filtering signal processing method, adaptive finite impulse response filter (FIR) combined with normalized least mean square (NLMS) and recursive least squares (RLS). These two algorithms are common methods to produce the signal of agro-products by filtering de-noising processing.

1. **Least Mean Square Adaptive Filter (LMS)**

   LMS is a linear adaptive filtering algorithm with a simple structure and good stability, one of the classical and effective adaptive filtering algorithms. In the filtering process, the mean square of the error between the expected response and the filter as the principle for LMS estimates the gradient vector according to the input signal in the iterative process and updates the weight coefficient to achieve

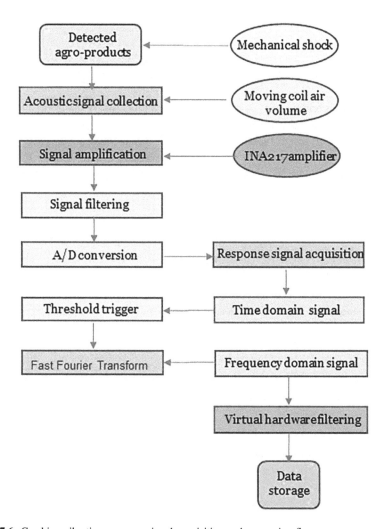

Fig. 7.6 Cracking vibration response signal acquisition and processing flow

the optimal adaptive iteration [4, 5]. However, there are contradictions in selecting step size (μ) in terms of convergence rate and steady-state imbalance by the LMS algorithm. Many improved LMS algorithms have been proposed in the research. NLMS has a simple structure and a large dynamic range of input, which is widely used.

In the basic LMS algorithm:

$$e(n) = d(n) - X^T(n)W(n)$$
$$W(n+1) = W(n) + 2ue(n)X(n) \tag{7.1}$$

In this algorithm, $d(n)$ is as the expected output, $e(n)$ is the error, $W(n)$ is the weight of node and u is the iteration step size. The eigenvalue has changed with the change of input signal change, which affects the convergence rate and maladjustment. It may even destroy the convergence condition, so the NLMS algorithm is proposed.

$$e(n) = d(n) - X^T(n)W(n)$$
$$W(n+1) = W(n) + ue(n)X^T(n)X(n)^{-1}X^T(n) \tag{7.2}$$

$u(n) = u(X^T(n))X(n)^{-1} = \frac{u}{p_j}$ replaces u, p_j is the normalized value of the input power. Convergence time of mean square error is: $\tau = T_s/(4u\lambda_i/p_j)$, $M = (u/p_j)\text{tr}[R]$. Because λ_i is proportional to $\text{tr}[R]$ and p_j, p_j can keep the property of LMS stable and expand its dynamic range. To ensure the stable convergence of the adaptive filter, NLMS is used to normalize the convergence factor u to realize the adaptive filter.

Some researchers obtained the response signal before and after the adaptive filtering by tapping eggs. After processing with the NLMS algorithm, they found that this method significantly affected by filtering and denoising. After adaptive filtering, the noise of the vibration response signal was significantly reduced, and the resolution and sensitivity of the signal were improved, which provide the possibility to judge the effective recognition of the vibration response signal. Figure 7.7 shows the response signal diagram before and after NLMS adaptive filtering.

2. **Adaptive RLS Filtering**

   RLS is an adaptive filtering algorithm that is a recursive method with a partial least square of the adaptive transversal filter. In this algorithm, the least square estimation of filter tap weight vector is used to recurse the latest estimation from $n - 1$ to n. In RLS, the relationship between expected response $d(n)$ and tapped input vector $u(n)$ can be represented by a multiple linear regression model:

$$d(n) = W_0^T u(n) + e_0(n) \tag{7.3}$$

In this formula, $W0 = [W0, 0, W0,1 ..., W0, m-1] T$ is the regression parameter vector, the mean value of $e_0(n)$ is 0. The tap input vector $u(n)$ is generated randomly, and its auto-correlation function has ergodicity. Therefore, the statistical average can be replaced by the time average. Like NLMS, the RLS is implemented in a 32-bit adaptive filter by setting the forgetting factor to 1 and the initial filtering factor to 0. Figure 7.8 shows the corresponding signal graphs of cracked eggs before and after the RLS algorithm's adaptive filtering.

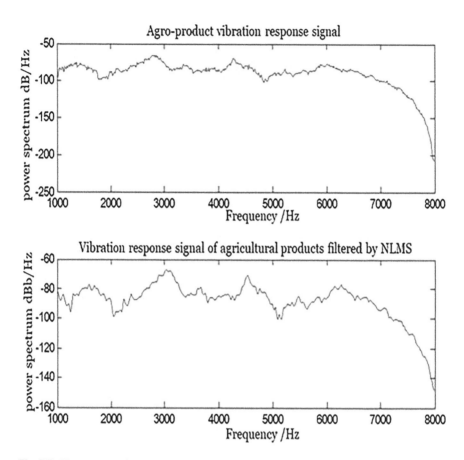

Fig. 7.7 The response signal diagram before and after NLMS adaptive filtering

3. **Independent Component Analysis (ICA)**

Recently, ICA is a multivariate statistical data analysis technique to extract non-Gaussian and independent source signals. It has been concerned in the field of blind de-convolution to the low-resolution image. Requiring before no information about the blur type and size is one of the advantages of ICA. Besides, the computational load can be saved because only the mixing matrix absorbing blurring process needs to be searched rather than the whole image data.

7.3.2 Feature Extraction and Analysis

1. **Feature Information Extraction**

Vibration mechanics response signal power spectrum contains more information, and its characteristic information needs to be extracted for further analysis [6]. The following introduces several commonly used analysis methods which

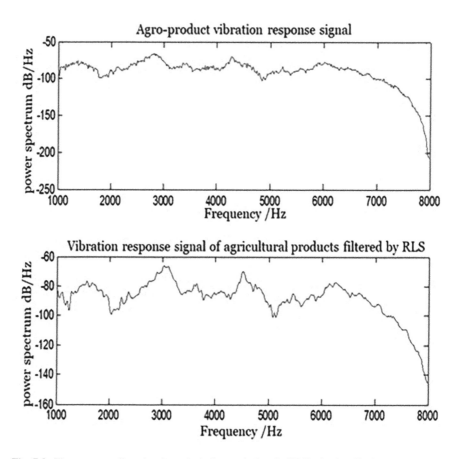

Fig. 7.8 The corresponding signal graphs before and after the RLS adaptive filtering

is for vibration mechanics response signal: characteristic power spectrum area, power spectrum standard deviation, frequency band corresponding to the first formant, the variance of the power spectrum of the first three formants, and power spectrum energy and variation in the middle and low-frequency bands—calculation of coefficients, quotients, zero-crossing points, etc.

(a) The power spectrum area (area)

The power spectrum area is the sum of the amplitude of the power spectrum of each point. By calculating and comparing the test data, the food and agro-products with different structural stiffness have different power spectrum areas. Therefore, the power spectrum can be used as a distinguishing factor for discrimination. Its calculation method is as follows:

$$\text{Area} = \sum_{i=0}^{512} P_i \tag{7.4}$$

P_i is the power spectrum amplitude corresponding to each sampling point.

(b) The standard deviation of the power spectrum (σ)

The standard deviation of the power spectrum can reflect the data dispersion for the sampling points. Generally speaking, the peak value of the power spectrum of agro-products with good structural stiffness is relatively prominent, and the difference between the amplitude is obvious. Its calculation formula is as follows:

$$\sigma = \sqrt{\frac{\sum_{i=0}^{n} P_i^2 - nP^2}{n-1}} \tag{7.5}$$

In this formula, n is the sampling point, P_i is the power spectrum amplitude corresponding to each sampling point, P is the mean of the amplitude of the power spectrum.

(c) The frequency band corresponding to the first formant (f_1)

The frequency band corresponding to the first formant is the frequency point corresponding to the maximum amplitude. f_1 can be described by this formula:

$$f_1 = \text{Index}_{\max}(P_i) \tag{7.6}$$

(d) The variance of the power spectrum of the first three formant (D)

When an agro-products with high structural stiffness is stimulated, its power spectrum in frequency domain usually has an obvious peak, and with large variance among formants. The power spectrum variance of the first three formants is relatively large. The power spectrum variance D of the first three formants is described by the formula as follows:

$$D = \text{Max}_{1:3}(P_i/3) \tag{7.7}$$

(e) The ratio of mean power spectrum energy of middle-frequency bands to low-frequency bands (M)

Generally, agro-products' large structure stiffness has more power spectrum in the middle-frequency band (3500–6000 Hz) than the low-frequency band (1000–3500 Hz) by the extra external cracking. However, the agro-products with small structure stiffness have little difference, so it can distinguish the structural stiffness of agro-products. The ratio of mean power spectrum energy of middle frequency bands to low frequency bands is calculated as follows:

$$M = \frac{\left(\sum\limits_{i=1}^{200} P_i\right)}{\left(\sum\limits_{i=201}^{400} P_i\right)} / 200 \tag{7.8}$$

(f) Coefficient of variation

The coefficient of variation is an index reflecting the difference degree or dispersion degree of each unit's marker value in the whole population. It is also one of the indexes reflecting the data distribution. Compared with standard deviation, the coefficient of variation can be more comprehensive to reflect the data volatility; the structure of agro-products is relatively complete [7, 8]. The incentive signals emitted from different regions of the variation coefficient in its surface is small. Still, when there is damage or crack of agro-products, the signals emitted from different regions of the variation by external excitation will get bigger, its formula is as follows:

$$CV == \sqrt{\frac{1}{N}\sum_{i=1}^{N}(x_i - \mu)^2} / \mu \tag{7.9}$$

$\mu = \frac{1}{N}\sum\limits_{i=1}^{N} x_i$ $(i = 1, 2, 3, \ldots, 512)$ is the arithmetic mean.

- Difference quotient

  The ratio difference between maximum and minimum values set data, which represents the relative fluctuation amplitude of data, and the formula is as follows:
- Number of passing zero

  When two continuous amplitudes are one positive and one negative, the zero-crossing number of the waveform is calculated by the self-programmed program. The agro-products with different structural stiffness also have a great difference in zero passing number.

2. **Optimization of Eigenvalues**

In agro-products, the external eigenvalue is used to eliminate redundant data and simplify the modeling process. However, the number of characteristic parameters is not the better. Some eigenvalues could not reflect the characteristics of the global data well, and its stability is poor, which cannot represent the excitation signal of agro-products with different structural stiffness. Therefore, it is necessary to screen the extracted feature variables further, select more reasonable feature parameters, remove the parameters with a small contribution, and enhance the model's stability. "Goodness" has more advantages for the screening of fewer eigenvalues. This method considers the stability and recognition of waveform eigenvalues. The judging standard of excellence is parameter D.I. (distinguished index), which is calculated as follows:

$$D.I = \frac{|\overline{X}_2 - \overline{X}_1|}{\sigma_2 + \sigma_1} \tag{7.10}$$

In this formula, $\overline{X}_1, \overline{X}_2$ are the mean values of agro-products with different structural stiffness under a certain characteristic value. σ_1, σ_2 are the standard deviations of the corresponding signal under the eigenvalue. The higher the value, the higher the resolution, and the recognition ability on characteristic parameters is strong. When the value of a parameter is ≤ 0.85, it is considered that its goodness is not enough to be suitable for further modeling and recognition, so it should be removed.

7.4 Application in Detection of Agro-Products Quality

7.4.1 Detection of Poultry Eggshell Strength

In commercial egg production, one of the main concerns is to avoid the cracks in eggshells, because cracked eggs are very vulnerable to bacterial infections leading to health hazards. Consumers may perceive eggs with cracked shell as lack of quality. In such case, it is essential to measure the strength of eggshell, avoiding overload for eggshell in the processing of collection, sorting, and transportation. The strength difference of eggshell would be reflected in their acoustic response signals. Lin et al. presented a nondestructive method for eggshell strength measurement by using acoustic response analysis [9].

The samples of this experiment were purchased from Zhenjiang Farm, Jiangsu Province in China. The eggs were cultured with intact brown shells 1–2 days after production. They have regular shapes and uniform sizes. The egg sample was tapped six times near the equator, and the average value of the vibration response signals of the six taps was collected as the original signal of the sample. After each sample was tapped to collect the tapping vibration response signal, a quasi-static compression test by a physical property instrument (Stable Mcro System, UK, model TA-XT2i) was performed to test the compression stress of the eggshell. The maximum compressive stress of the eggshell was used as the criterion of the eggshell strength. Figure 7.9 showed the stress distribution of eggs.

Quasi-static compression could only detect the static and continuous egg load, and it is difficult to detect the dynamic and instantaneous egg load. Therefore, the vibrational response signals were also used for the eggshell strength measurement [10–12]. The frequency range of the vibration response signal of egg cracking was 1000–8000 Hz. The vibration response signal of egg cracking was adaptively filtered by RLS. Figure 7.10 shows the power spectrum of the vibration response signal of egg cracking with different eggshell strengths after denoising by RLS filter. Figure 7.10 (1–3) shows the response diagram of the signal with higher eggshell strength (higher than 40 N). Figure 7.11 (4–6) showed the eggshell strength is lower (less than 40 N) vibration response signal diagram of acoustic. It could be

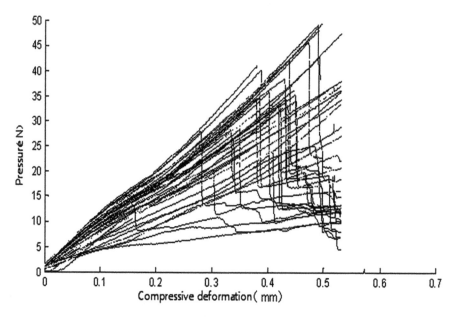

Fig. 7.9 Stress distribution curve of eggs

observed from the figure, the vibration response signal of an egg with a larger shell strength (maximum bearing stress) had a prominent resonance peak, which did not present in the lower shell strength eggs. The reason may be eggs with stronger eggshells would receive sharper pulse signals when impacted, the corresponding resonance peaks may be more prominent. In contrast, eggs with lower eggshell strength will have relatively gentle pulse signals.

This study used genetic partial least squares (GA-PLS) to establish an eggshell of strength test model. The GA-PLS model results show that when 39 frequency variables are used, the lowest RMSECV value of 3.599 could be obtained, and the genetic algorithm synergy interval partial least squares (GA-SiPLS) model results show that when 29 frequency variables are used, the lowest RMSECV value of 3.608 can be obtained (Fig. 7.11).

7.4.2 Detection of Cracking Poultry Eggs

After the production and processing, it is inevitable for eggshell to be cracked. Large cracks are easy to observe with the naked eye and are convenient for screening. However, many of them are small cracks. Damage is not easy to be found [13]. Therefore, the rapid detection of surface cracks on poultry eggs needs to be solved urgently. When the eggshells appear cracks, the surface characteristics of eggs have changed, such as damping coefficient and stiffness [14–17]. Lin et al. presented a novel method for egg crack detection using multiple vibration sensor analysis.

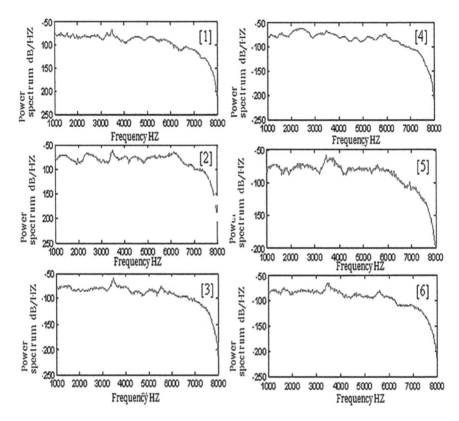

Fig. 7.10 The power spectrum of the vibration response signal of eggs

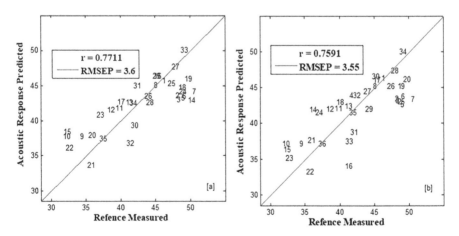

Fig. 7.11 The correlation diagram between the predicted value and the measured value of the cracking egg vibration response signal of GA-PLS (**a**) and GA-SiPLS (**b**) models

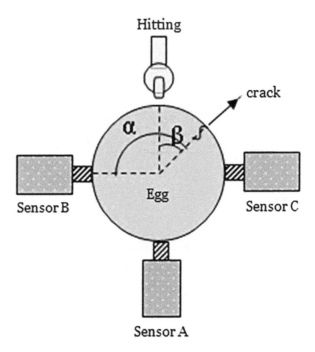

Fig. 7.12 Side view of the multi-sensor egg crack detection platform (α represents the angle between the crack and the sensor, β represents the angle between the striking point and the crack)

One hundred fresh and well-wiped eggs from the farm were selected and labeled in order. In the work, the eggs with intact shells were first tapped. Each tapping point was repeated three times. Eight tapping points were selected for each egg. After completing the sequence, tap the eggshell with a metal rod until the shell tapping sound was dull. Then the same acoustic experiment was performed on such eggs, and each hitting point was repeated three times, with eight points per egg. The position of sensors and the integrity of the eggshells would affect the signal. The detection platform structure of eggs was shown in Fig. 7.12. For eggs, the maximum amplitude of intact eggs was larger than cracked eggs. The possible reason was that the vibration wave on the eggshell surface changed when the egg was excited. The integrity of the intact eggshell ensured the smooth propagation of the vibration wave.

The crack could reduce the vibration wave, so the amplitude of the cracked egg would be corresponding decrease [18]. Figure 7.13 shows the comparison of the waveforms of cracked and intact eggs.

LDA was used to identify the quality of samples. The sensors from three groups obtained satisfied performances, the optimal result was achieved when sensor A was used, where identification rates for crack eggs were 97.50% in training set and 93.75% in prediction set, respectively. Table 7.1 describes the joint linear discriminant analysis results of sensors. The data showed that the standard deviation of sensor A to sensor C and the ratio of the sum of the top ten maximum values had a better recognition effect on cracked eggs. At the knocking point 5, the number of higher resolution index was more than other tapping points which further illustrated the advantage of discrimination of the bottom sensor.

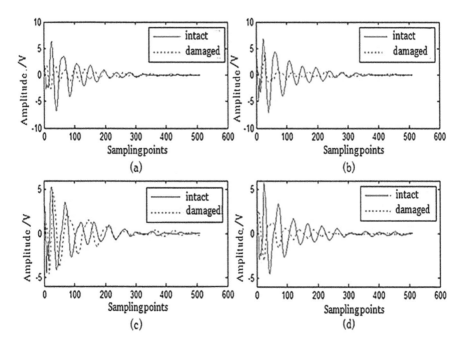

Fig. 7.13 Comparison of time domain waveforms of intact cracks collected by sensor B

Table 7.1 Joint linear discriminant analysis results of sensors

Hitting point	Sensor A		Sensor B		Sensor C	
	Test (%)	Train (%)	Test (%)	Train (%)	Test (%)	Train (%)
1	**91.52**	**95.83**	**81.25**	**85.83**	**93.75**	**95.83**
2	61.52	58.33	68.75	72.50	52.50	62.50
3	**80.00**	**86.67**	**83.75**	**90.83**	**83.75**	**89.17**
4	60.00	59.17	61.25	69.17	55.00	59.17
5	**93.75**	**97.50**	**86.25**	**95.00**	**80.00**	**95.83**
6	63.75	59.17	60.00	56.67	58.75	56.67
7	**87.50**	**90.83**	**91.25**	**83.33**	**92.50**	**92.50**
8	65.00	61.67	61.25	65.83	63.75	62.50

The bold data indicate the sensors (placed in the corresponding hitting points) that have better performances for the identification of cracked eggs

7.4.3 Detection of Fruit

The surface structure and internal quality of agro-products are different, and when they vibrate freely under external excitation, the sound response signal produced would show different characteristics. Based on the characteristic differences of acoustic impedance and natural frequencies in agro-products, their vibration response signals could be used to analyze the structural quality of fruit. Damaged and intact fruit would produce different sound signals under external excitation.

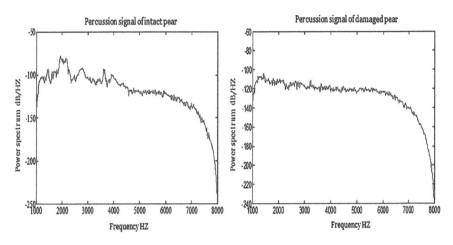

Fig. 7.14 Frequency domain signal of pear hitting response

1. **Application of Acoustic Vibration in Detection of Bruises of Pears**

The samples were purchased from the Zhenjiang Supermarket in April 2011, a total of 105 pieces. The experimental crystal pear surface damage was man-made with a diameter of about 1 cm. The cracking position was selected at the marked point, and through many previous tests, it was guaranteed that the cracking strength would not damage the pear tissue damage. Each marked position was tapped three times and the average value of the data taken as the original signal.

The time domain diagram of intact and damaged pears was similar and difficult to distinguish. This work used FFT to transform the pear vibration response signal from the time domain space to the frequency domain space. In the frequency domain space, the information difference between the two types of pears was discussed. The number of sampling points of FFT was 512, and the vibration response signal within 8000 Hz was analyzed through preliminary tests. The signal at high frequency was weaker, and the signal-to-noise was relatively low. Figure 7.14 shows the frequency domain power spectrum signal of the pear tapping response after bandpass filtering. The power spectrum difference between intact pears was not obvious. The resonance peak of the power spectrum among the most pears was relatively stable. When the structural rigidity of the damaged part of the pear was destroyed, the damping coefficient was increasing.

LDA was used to establish a discriminant model for intact and damaged pears. The 1-4d discriminant model had a recognition rate of more than 82% for intact and damaged pear samples, which could identify most experimental pear samples. By comparing the discrimination rate of the 1-4d training set and the prediction set, it was found that the overall discrimination rate showed a downward trend with time. As the storage time increased, the quality of intact pear gradually decreased, and the surface structure became soft, so the discrimination rate gradually decreased.

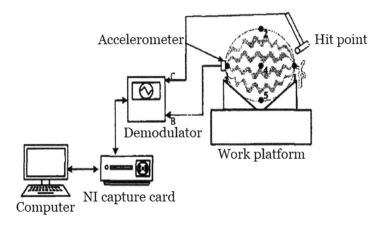

Fig. 7.15 Watermelon tapping vibration monitoring system

2. **Application in Ripeness Evaluation of Watermelon**

Maturity is an important indicator of watermelon quality. Generally, the maturity of watermelon could be divided as immature, mature, and overripe. At present, the maturity identification of watermelon is mainly through manual picking and experience, which is time-consuming and low inefficiency.

The acoustic vibration method is a simple and effective method for the detection of watermelon maturity. Lu et al. established a self-made acoustic vibration detection platform to knock the surface of watermelon for acoustic detection [19]. The sample was artificially ripened and divided into watermelons of different maturity levels. The stalk, the watermelon equator, and the pedicle were selected as the hitting points. In this part, FRF analysis was used to calculate the response function. A nonlinear functional relationship and a multiple linear regression model were used to describe the storage time by multiple vibration characteristic values. The coefficient of determination of the multiple linear regression model was $R^2 = 0.931$, and the maximum standard error of the model was 3.87. Yuan et al. used the similar way to detect the quality of watermelon [20]. The study used PCA to extract the response signal characteristics and constructed a model of watermelon maturity classification and watermelon sugar content detection by extreme learning machine. In the machine learning classification algorithms, ELM had the best discrimination effect and the fastest speed. However, there are still many problems in actual production and application.

Figure 7.15 shows the tapping vibration monitoring system for the watermelon. Li et al. used the vibro-acoustic method as the principle, through the transmission of signals, FFT could obtain the natural frequency of watermelon, and used the relationship between natural frequency and maturity to make a judgment [21]. Figure 7.16 shows the signal waveform of the watermelon. This fruit testing equipment was set up successfully, which is suitable not only for

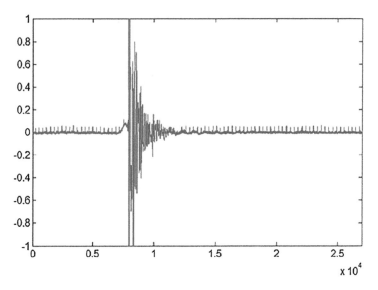

Fig. 7.16 Watermelon audio signal waveform

individual users such as watermelon salespersons and melon farmers but also for large-scale engineering applications.

3. **Application of Acoustic Vibration on Other Fruits**

    Generally, the detection for banana quality in the market is carried out by using puncture test. Jagannath [22] judged the quality of bananas by collecting the acoustic response spectrum before and after the banana damage. The experimental device was similar to the structure set for eggs. It found that parts of bananas were cracked when these samples received dynamic impact from the external.

    As the banana matured, the response resonance frequency decreased and the damping coefficient increased at the same time. The resonance frequency and damping coefficient were related to the color change of bananas, and this color change was due to the increase in respiration rate caused by mechanical or temperature absorption.

    Fruit damping is an important parameter, which is related to fruit maturity and fruit tissue. There was a significant correlation among the three parameters of color index, frequency decrease, and damping ratio increase. It proved that the above three parameters were related to the physical damage caused by fruit bumps, indicating that frequency and damping ratio provided an indication of fruit freshness.

7.4.4   Other Food Inspection

The acoustic detection method suffers from less interference, and the detection device has a lower cost and was easy to realize intelligence. The researchers used

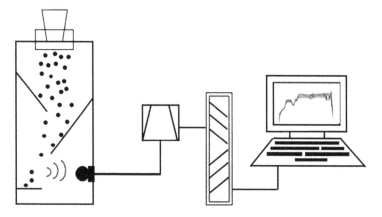

Fig. 7.17 Wheat acoustic signal collection device

acoustic characteristics of fruits and vegetables to reflect basic law relationship between signal and object.

The development of agriculture has always been the guarantee of the national economy. The quality of agro-products directly affected its competitiveness in the market. In the past, some works studied the quality of rice based on the specificity of the color. However, Li et al. used acoustic measurement of wheat hardness to analyze the signal obtained from wheat and studied the characteristics of wheat in the time domain and frequency domain. Figure 7.17 shows the device for collecting the acoustic signal obtained from wheat. The study established a BP neural network prediction model based on the characteristic parameter WFI to predict wheat hardness. The maximum relative error of this model's predicted value was −2.24%, and the average relative error was about 0.15%. It illustrated that the wheat hardness could be measured by the acoustic detection.

Li et al. designed a system for measuring wheat moisture according to the characteristics of wheat grains and the principle of acoustic properties, which included the generation, collection, and later signal analysis of wheat grain sound signals [23]. The samples in this work were carried out in a self-made moisture balance box, and the moisture content of wheat could be controlled by adding moisture. When the samples put into the feeder and collided with the bottom, the computer could collect the acoustic signal produced by impact. Compared with the acoustic signals obtained from the wheat gravity impacting on the target, this work added a sound insulation part to the device, which reduced external noise interference to a certain extent and increased the accuracy of the results.

The use of acoustic for nondestructive testing of grain was a new research idea, but there were many problems in large-scale production. The employment of acoustic characteristics to detect the moisture content of wheat in the pretreatment was more troublesome, it required manual impurity removal and the removal of small pieces of grain. Also, the number of wheat samples was small, and the result was not universal, so this method needed to be improved in the future study.

7.5 Conclusion and Future Trends

The acoustic experiment could determine whether the structure of the test sample was damaged by testing the indicators, such as the reflection characteristics, propagation speeds of sound waves, attenuation coefficients, and scattering characteristics. This method was simple and convenient, especially in agro-products and poultry products. It is possible to establish models related to product characteristics and acoustic characteristics. If the detection method could be optimized, the relevant measurement instruments could be developed to reduce the error caused during the experiment.

With the development of electronic communication science and technology, various new nondestructive testing technologies have gradually been involved in all fields of agricultural and poultry product quality testing. Machine vision and near-infrared spectroscopy technology were used to detect the integrated quality of eggs by the vibration response signal. These studies are aimed to promote these new methods from the laboratory to the practical application and provide the possibility for the automation and industrialization of quality inspection.

References

1. Ali MM, Hashim N, Bejoa SK et al (2017) Rapid and nondestructive techniques for internal and external quality evaluation of watermelons: a review. Sci Hortic 225:689–699
2. Jie DF, Zhou WH, Wei X (2019) Nondestructive detection of maturity of watermelon by spectral characteristic using NIR diffuse transmittance technique. Sci Hortic 257:108718
3. Abdullah MH, Nashat S, Anwar SA et al (2017) A framework for crack detection of fresh poultry eggs at visible radiation. Comput Electron Agric 141:81–95
4. Li PC, Wang QH, Zhang QL et al (2012) Non-destructive detection on the egg crack based on wavelet transform. IERI Procedia 2:372–382
5. Baltazar A, Arandab JI, Aguilar GG (2008) Bayesian classification of ripening stages of tomato fruit using acoustic impact and colorimeter sensor data. Comput Electron Agric 60:113–121
6. Craessaerts G, Saeys W, Missotten B et al (2007) A genetic input selection methodology for identifying the cleaning process on a combine harvester, part II: selection of relevant input variables for identification of material other than grain (MOG) content in the grain bin. Biosyst Eng 98:297–303
7. Lin H, Zhao JW, Sun L et al (2011) Stiffness measurement of eggshell by acoustic resonance and PLS models. J Food Eng 103:351–356
8. Sun L, Bi XK, Lin H et al (2013) On-line detection of eggshell crack based on acoustic resonance analysis. J Food Eng 116:240–245
9. De Ketelaere B, Coucke P, De Baerdemaeker J (2000) Eggshell crack detection based on acoustic resonance frequency analysis. J Agric Eng Res 76:157–163
10. De Ketelaere B, Vanhoutte H, De Baerdemaeker J (2003) Parameter estimation and multivariable model building for the non-destructive, on-line determination of eggshell strength. J Sound Vib 266:699–709
11. Yang ZJ, Mao L, Yan B et al (2020) Performance analysis and prediction of asymmetric two-level priority polling system based on BP neural network. Appl Soft Comput 99:106880
12. Wen F, Jing FS, Zhao WH et al (2020) Research on optimal receiver radius of wireless power transfer system based on BP neural network. Energy Rep 6:1450–1455

13. Wang HJ, Mao JH, Zhang JY et al (2016) Acoustic feature extraction and optimization of crack detection for eggshell. J Food Eng 171:240–247
14. Sun L, Cai JR, Lin H et al (2013) On-line estimation of eggshell strength based on acoustic impulise response analysis. Innov Food Sci Emerg Technol 18:220–225
15. Zhao Y, Wang J, Lu QJ et al (2010) Pattern recognition of eggshell crack using PCA and LDA. Innov Food Sci Emerg Technol 11:520–525
16. Deng XY, Wang QH, Chen H et al (2010) Eggshell crack detection using a wavelet-based support vector machine. Comput Electron Eng 70:135–143
17. Kemps B, De Ketelaere B, Bamelis F et al (2006) Development and validation of a methodology for dose calculation in electron beam irradiation of complex-shaped foods. J Food Eng 89:215–221
18. Pan LQ, Zhan G, Tu K et al (2011) Eggshell crack detection based on computer vision and acoustic response by means of back-propagation artificial neural network. Eur Food Res Technol 233:457–463
19. Lu Y, Pu HJ, Wang DS et al (2015) Research on non-destructive testing of watermelon storage time by acoustic vibration method (Chinese)
20. Yuan PP (2017) Research on extreme learning model and application of non-destructive testing of watermelon maturity (Chinese)
21. Li N, Zhang YG, Hao YL et al (2008) A new variable step-size NLMS algorithm designed for applications with exponential decay impulse responses. Signal Process 88:2346–2349
22. Jangannath JH, Dasgupta DK, Bawa AS et al (2005) Assessment of ripeness/damage in banana (musa paradisiacal) by acoustic resonance spectroscopy. J Food Qual 28:267–278
23. Li GC (2012) Research on the acoustic measurement method of wheat moisture (Chinese)

Chapter 8
Multi-sensor Data Fusion Technologies in Food

What Is Detection Technology of Multi-sensor Data Fusion?

The food quality is evaluated by external appearance and interior composition, while a single detection technology has limitations for achieving highly accurate and stable results. In comparison with single-sensor, multi-sensor data ffusion (MSDF) has several merits offering detailed information, medium tolerance human cognitive level ability to distinguish food comprehensively, since it incorporates food quality and safety information collected by multiple sensors. Briefly, the purpose of this chapter is to introduce the principle of MSDF detection technology, the data processing of MSDF, as well as its application in food quality discrimination. We will herein provide guidelines and reference for its practical application in food quality and safety (Fig. 8.1).

Abbreviation	Full name
BPANN	Back propagation artificial neural network
CCD	Charge-coupled device
CV	Computer vision
D-S	Dempster–Shafer
E-nose	Electronic nose
HLA	High-level abstraction
ILA	Intermediate-level abstraction
LLA	Low-level abstraction
NIR	Near-infrared
NIRs	Near-infrared spectrometer
NMR	Nuclear magnetic resonance
MSDF	Multi-sensor data fusion
PCA	Principal component analysis
PCs	Principal components
RMSEP	Root mean square error of the prediction
RMSEC	Root mean square error of calibration

(continued)

Abbreviation	Full name
SVM	Support vector machine
TVB-N	Total volatile basic nitrogen

8.1   Introduction

8.1.1   General Concept

The process of MSDF involves the simulation of synthesizing information in human brain and utilizes all the multi-sensor materials. By virtue of rational control and utilization of various sensors and their observation results, it combines each time complementation and redundancy by the optimization criteria, in order to make a consistent interpretation and description of the targeted environment. To date, in comparison with the single sensor, MSDF has drawn increasing attention and application in food, environment, military, and other fields, due to its provision of detailed information, reasonable degree of tolerance and cognitive processing ability equivalent to humans, etc. MSDF is usually carried out using three methods, namely data fusion (low-level fusion), feature-level fusion (intermediate-level fusion), and decision-level fusion (high-level fusion). Data fusion is the fusion of the original data of the sensors retained to the greatest extent. However, on the account of the hosts of data processing information, it has the demerits of the elongated operation time requirements and less-efficient real-time performance. Additionally, feature-level fusion is to extract the characteristic information from the original data provided by each sensor, and fuse them. Moreover, in decision-level fusion, various types of sensors form their discriminant estimates respectively. Then the conclusion is formed by that basis. Decision-level fusion has high flexibility and strong anti-interference ability, but causes loss of information, and thereby discrimination accuracy is influenced to some extent. In contrast, the information extracted by feature-level fusion is directly related to decision analysis. Additionally, the consequence of fusion can provide the feature necessities for decision analysis to the maximum extent. Furthermore, it is conducive to real-time process, for attaining considerable information. Indeed, the temporal or spatial resolution and monitoring

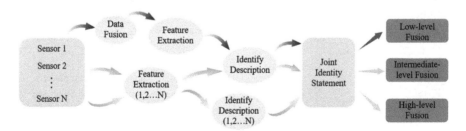

Fig. 8.1 Overview of multi-sensor data fusion

scope could be extended by MSDF. Additionally, multi-dimensional information increases the dimension of target feature vector and reduces the uncertainty. Above all, MSDF enhances the tolerance and self-adaption of the system, reduces the vagueness of reasoning, and improves the reliability and robustness of the system [1–3].

To sum up, the essence of MSDF is to process the information and data from multiple sensors or multiple sources comprehensively to obtain accurate and reliable conclusions. In fact, MSDF is ubiquitous in nature. Take human brain as an example, MSDF is one of its basic functions in way that it uses each physical sensation (such as eye, ear, nose, skin) from daily life, and integrates the interaction between them. Simultaneously, the brain uses prior knowledge that it has acquired through experience to estimate and comprehend the external world and the current events. During handling problems, the brain takes full advantage of the various characteristics of the multi-sensors, such as real-time process, changing velocity, the relationship between information, like complementary, mutual supporting, contradictory, or competing relate. Concept behind MSDF is analogous to the human brain, making full use of the redundant or complementary multi-sensors information to carry out synthesis using certain criteria to consistently explain or describe the objects [4].

8.1.2 Principles and Methodology

Multi-sensor information fusion allows the maximum use of data from different sensors to explain the varying quality features of the same object. The data is analyzed, synthesized, and balanced, on the basis of the criterion to obtain hosts of optimal simplified synthetic variables. The ultimate goal is to find a comprehensive statement based on two or more sensing information [5].

As a result of the interaction that may have occurred during the presentation of information resources for different sensors, the selection of the fusion for multi-sensing information technology is important. It can be classified as three methods: data fusion (low-level fusion), feature-level fusion (intermediate-level fusion), and decision-level fusion (high-level fusion) [2].

Data fusion refers to making correlation of the original data of each sensor directly and sending them to the fusion center to complete the comprehensive evaluation of the tested object, which belongs to sensor-level fusion, and its structural principle is shown in Fig. 8.2. Its advantages are that it keeps as much original

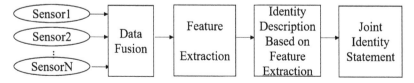

Fig. 8.2 Low-level fusion

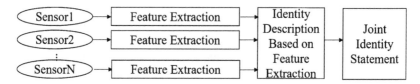

Fig. 8.3 Intermediate-level fusion

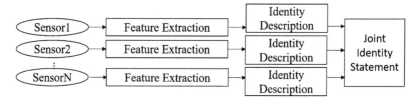

Fig. 8.4 High-level fusion

data information as possible, while containing large amount of information, poor real-time performance and elongation times are its linked shortcomings. Moreover, without the appropriate methodology to check consistency for the character contained in the original data, data fusion would be blind for inconsistency of various types of sensors [6].

Additionally, in feature-level fusion, the composition, association, normalization of the characteristic extraction for original data would be transformed into fusion center to analyze and synthesize for the comprehensive evaluation of the measured objects [3]. The structural principle is shown in Fig. 8.3. Recently, feature-level fusion has been widely applied to preserve sufficient original information and compress it for real-time processing. Besides, its enhanced feature extraction ability and its robustness in complex environment needs further improvement [7].

Moreover, decision-level fusion means that the signals of each sensor are locally processed before fusion. That is to say, before transmitting to the fusion center to build the correlation, the feature extraction and decision would be completed by the corresponding processing units respectively. Therefore, the essence of decision-level fusion is to make the optimal decision according to certain criteria and the credibility of each decision, which belongs to the highest-level fusion. Its structure is shown in Fig. 8.4. Small volume of data communication, high real-time control, and asynchronous information can be processed, and different types can be effectively integrated. The system can still work when one or more sensors fail and has good fault tolerance, but it also has some shortcomings, such as easy loss of original information, the challenge of gaining prior knowledge of each index of the item being examined, and the enormous knowledge base [6, 7].

In the system, the two detection devices, near-infrared spectroscopic (NIRs) instrument and machine vision, are abstracted into two different types of sensors, with inconsistent characteristics. Therefore, various detection modes, large intervals, and the difficulty of having direct contact add to the difficulty of data fusion. In addition, in a real applications, it is difficult to obtain several quality metrics for tea, and decision-level fusion requires repository consisting of indicators that reflect tea quality, such as each sensory index and physiochemical index obtained by the tea experts for sensory evaluation and chemical methods. Consequently, a more realistic approach is to adopt the feature layer fusion. Therefore, not only enough original information is retained, but also a certain degree of data compression is realized, which is useful in real-time processing.

The comprehensive and complete information about the object and environment obtained by using multiple sensors is mainly reflected in the fusion algorithm. With proper use of the fusion algorithm, its essence selecting suitable selection algorithm with basic specifications as robustness and parallelism to information diversity and complexity. Similarly, it is indispensable for the speed and accuracy of the method, interface performance with the preprocessing and the subsequent recognition system, ability to coordinate with various technologies and methods as well as requirements for information samples. Nonlinear mathematical methods are typically fault-tolerant, adaptive, associative, and parallel, and can be implanted in data fusion algorithms.

8.2 Data Processing and Analysis

MSDF analyze and synthesize data of various sensors to make a unified and optimal estimation of the object, thus it can be used in generating high accuracy, reliability, and integrity, which is significantly improved as compared to individual sensor. As a result of the uncertainty in a degree provided by the sensor, the essence of data fusion is a process of uncertain reasoning and decision-making.

Recently, MSDF has proposed several data fusion methods to solve uncertainty. The following is a brief introduction of Bayesian method, evidence theory method, fuzzy set method and artificial neural network which are the most commonly used data fusion methods.

8.2.1 Bayes

Bayes, the earliest method for processing uncertainty, proposed by *Duda* in 1976, is a common method to fuse low-level data of multiple sensors in a static environment. The evidence or information was described via probability distribution in Bayesian estimation, which is applicable to the uncertainty with Gaussian noise. When using Bayesian method to deal with problems from MSDF, it needs hosts of prior knowledge to express the uncertainty provided by sensors as probability. Simultaneously, Bayesian conditional probability formula is used to handle these problems. The main issues are the acquisition of prior knowledge, accurate probability

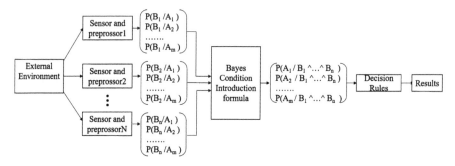

Fig. 8.5 The process diagram of multi-sensor information fusion

expression, restricting its application. In addition, Bayesian methods could not distinguish between the two recognition states of uncertainty and ignorance.

In multi-sensor information fusion, Bayesian method represents various uncertainties as probability, processed by Bayer conditional probability formula in probability theory. The conditions of using Bayer method include the possible independent decisions of the system, in which decisions can be considered as a partition of the sample space, and the policy problems can be solved via Bayesian conditional probability formula.

Suppose that the possible decision of the system is $A_1, A_2 \ldots A_m$, and observation results is B. Only if the results are the prior probability $P(A_i)$ and conditional probability $P(B/A_i)$ by the system of prior knowledge and the characteristics of the sensor, Bayes conditional probability formula is employed. According to observation of the sensors, prior probability $P(A_i)$ will be updated to posterior probability $P(A_i/B)$. Moreover, the observation results are $B_1, B_2, \ldots, B_n$ when there are sensors. If they are independent of each other and independent of the observed object conditions, the total posterior probability of each decision can be obtained as follows:

$$P(A_i/B_1 \wedge B_2 \wedge \ldots B_n) = \frac{\prod_{k=1}^{n} P(B_k/A_i)P(A_i)}{\sum_{j=1}^{m} \prod_{k=1}^{n} P(B_k/A_j)P(A_j)} \qquad (8.1)$$

$j, i = 1, \ldots, m, k = 1, \ldots, n$. Finally, the decision of the system can be drawn by rules, such as taking the decision with the maximum posterior probability as the final decision of the system. The information fusion process of multiple sensors in Bayes method can be shown in Fig. 8.5. The main steps of Bayes information fusion algorithm are summarized as follows:

1. Transform the observation of each sensor about the target into the classification and description of the target identity $B_1, B_2, \ldots, B_n$.

2. Calculate the uncertainty of each sensor about the target identity or decision, that is $P(B_i/A_j)$, $j = 1,2,...m$; $i = 1,2..., n$.
3. Calculate the fusion probability of the target identity:

$$P(A_j|B_1, B_2, \ldots, B_n) = \frac{P(B_1, B_2, \ldots, B_n|A_j)P(A_i)}{P(B_1, B_2, \ldots, B_n)} \tag{8.2}$$

If B_1, B_2, ..., B_n are independent to each other, that is

$$P(B_1, B_2, \ldots, B_n|A_j) = P(B_1|A_j)P(B_2|A_j) \ldots P(B_n|A_j) \tag{8.3}$$

Bayes reasoning has a wide range of applications in various fields; however, there are mainly two difficulties while using probability calculation formula directly in identity recognition: first, the probability of evidence A is based on numerous statistical data, and enormous statistical workload is required to confront complicate problems, making it exceedingly difficult to define the prior function; additionally, multiple possible hypotheses and multi-conditional correlated events, caused by necessary incompatibility or independence of each other, enhance the computational complexity rapidly; another demerit in Bayes' reasoning is the deficiency of capacity for total distribution of uncertainty. Above all, these defects limit the application of the Bayes method.

8.2.2   D-S Evidential Reasoning

D-S (Dempster–Shafer) evidential reasoning, a method for the expression and processing of uncertain information, was introduced by Dempster primitively and further improved and developed by Shafer. As an extension of the Bayesian method, D-S theory distinguishes general probability theory (axiom stronger than probability theory) from conditional probability theory (axiom weaker than probability theory). In D-S theory, the probability of several events is constrained by basic probability assignments to establish belief function, instead of accurate and difficult-to-obtain probabilities to distinguish between "uncertain" and "unknown." In addition, in the application of D-S theory, for a certain sensor, it can not only affect a single hypothesis but also impact a more general ambiguous hypothesis. Therefore, the D-S theory can collect and process information at different details and levels.

D-S evidence reasoning consists of three basic points: basic probability assignment function mi, trust function $Beli$, and likelihood function $Plsi$, which are divided into three levels from top to bottom. And the structure is shown in Fig. 8.6.

Primary level is the target synthesis to comprise the observations from independent sensors into a total output (ID). Second-level reasoning, based on the authoritative and logical target report generated by the sensor report, obtains, and deduces the sensor observation results, and expands it into the target report. Generally,

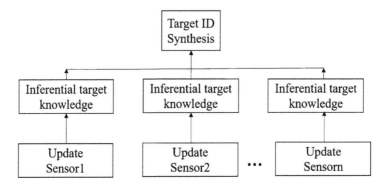

Fig. 8.6 D-S inference structure

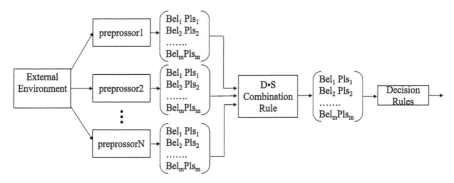

Fig. 8.7 The application of evidential reasoning in multi-sensor data fusion

random errors exist in sensors, it is a group of continuous reports from the same sensor in sufficiently independent time compared to any single report. Tertiary level is the update in an attempt to combine (update) the observation data of the sensor prior to reasoning and multi-sensor synthesis.

The basic application process of D-S evidential reasoning in MSDF is shown in Fig. 8.7. First, there is calculation for basic probability assignment function mi, trust function $Beli$, and likelihood function $Plsi$ of each evidence. Subsequently, with D-S combination rule, calculate them under the cooperative action of all evidences. Finally, by the decision rules, the hypothesis with the maximum support under combined effect is selected.

Despite the broad variety of data fusion applications, D-S theory is required for the evidence independence in the application. In addition, the strong independent conditions are such that they can hardly be coped with in different situations. Even though researchers assume independence to the approximation and apply D-S theory directly, undoubtedly, the conclusion is beyond the estimate.

The progress of artificial intelligence, fuzzy logic, neural network, and other disciplines has greatly provided technologies that provided the uncertain reasoning process of information in a simple and effective way, and solved the uncertainty problem in the information fusion algorithm by means of human identification mechanism, exploration, and simulation [8].

8.2.3 Fuzzy Set Theory

The concept of fuzzy set was first proposed by *L.A.Zadeh* in 1965. The basic idea is to flexibly extend the absolute membership relationship in the ordinary set such that the membership degree of the element in the set can be extended from the original value of $\{0, 1\}$ to any value in the $[0,1]$ interval, suitable for defining and processing sensor information uncertainty. In the application of MSDF, A is regarded as a set of possible decisions of the system, and B as a set of sensors. The element μ_{ij} in the relation matrix $R_{A \times B}$ represents the possibility, the decision inferred from sensor i to j. X is taken as the credibility of each sensor judgment, and Y obtained by fuzzy transformation is the possibility of each decision.

Specifically, given that there are m sensors to observe the system, and the system has n possible decisions, then:

$A = \{y_1/\text{decision}_1, y_2/\text{decision}_2, \ldots, y_n/\text{decisions}_n\}$

$B = \{x_1/\text{sensor}_1, x_2/\text{sensor}_2, \ldots, x_m/\text{sensors}_m\}$

The sensor's judgment of each possible decision is expressed by the membership function defined on A, shown as follows:

$$\left[\mu_{ij}/\text{decision}_1, \mu_{ij}/\text{decision}_2, \ldots, \mu_{ij}/\text{decisions}_n\right], \quad 0 \le \mu_{ij} \le 1$$

It is considered that the possibility that the result is a decision j is written as a vector μ_{ij} ($\mu_{i1}, \mu_{i2}, \ldots, \mu_{in}$), the relation matrix of m sensors to form $A \times B$ is as follows:

$$R_{A \times B} = \begin{bmatrix} \mu_{11} \ \mu_{12} \ldots \mu_{1n} \\ \mu_{21} \ \mu_{22} \ldots \mu_{2n} \\ \mu_{31} \ \mu_{32} \ldots \mu_{3n} \end{bmatrix}$$

The credibility of each sensor is determined by the membership number on B, $X = \{x_1/\text{sensor}_1, x_2/\text{sensor}_2, \ldots, x_m/\text{sensor}_m\}$, then, according to the fuzzy transformation of $Y = X \cdot R_{A \times B}$, we can get $Y = (y_1, y_2, \ldots, y_n)$, the possibility y_i of each decision after comprehensive judgment.

8.2.4 Extraction and Analysis of Multi-sensor Interactive Induction Information

Obtaining information in a particular sensory organ could be vulnerable to other senses, consciously or unconsciously. Examples include the taste of yellow rice wine. Undoubtedly, it is the tongue that occupy the dominate position, but the result

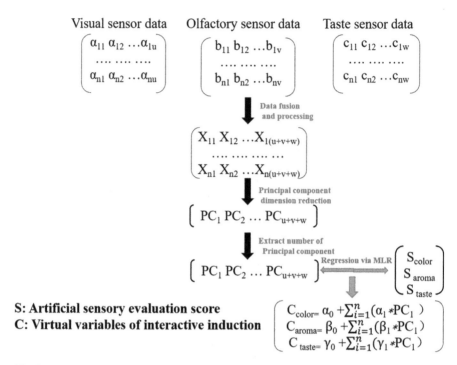

Fig. 8.8 The steps for extracting feature variables with interactive induction

could be affected by the color (vision) and odor (smell). In other words, there are objective interaction between the senses. Additionally, the description of the style sensory characteristics of yellow rice wine is also the result of the comprehensive discrimination of each sensory organ by brain, integrating each. Therefore, the interaction between various sensor data, that is, the interaction between human senses, is not considered in the feature variables obtained by traditional data layer fusion and feature layer fusion.

The mathematical bionic theory of interactive induction was introduced during the smart evaluation of food, aiming at the problem of interactive induction with neglect between different organs. Take yellow rice wine as an experimental object. As shown in Fig. 8.8, first, combine visual, smell, and taste sensor data. Second, 99% of sensory scores and principal component scores of color, aroma, and taste were fitted by multiple linear regression (MLR), and virtual visual, olfactory, and gustatory variables reflecting interactive induction were obtained. Finally, the correlation between virtual vision, smell, and taste and the total sensory score of yellow rice wine collected by the visual, smell, and taste sensor systems were combined through linear and nonlinear models to obtain a total of 29 sensor variables, and the sensor data set would be pretreated by the Z-Score function. Then, the crucial information of the sensor data set is extracted through principal component analysis (PCA). Subsequently, via MLR, the color, aroma, and taste of yellow rice wine and the 99% sensory information extracted from PCA, a series of regression coefficients (weights) were obtained. Clearly, the combination of principal components and their respective regression coefficients constitutes virtual visual, smell, and taste variables with interactive induction.

8.2.5 Neural Network Method

When statistical models of the system are not available, neural networks provide a way to fuse information from different sensors into a unified framework. The rationale is to get the corresponding system control output directly by training the pretreated sensor information set. The essence of multi-sensor information integration and fusion is set for nonlinear reflection from data space to command space under the constraints. Simultaneously, the nonlinear function mapping expressed by the neural network meets the need, with its suitable inherent structure and learning method for parallel process of multi-sensor fusion.

Neural network composed of abundant simple processing units similar to neurons has the following three elements:

1. Neuron characteristics. It defines a function to merge each single input into overall value to map to another function.
2. Rulemaking. It gives a set of initial weights with the way to change the weights to improve performance.
3. The topology of the network. It describes the connection between neural units in the network.

Neural networks with various topologies can be constructed by neurons, with two typical structural models: feedforward and feedback. Recently, representative neural networks include BP network, Hopfield network, Boltzman machine, Koheon network, and so on.

In the application of the neural network for the fusion of sensor information, the initial step is to select the appropriate neural network model according to the requirements of the system and the characteristics of the sensor, including the network topology, neuron characteristics, and learning rules. Simultaneously, it is necessary to establish the mapping relationship between the input and sensor information, the output, and the system decision. Next, it could be learnt according to the existing sensor and the corresponding system decision, to determine the weight distribution and complete the training of the network. The trained neural network participates in the actual fusion process, as shown in Fig. 8.9. The information obtained by the sensor first goes through the appropriate in process 1. Then, the neural network processes it and generates the related results as the neural network data, and then the processing process 2 interprets it as the system's basic decision-making.

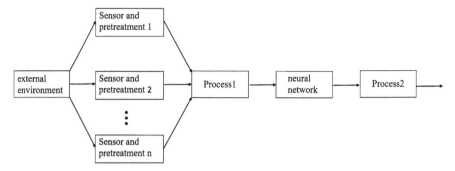

Fig. 8.9 Fusion based on neural network

8.2.6   General Steps in Nondestructive Testing of Food Quality and Safety

The applications of multi-sensor information fusion method in nondestructive testing of food quality and safety is generally divided into eight steps, shown in Fig. 8.10:

1. Identify the characteristics of the item significant to the quality characteristic.
2. Determine the auxiliary methods (qualitative or quantitative methods) for the evaluation of the quality of the tested items.
3. Determine the nondestructive testing method detecting the characteristics of the selected items.
4. Obtain data via the auxiliary method and then select sensor nondestructive testing method.
5. Evaluate whether the sensors used in nondestructive testing are redundant or need to be supplemented with new sensors.
6. Select and apply appropriate multi-sensor fusion methods.
7. Evaluate and improve the sensor fusion system via comparison with the results obtained by the auxiliary method.
8. Accept, reject, or continue to improve the proposed sensor fusion method.

   Step 1: Attributes related to quality characteristics.
   In this step, indistinct as these relationships are, it is important to understand the possible counterparts between different properties in items initially, such as the relationship between the color of a fruit and the sugar composition. Second, it requires full friendly cooperation between sensor fusion designers and experts, in the field of production or agricultural products (such as experts in the field of food science). The result of this step is a series of physical or chemical properties related to the sensory properties of the product, and a series of relationships between these properties.
   Step 2: Determine the auxiliary method for product quality evaluation.

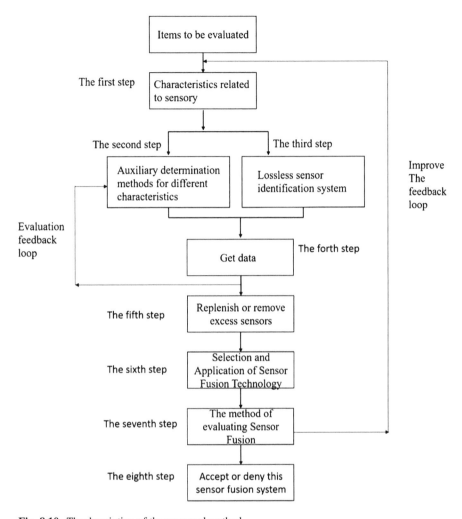

Fig. 8.10 The description of the proposed method

The auxiliary method can be defined as existing analytical procedures to detect quality index, or qualitative information provided by expert evaluation. In the former case, the test results can be utilized to build an automatic classification library. Or it would be used to classify it in the latter case, the selection of auxiliary methods will provide more useful information and knowledge for the third step of sensor nonde-structive testing. Remarkably, the accomplishment of the step depends on the cooperation of designers of sensor fusion, technical experts in the field of production and sensory analysis, and cooperation with a group of professionals in chemical and physical measurements. Hence, a series of lossy test information could be obtained, related to the characteristics selected in the first step or to evaluate the results of the expert evaluation.

Step 3: Nondestructive testing method for determining product quality.

Nondestructive detection is applied to various models in measuring physical or chemical properties. For example, the theory of elasticity provides a model for measuring hardness with sensors, while the diffusion model of light in fruit gives a theoretical basis for the detection of sugar. The signals and features proposed by each sensor and their relationship with the characteristic metrics determined in the first step must be confirmed here. Furthermore, both the real-time and economic cost (economic strength) must be taken into account. Several sensors are capable of measuring an object's characteristics with long time, low speed, and uneconomical cost, causing it seldom to sift out. For instance, nuclear magnetic resonance (NMR) has been utilized in the study of internal damage detection for fruits in 1988, but to date, it is still too expensive to be used for on-site integrated detection. The selection of sensors often depends on the factors for replenishing or removal of excess sensors and the analysis problem itself.

The achievement of this step is bound to the researchers' knowledge of sensors and chemical dosimetry. Consequently, a set of sensors concerning the characteristics proposed in the first step could be obtained. This step can be performed along with the second step, rather than following it, as shown in Fig. 8.10.

Step 4: Acquisition of data.

The data would be gathered after the selection of the appropriate device (sensor) and auxiliary method. There could be an experimental design phase: the properties of the different components of the test object, such as moisture and flour in the dough process and the optimized measurement parameters (frequency, acquisition time, temperature) should be emphasized. Tough control of experimental conditions also challenges the design. For example, the quality of the fruit in the garden is closely linked with the climate, uncontrollable natural climatic conditions, unless the greenhouse is applied to simulate the natural conditions. The selection of auxiliary methods is mainly based on the analysis of the results classified by experts and the results of lossy detection. Actually, there is a cycle between step 4 and step 2.

Step 5: Determine the criteria for redundancy and supplement in nondestructive testing sensors.

It is crucial to determine that the information obtained from different sensors is equal. When the different characteristics of the tested samples are from various sensors, counterparts complement each other, or if the sensors are superfluous, it is feasible to remove them. The criterion for defining the redundancy of the sensor in the array used to select the sensor should be taken into account in order to determine the characteristics of the tested object in the first step. This step is to determine the validity of the sensors, that is, whether they are complementary or redundant. The result is to leave available sensors and remove the excess ones by calculating statistical indicators, such as correlation coefficient, similarity, linear correlation between different sensors principle, allowed by component analysis and correlation loop, etc. The description could remove those redundant sensors and useless features, but it is necessary to retain a certain degree of redundancy to improve the robustness of the system. Thus, the system still maintains stability in the event of a sensor failure.

Step 6: Select and use appropriate sensor fusion technology.

This step is based on the understanding of the biological model, by the selection of appropriate sensor fusion level and fusion technology. The following describes the biological model, the selection of fusion level, and the choice of fusion technology one by one.

1. **Biological Model**

   There are five different senses, namely taste, smell, vision, audio, and touch, tools for our consciousness to connect with the outside world. Additionally, scientists have studied the interaction between senses. Examples include that *Gree* has revealed a strong interaction between taste and smell by testing people's responses to ethylbutyrate and saccharin. Similarly, *Mark* has studied the counterparts between hearing and vision, demonstrating that both senses are involved in receiving information from food. *Stillman* have also shown the effect of color on aroma recognition, while aroma is defined as information relevant to taste and smell. Understood as a tool for observing and transmitting information, each sense has a single expression or meaning, and can be biased therefore depending on the interaction of each of the human senses. In fact, senses may be involved in the transmission of the same information simultaneously with numerous characteristics sharing in the process. However, it is indistinct that the physiological mechanism of sensory interaction and the fusion in the human brain increases the difficulty of human beings' overall perception. The sensory interaction in food assessment by artificial neural networks, commonly used in food processing and agriculture, is particularly difficult to resolve. To summarize, the model of sensor fusion is inspired by biological model, such as artificial neural network. Nevertheless, sensor fusion technology is not merely a simple simulation of the human brain.

2. **Selection of Sensor Fusion Layer**

   Theoretically, the sensor fusion process would be most effective in the original data level, while the level is subject to several limitations in practice. First, the same sensor or the identical measurement sensor is rarely used in a system. Second, it is important to note that the ability of high-speed storage to-date is impractical. The information used in the feature- and decision-level fusion technology is less than the original signal provided by the sensor. Accordingly, error noises are included in the information transmission. The sensor fusion technology at the feature and decision level could be adapted to the actual situation containing various types of sensors, without fast data processing. Additionally, the choice of fusion level is related to the selection of fusion technology.

3. **Selection of Sensor Fusion Technology**

   Hosts of effective algorithms have been developed under diverse application backgrounds with various mathematical tools utilized in fusion. Particularly, the non-modeling intelligent algorithm, the main method used in fusion, represented by fuzzy theory, neural network theory, evidence theory, and inference network, with superiorities of language, is liable to reflect and cope with the problems. Moreover, random mathematics methods, statistical methods, wavelet analysis, set theory, decision theory, and algorithms are widely used in fusion. However, these methods should be selected as appropriate in practical application.

Step 7: Evaluate the results of multi-sensor information fusion.

In accordance with the final fusion detection results (test accuracy, correlation coefficient, etc.), whether the accuracy of MSDF meets the requirements could be determined. If not, it is necessary to re-optimize the entire system from the second step, as shown in Fig. 8.9.

Step 8: Identify the system of MSDF.

Determine whether the established multi-sensor system is satisfied, and if so, retain it, otherwise, abandon the whole system for redesign.

8.3   Application in Food Quality and Safety Detection

MSDF, utilized to increase the accuracy of classification, is based on the data from multiple sensors. Therefore, characteristic data on various aspects of a sample could be obtained [8]. On the edge of efficiency of processing data, the fusion technology is increasingly wide applied rather than the single sensor [9–14].

8.3.1   Application in Detection of Yellow Rice Wine

The sensory quality of yellow rice wine for example is an important standard to measure its quality, including four aspects such as color, aroma, taste, and characteristics, by which sensors score its quality. The visual, smell, and taste sensors, together with appropriate chemometrics, showed an excellent correlation with the scores of yellow rice wine [15].

It is a limitation that a single biomimetic sensor detection method only describes a certain aspect of the sensory quality of yellow rice wine, but cannot comprehensively describe the overall sensory quality information of yellow rice wine. During intelligent evaluation of food, Ouyang et al. (2013) use the mathematical bionic theory of interactive induction was aiming at the problem of interactive induction between different organs with neglecting [15]. Take yellow rice wine as the experimental object. First, visual, smell, and taste grades evaluated by artificial senses were obtained and combined. Second, after the standard pretreatment on function of Z-score, the sensory scores of visual, aroma, taste, and 99% of principal components (PCs) were fitted respectively by MLR to obtain these variables with reflection of interactive induction as the output of model. Finally, the correlation between variables including virtual visual, olfactory, and taste and the total sensory score of yellow rice wine was established through linear MLR, nonlinear model of back propagation artificial neural network (BPANN) and support vector machine (SVM)

Table 8.1 Results from different models of different sensors for predicting sensory scores of rice wine [16]

Fusion method	Model	Number of variables	Calibration set		Prediction set	
			R_c	RMSEC	R_p	RESEP
Traditional fusion	MLR	3	0.7882	3.30	0.7467	3.54
	BPANN	3	0.8498	2.83	0.7979	3.17
	SVM	3	0.8897	2.46	0.7729	3.33
Interactive fusion	MLR	3	0.9303	1.97	0.9064	2.31
	BPANN	3	0.9392	1.88	0.9060	2.27
	SVM	3	0.9315	1.96	0.9071	2.31

to realize the comprehensive intelligent evaluation of the sensory quality of yellow rice wine.

As input of the model for virtual vision, smell, and taste variables with interactive induction, with linear MLR, nonlinear BPANN and SVM, respectively, the prediction model of the total sensory score of yellow rice wine was established. Establishing the model with MLR, the calibration coefficient R_p between the predicted total sensory score of the model prediction set and the actual score value was 0.9064, and root mean square error of the prediction (RMSEP) was 2.31. With BPANN, R_p was 0.9060, and RMSEP was 2.27. With SVM, R_p was 0.9071, and RMSEP was 2.31. To compare with the traditional fusion technology which is regardless of the interactive induction, Ouyang et al. [16] extracted the first PCs on sensory data of visual, aroma, taste as the input of model, and constructed the model of MLR, BPANN, SVM, and R_p was 0.7467, and RMSEP was 3.54 with MLR; R_p was 0.7979, and RMSEP was 3.17 with BPANN; R_p was 0.7729, and RMSEP was 3.33 with SVM. As is shown in Table 8.1, compared with the traditional one, feature variables extracted by MSDF with interactive induction are closer to the real visual, olfactory, and taste information. Additionally, rice wine quality evaluation model established by MSDF has better predictive ability, and the research results attach great importance to intelligent bionic evaluation for food.

8.3.2 Application in Detection of Tea

There are various indexes to reflect tea quality, including external quality indexes such as color and shape, and internal quality indexes such as taste and aroma of tea. However, a single detection cannot comprehensively describe the object, for it can only describe one or several aspects. For example, NIRs characterizes the internal quality of tea (including the taste and aroma of tea), neglecting the external characteristics of tea (including the color and shape of tea), which can be represented by machine vision technology [17, 18]. Clearly, the limitation caused by the diverse emphasis of acquisition, inevitably, affects the accuracy and stability of the detection. Sequentially, it has been a new trend in rapid nondestructive testing research for

Table 8.2 Comparison of recognition results of three modes

Model	Recognition results		Comparison
	Training set (%)	Prediction set (%)	
Image model	90	74	General
Spectral model	90	87	Good
Fusion model	98	89	Best

tea that utilizing the merits of these two detection methods and combining the strengths to complement each other to improve reliability and sensitivity comprehensively.

1. **Detection of Green Tea Quality**

   To detect the quality of green tea, Chen et al. [19] combined multiple sensors information from NIR and computer vision (CV). Take four grades of green tea as experimental target. The score vectors were the input of pattern recognition after PCA, on the basis of imaging information and spectral information. Next, the fusion model to detect the comprehensive quality of tea was constructed by BPANN with optimization of PCs from two sensors. As shown in Table 8.2, the recognition rate of the model built on image and spectral information fusion is as high as 99% in training set and 89% in prediction set. Be the training or prediction set, the results preceded the recognition results of a single information model. Therefore, it was feasible to discriminate tea comprehensive quality by fusion of NIR and CV, rather than the single one, enhancing the accuracy and stability of data. In conclusion, the results offered a reference that MSDF could be applied to the detection of quality on green tea [18].

2. **Classification of Oolong Tea Varieties**

   To date, gustatory or olfactory sensors are widely applied to detect the type of oolong tea and other foods, imitating animal's recognition organs. Therefore, sensors respond a school of correlative chemical species rather than selecting the specific one, since the overall data are fused into a group, and they could be analyzed by a certain recognition program.

   Oolong tea is a wide concept and tea flavor is the most important characteristic, varying frequently. Hence, detecting the quality of oolong tea in a single sensor would not be accurate. Chen et al. [10] used the integration of gustatory and olfactory sensors to obtain plentiful information from the objects to classify the abundant varieties of oolong tea. Herein, four electrodes, including gold, copper, platinum, and glassy carbon, in a standard three-electrode configuration were utilized to complete the gustatory sensors system, and a colorimetric sensors array was used to develop the olfactory sensors system. A functional prototype of the olfactory sensors system was constructed, and its schematic diagram is shown in Fig. 8.11. Specifically, first, the data were obtained from two senses and then analyzed separately. Next, it was analyzed that the potential of the combination of the two sensors system. PCA and LDA were used to analyze and separate that data. In conclusion, the capability of recognition on the fusion system performed

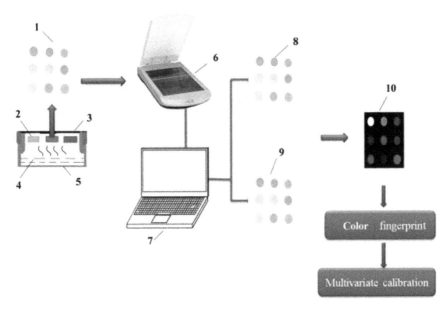

Fig. 8.11 Schematic diagram of olfactory sensors system based on: 1 Odor imaging sensors array; 2 Sensor array; 3 Lid; 4 Oolong tea infusion; 5 Plastic dish; 6 Scanner; 7 Computer; 8 Original image; 9 Final image; 10 Different image [10]

better than each single one, and LDA achieved 100% classification rate by cross-validation [10]. Hence, the combination of gustatory and olfactory sensors is an effective tool to classify oolong tea.

8.3.3   Application in the Detection of Meat

1. **Detection of Pork Quality**
   The spoilage of pork is mainly caused by microbial contamination, with the change of color, texture, tissue structure, chemical composition, and smell of deteriorated pork, as well as the decrease of meat quality and nutritional value. The relationships between the proper sensing techniques (NIRs, CV, and E-nose) and the changes in chemical composition, color, texture, and/or gases in meat samples are extremely intricate with a likely nonlinear nature [13].

   Huang et al. [20] utilized NIRs, CV, and E-nose techniques were applied to obtain internal and external characteristic information such as chemical composition, color, texture, and gases released during pork deterioration, to evaluate pork freshness comprehensively, which is shown in Fig. 8.12. Next, these individual characteristic variables were fused in feature-level fusion, and the top PCs were extracted through PCA as input of model to construct total volatile basic nitrogen (TVB-N) model with BPANN. As is shown in Table 8.3, the prediction performance of the BP-ANN model on the grounds of single-sensor

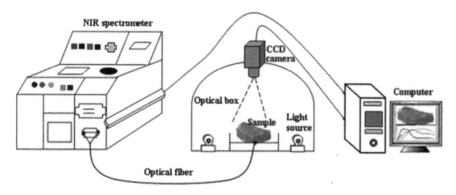

Fig. 8.12 Information fusion detection system based on NIR and CV technology [21]

Table 8.3 The modeling results of five times for GA-siPLS operation [21]

Model number	Spectral variables	The best number of PCs	Training set		Prediction set	
			RMSEC (mg/100 g)	R_c^2	RMSEP (mg/100 g)	R_p^2
1	76	7	4.62	0.785	6.84	0.543
2	**65**	**7**	**4.45**	**0.805**	**5.88**	**0.682**
3	69	8	4.59	0.788	6.33	0.624
4	60	8	4.76	0.774	6.01	0.654
5	63	7	4.56	0.793	5.92	0.681

technology data, CV image, and electronic nose (E-nose) data performed similarly and data fusion model by any two kinds of sensor technology was superior
to the corresponding single sensor, which was achieved with root mean square
error of calibration (RMSEC) was 1.46 mg/100 g and R_c^2 was 0.984 in the training
set, root mean square error of the prediction (RMSEP) was 2.73 mg/100 g and R_p^2
was 0.984 in the prediction set. Therefore, compared with the model from single
sensors or two sensor information, the accuracy and robustness of the model from
three sensor information fusion enhanced obviously. In conclusion, the study
offered a reference that multi-sensor information could be applied to evaluate
comprehensive pork quality and safety [20, 21].

2. **Detection of Chicken Meat Quality**

Chicken meat occupies 17% of total meat consumption in China for it provides
fast-growing protein. MSDF is a technique that combines the multiple data from
various sources, to achieve the high-efficient representation of the data.
According to abundant literatures, data fusion based on data abstraction could
be classified into three types, including low-level abstraction (LLA) or original
data fusion, intermediate-level abstraction (ILA) or medium-level or feature-level
data fusion, and high-level abstraction (HLA) or decision fusion [21]. In fact, it
could be classified into four levels according to the level of data abstraction or the
way how data from different instruments derived one to another. The fourth level

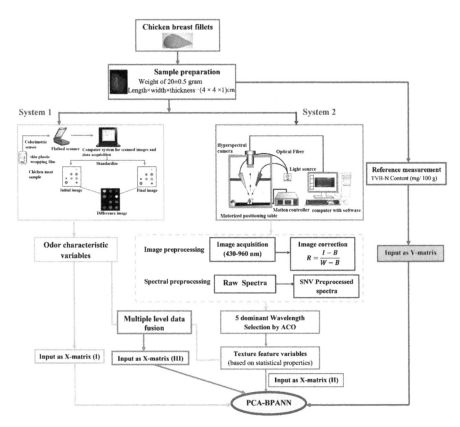

Fig. 8.13 Schematic diagram of integrating two nondestructive analytical systems for multiple sensor fusion, E-nose/olfactory system based on low-cost colorimetric sensors array (system 1), and hyperspectral imaging system (system 2) [11]

is called multiple level data fusion (MLF). MLF could be used to process the data from various abstraction level during the combination with a feature to obtain a decision. Urmila et al. [22] proposed a fusion technique based on LLA of colorimetric sensors data and ILA of HIS optical sensors to evaluate total volatile basic nitrogen (TVB-N) content in chicken meat, which is shown in Fig. 8.13. Compared with each individual system, MLF produced better prediction results of the combination PCA and BPANN with RMSEP of 4.587 mg/100 g and R_p of 0.8659. Additionally, after applying Person's correlation analysis, to reduce the data variables rather than the original information, the prediction level was promoted with RMSEP of 4.3137 mg/100 g and R_p of 0.8819. In conclusion, the results promoted the scale of MSDF and improved MLF technique for better model performance to detect the quality of chicken meat [11, 22].

8.4   Conclusion and Future Trends

In recent years, MSDF has received considerable interest and rapidly evolved as a novel interdisciplinary technique. It is a comprehensive processing and optimization technology for the acquisition, representation, and internal relationship of various kinds of information. It also provides a new concept for the research of intelligent information processing technology. With the advancement of sensing and detection technology, food is no longer classified according to a single quality. As a multidisciplinary method, MSDF is applied to food quality evaluation, which will promote the development of food online classification toward fast, nondestructive, accurate, and multi-index.

References

1. Paulus I, Busscher RD, Schrevens E (1997) Use of image analysis to investigate human quality classification of apples. J Agric Eng Res 68:341–353
2. Martens H, Nielsen JP, Engelsen SB (2003) Light scattering and light absorbance separated by extended multiplicative signal correction. Application to near-infrared transmission analysis of powder mixtures. Anal Chem 75:394–404
3. Nakano K (1997) Application of neural networks to the color grading of apples. Comput Electron Agric 18:105–116
4. Petersen D, Xiao-Bo Z, Shou-Yi W (2002) Evaluating the quality of cigarettes by an electronic nose system. J Test Eval 30:11063
5. Steinmetz V, Roger J, Moltó E, Blasco J (1999) On-line fusion of colour camera and spectro-photometer for sugar content prediction of apples. J Agric Eng Res 73:207–216
6. Luzupiaga IA (1999) Application of computer vision and electronic nose technologies for quality assessment of color and odor of shrimp and salmon. University of Florida, Gainesville
7. Martin H et al (1998) Bacteria classification based on feature extraction from sensor data. Biotechnol Tech 12:319–324
8. Penza M, Cassano G (2003) Application of principal component analysis and artificial neural networks to recognize the individual VOCs of methanol/2-propanol in a binary mixture by SAW multi-sensor array. Sens Actuators B Chem 89:269–284
9. Banerjee(Roy) R, Chattopadhyay P, Tudu B, Bhattacharyya N, Bandyopadhyay R (2014) Artificial flavor perception of black tea using fusion of electronic nose and tongue response: a Bayesian statistical approach. J Food Eng 142:87–93
10. Chen Q, Sun C, Ouyang Q, Liu A, Li H, Zhao J (2014) Classification of vinegar with different marked ages using olfactory sensors and gustatory sensors. Anal Methods 6:9783–9790
11. Khulal U, Zhao J, Hu W, Chen Q (2017) Intelligent evaluation of total volatile basic nitrogen (TVB-N) content in chicken meat by an improved multiple level data fusion model. Sens Actuators B Chem 238:337–345
12. Li H, Chen Q, Zhao J, Wu M (2015) Nondestructive detection of total volatile basic nitrogen (TVB-N) content in pork meat by integrating hyperspectral imaging and colorimetric sensor combined with a nonlinear data fusion. LWT Food Sci Technol 63:268–274
13. Li H, Kutsanedzie F, Zhao J, Chen Q (2016) Quantifying total viable count in pork meat using combined hyperspectral imaging and artificial olfaction techniques. Food Anal Methods 9:3015–3024

14. Xu Y, Kutsanedzie FY, Sun H, Wang M, Chen Q, Guo Z, Wu J (2017) Rapid Pseudomonas species identification from chicken by integrating colorimetric sensors with near-infrared spectroscopy. Food Anal Methods 11:1199–1208

15. Ouyang Q, Zhao J, Chen Q, Lin H (2013) Classification of rice wine according to different marked ages using a novel artificial olfactory technique based on colorimetric sensor array. Food Chem 138:1320–1324

16. Ouyang Q, Zhao J, Chen Q (2014) Instrumental intelligent test of food sensory quality as mimic of human panel test combining multiple cross-perception sensors and data fusion. Anal Chim Acta 841:68–76

17. Yud-Ren C, Kuanglin C, Moon SK (2002) Machine vision technology for agricultural applications. Comput Electron Agric 36:173–191

18. Xiaobo Z, Jiewen Z, Yanxiao L (2010) Objective quality assessment of apples using machine vision, NIR spectrophotometer, and electronic nose. Trans ASABE 53:1351–1358

19. Chen Q, Zhao J, Cai J (2008) Inspection of tea quality by using multi-sensor information fusion based on NIR spectroscopy and machine vision. Trans Chinese Soc Agric Eng 24:5–10

20. Huang L, Zhao J, Chen Q, Zhang Y (2014) Nondestructive measurement of total volatile basic nitrogen (TVB-N) in pork meat by integrating near infrared spectroscopy, computer vision and electronic nose techniques. Food Chem 145:228–236

21. Chen Q, Zhao J, Zhang H, Wang X (2006) Feasibility study on qualitative and quantitative analysis in tea by near infrared spectroscopy with multivariate calibration. Anal Chim Acta 572:77–84

22. Urmila K, Li H, Chen Q, Hui Z, Zhao J (2015) Quantifying of total volatile basic nitrogen (TVB-N) content in chicken using a colorimetric sensor array and nonlinear regression tool. Anal Methods 7:5682–5688

Chapter 9
Intelligent and Portable Equipment of Nondestructive Detection Technologies in Food

What Is Portable Evaluation Instrument for Quality of Agro-Product?

The quality evaluation of common agricultural products (agro-products) in the purchase and storage process is an important link in its quality and safety. The evaluation content of agricultural products generally involves fruit quality (sweetness, freshness), cereal aging, food flavor, product damage, and cracking. Research on portable rapid nondestructive instrument to evaluate agricultural products aims at different agricultural product quality indicators as mentioned above, using electrical, optical, bio-sensing, nuclear magnetic, and other scientific and technological means. Figure 9.1 illustrates some of the portable nondestructive detection instrument and their respective application areas.

Abbreviation	Full name
Agro-product	Agricultural product
AD	Analog–digital
BLE	Bluetooth low energy
DLP	Digital light processing
DMD	Digital micromirror devices
E-nose	Electronic nose
FFT	Fast Fourier transform
LDA	Linear discriminant analysis
M-porphyrin	Metalloporphyrin
MDA	Malondialdehyde
MAE	Mean absolute error
MRE	Mean relative error
NIR	Near-infrared
NIRS	Near-infrared spectrometer
PLC	Programmable logic controller
R&D	Research and development
TVB-N	Total volatile basic nitrogen
Tablet PC	Tablet personal computer

(continued)

Abbreviation	Full name
UAV	Unmanned aerial vehicle
UV-vis	Ultraviolet and visible

9.1   Introduction

During the purchase, storage, and transportation process of agro-products, the influence of adverse external conditions and the action of internal enzymes would result in certain change in the nutrients nature. For example, the color and smell of rice changes with extended storage time, thereby directly affecting its commercial value. Fresh rice is bright in color, translucent, and shiny. When the storage time increases, the rice grains become darker, rougher, its chalkiness increases, and vertical white streaks appears on it, and even the white rice grains turn into yellow rice grains. Fresh rice has a delicate and pure smell, while aged rice will deteriorate

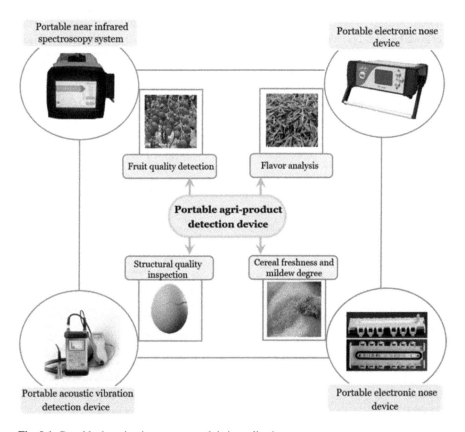

Fig. 9.1 Portable detecting instruments and their application areas

and give off hala taste. In addition, the changes in the chemical bonds, peptides of the protein during the storage of rice could reduce the solubility of the protein and the viscosity of the rice. Pork is very susceptible to microbial contamination during processing and storage. The proliferation of microbes and their catabolism can cause protein and fatty acid degradation and carbohydrate fermentation to cause spoilage. Owing to these sequence of chemical decomposition, there is a shift in the original color of the meat, formation mucus, unpleasant smell, and even harmful and toxic chemicals. In addition, in the process of agro-products sales, on-site evaluation and grading of the internal and external quality are key links in the industrialized processing and production. On the one hand, this link is related to the realization of the value-added of agricultural products and the increase of farmers' income. It also has influence on the quality and safety of agro-product, customer interests, and the realization of declining costs and rising profit for processing enterprises. This chapter first introduces the common on-site food evaluation indicators in the storage process of primary agricultural products and systematically elaborates the existing on-site portable instrument for these evaluation contents, including its vital features and specific configuration components. Finally, in the agro-product collection and storage process, the application methods, research methods, and evaluation methods of these tools are clarified with examples, and the basic content and research ideas are given to further promote the development of fast and nondestructive portable tools to detect the quality of agro-product collection and storage.

The appearance and internal composition of agricultural products will change with changes in external conditions, and under the action of enzymes present within products. External attributes such as size, color, shape, surface texture and external defects can be evaluated by acoustic vibration detection technology or imaging techniques. The chemical composition of agro-products such as moisture, fat, and protein contents are possible to be determined with spectral imaging. These substances are closely related to the edible value and storage period of agricultural products, therefore, on-site and real-time detection of these quality indicators is of great significance for ensuring the quality and safety of agricultural products. To ensure the safety of food at all levels, there needs to be high-level quality evaluation system of agro-products across the food processing industry. Therefore, the application of quality control regimes in food or agro-products encourage the development of nondestructive technologies.

9.1.1 Evaluation Indicators for Cereals and Oils

During the storage process of cereals and oils, the moisture, volumetric weight, fatty acid value, and protein content change with microbiol action as well as aging [1, 2]. The hydrolysis and oxidation of the lipid in the rice storage process would produce free fatty acids and a rancid taste. Among them, oxidation would cause the unsaturated fatty acids to be oxidized to hydroperoxides, and then produce carbonyl compounds such as aldehydes and ketones, which would give rise to the smell of hala and aging. Wheat has the characteristics of, long after ripening period, strong hygroscopicity and easy to be infected with mold. The newly harvested wheat has a physiological post-ripening period of about 2 months [3]. During the post-ripening period, wheat has a large respiration capacity and strong metabolism. It is extremely sweat and affects the storage stability. Wheat grains have the characteristics of thin skin, soft tissue, etc., and contain a lot of hydrophilic substances, so they have strong moisture absorption capacity. In the early stage of mildew, only dry-type mold such as *Aspergillus glaucus* can grow and carry out metabolic activities [4]. When the moisture content and the grain temperature of the wheat rise to a certain range, and the suitable environmental conditions for the reproduction of molds carried on the wheat grain reach, the nutrients can be decomposed and consumed for molds' growth and metabolism. Due to the strong infection ability and large amplification range of microorganism, experiencing local mildew, when the temperature and humidity of the entire grain pile meet the growth conditions of mold, the entire granary would be infected by mold. Applying electronic nose technology, olfactory visualization technology, and so on, it can realize rapid detection of the above-mentioned stored grain fungus and fungal contamination [5, 6].

9.1.2   Evaluation Indicators for Fruits and Vegetables

Although the harvested fruits and vegetables are separated from the plant matrix and stop assimilation, the respiration still continues. During the storage and transportation of fruits and vegetables, as the respiration progresses, the nutrient amount continuously increases decreasing the quality [7]. In addition, freshly picked fruits and vegetables have a high water content, and the water loss results in the color change and texture of the fruits and vegetables. Therefore, water loss is also one of the important reasons for the deterioration of fruits and vegetables quality during storage after being harvested. The spoilage and deterioration caused by microbial contamination in fruits and vegetables during storage cannot be ignored. The growth and reproduction of microorganisms such as *Penicillium* and *Bacillus* can cause fermentation, softening, mildew, rot, and discoloration of fruits and vegetables [8–10]. In summary, as the storage time of fruits and vegetables increase, the color, flavor, moisture, hardness, weight loss rate, soluble solid content, soluble sugar content, titratable acid content, *vitamin C* content, and malondialdehyde (MDA) content are all prone to occur changing. However, it is possible to reduce the respiration and water evaporation loss of fruits and vegetables by reducing the environmental temperature and with nitrogen-rich and low-oxygen method, so as to delay the quality decline and the aging process of fruits and vegetable after picking. At the same time, near-infrared spectrometer can be used to detect and monitor the moisture and soluble sugar content of fruits and vegetables [11–15].

9.1.3 Evaluation Indicators for Eggs

Eggs usually refer to hen eggs, duck eggs, quail eggs, and goose eggs. Among them, hen eggs (eggs) occupy more than 80% of the market supply according to China Agriculture Data. Eggs are composed of egg shell, egg white, and egg yolk. Air chambers are often formed inside the blunt end of eggs. There is a layer of outer eggshell membrane on the eggshell, which forms protective clothing for preventing external microorganisms from entering the eggshell causing the eggs to spoil and deteriorate [16]. The eggshell helps fix the egg white and egg yolk shape, and further stops the invasion of microorganisms. The cracks and strength of eggshell are the main evaluation standards for its quality as they are easily damaged during storage and transportation. Eggshell cracks are a form of minor scratches that is often difficult to discern, but they have a significant effect on evaluating its quality. The thick protein content in the egg white is closely related to the quality of eggs and is affected by poultry species, season, feed, storage environment, and time [17]. The bactericidal and antibacterial lysozyme gradually lose their activity. Therefore, the reduction of thick protein is one of the important manifestations of aging and deterioration of eggs. With the increase in storage, the water in the egg white gradually penetrates into the egg yolk over time to increase the yolk volume and sometimes large enough that it penetrates through the yolk membrane and overflow to form loose yolk eggs. The air chamber height is one of the important manifestations of the degree of changes in the internal chemical composition of the eggs, and also one of the indicators of the freshness; with the increase of the storage days, the height increases. Hyperspectral technology, percussion vibration signal analysis technology, etc. have been widely studied and applied to detect egg gas chambers, cracks, and shell strength [18–20].

9.1.4 Evaluation Indicators for Meat

Meat is currently being stored and processed using cooling and freezing methods. Owing to its nutrient-rich nature, it is very susceptible to microorganisms contamination during the storage and transportation, such as *Salmonella, Bacillus cereus, Staphylococcus aureus, Escherichia coli, Pseudomonas fluorescens*, etc. [21]. Spoilage is mainly manifested in the proliferation of microorganisms, metabolism and decomposition of meat substrates, such as protein degradation, fatty acid degradation, and glycolysis of carbohydrates. Protein is decomposed to produce ammonia and amines and other basic nitrogen-containing substances (histamine, putrescine, etc.). Carbohydrates are decomposed to produce organic acids and carbon dioxide, and fats are decomposed to produce compounds such as fatty acids, glycerol, aldehyde and ketones. These catabolism changes the appearance and sensory indicators such as meat color, texture, smell, elasticity and viscosity, and cause significant changes in physical and chemical indicators such as total volatile basic nitrogen

(TVB-N), ammonia value, hydrogen sulfide, peroxidase, acidity, oxidation, and pH. TVB-N, protein, total sugar, total fat, total number of bacteria, etc. have their specific absorption peaks in the NIR region of the electromagnetic spectrum. Therefore, portable NIR and hyperspectral technologies are used to improve the detection capability of meat quality during storage and transportation [12, 22].

9.1.5 Evaluation Indicators for Tea

Tea has been known to be vulnerable to external environmental stimulus during the storage process. The external factors such as temperature, humidity, oxygen, and light can influence the free amino acids, polyphenols, and polysaccharides contents present, especially high temperature, high humidity, rich oxygen, and strong light would accelerate the aging and deterioration process [23, 24]. With the increase in storage time, the tea polyphenol content, caffeine, free amino acids, soluble sugar contents, and antioxidant capacity of *Pu-erh* tea gradually decrease, while the flavonoids level increases. Besides, the soup of *Pu-erh* tea which gradually changes to orange-yellow color, the aging degree in the tea aroma increases gradually with a thick and slightly astringent taste [25–27]. For white tea, the total amount of water extract, catechins and amino acids decrease with the increase of storage time, while the content of gallic acid, alkaloids and some amino acids show a significant increase. Among them, the water extract content of white tea, the total amount of catechins, ester catechins, non-ester catechins, and three alkaloids including caffeine, theobromine, and theophylline are all important indicators of Baichading [28, 29]. The quality evaluation in the process of tea storage and transportation generally adopts a combination of sensory assessment with physical and chemical analyses. Sensory assessment is easily affected by the subjective influence of the reviewers and objective influences of environmental factors. The data has low accuracy and does not meet criteria of physical and chemical assessment, and these are time-consuming and labor-intensive. Visible/near infrared (Vis-NIR) spectroscopy detection technology has been widely developed to portable devices and applied for tea quality evaluation, involving quantification of tea polyphenols, caffeine, catechins, etc. [11, 24, 30, 31].

9.2 Instrumentation

9.2.1 NIR Spectroscopy

Near-infrared analysis technology is simple to operate, easy to use, fast to measure with capability to obtain a wealth of molecular information, making it an ideal online evaluation technology. At the same time, various types of NIR instruments, complete measurement accessories, and high portability are other features in selecting

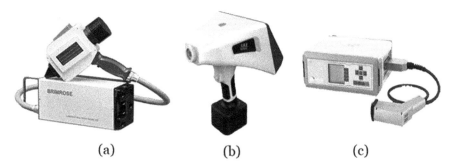

(a)   (b)   (c)

Fig. 9.2 Portable NIRS. (**a**) Luminar5030, (**b**) IAS-2000, (**c**) IAS-8100

NIR technology for online evaluation [32]. With the development of digital information technology, portable NIR instruments have developed rapidly in agro-product detection, and their detection accuracy has also been greatly improved. Compared with traditional general-purpose laboratory instruments, NIR spectrometer (NIRS) has the advantages of compact instrument, and easy secondary development according to user needs [33, 34].

Some portable NIRS that can be used for food and agro-product quality evaluation include the Luminar 5030 Mini-AOTF/NIR which is a portable spectroscope designed by Brimrose, USA, as shown in Fig. 9.2a. It is light in weight, small in size and convenient to carry and can be used not only in the laboratory but also in the production site and industrial online setups. The spectral resolution of the Luminar 5030 NIR spectrometer is 2–10 nm, wavelength repeatability is ±0.01 nm, and the spectral range is divided into two types: 1100–2300 and 1200–2400 nm. The wavelength range detected by *Intelligent Analysis Service Co. Ltd* is 950–1650 nm [35]. IAS-2000 portable NIRS is a special analytical instrument developed for grain detection and analysis, which is rapid, accurate, and nondestructive in detecting protein, fat, moisture, and other indicators of cereal samples. The instrument is small in size and light in weight and could be easily carried to the field, storage workshop, and truck for cereal detection and analysis, as shown in Fig. 9.2b. The IAS-8100 handheld NIRS (Fig. 9.2c) is an analytical device for rapid on-site detection. It could be used for rapid detection and identification of the quality of raw materials and auxiliary materials in the fields of fruit, tobacco, wood, textiles, and pharmaceuticals. The instrument is easy to carry, equipped with a touch screen, and easy to interact.

Figure 9.3 is a series of portable NIR spectroscope developed by domestic universities and have been used for rapid detection of some internal components such as soluble solids in fruits, caffeine, polyphenols in tea, etc., and moisture in cereals. It could attain on-site, rapid, and intelligent detection of food and agro-product. NIR fully exploits the progress made in the microelectronics and micro-processing technology. It uses sensor arrays or micro-electromechanical system instrument as the intermediate spectrum splitting element, so that the optical device not only greatly reduces the volume but also guarantees measurement accuracy. NIR instrument is battery-powered, and small light sources are embedded inside, so there

Fig. 9.3 Portable NIR detecting instrument

is no need to consider the use of on-site power and radiation sources. Some NIRS are combined with modern wireless communication technologies such as such as wireless transmission via Bluetooth, to facilitate the rapid collection and transmission of field data.

A miniature mode NIRS of the model digital light processing (DLP) NIR scan Nano produced by *American Texas Instruments (TI) Company* was selected. Compared with traditional spectrometers, it uses DLP technology to realize a low-cost, small-size, high-performance module. The physical map is shown in Fig. 9.4a. The wavelength scanning range of the DLP NIR scan Nano is 900–1700 nm, and it is integrated with a light source module for offering a broadband optical signal. The spectrometer microprocessor board is integrated with a BLE module, which is used for the control of the spectrometer by the mobile terminal and the long-distance wireless transmission of data. The DLP NIR scan Nano spectroscopic system uses diffraction gratings and uses digital micromirror devices (DMD) and single-point InGaAs detectors to replace traditional linear array detectors. The internal optical path and signal transmission diagram of the spectrometer is shown in Fig. 9.4b. The light diffusely reflected from the sample is collected by the collection lens and focused on the light engine through the input slit, and the light passing through the slit is calibrated by the collimating lens. After that, the light passes through an 885-nm-long wave-pass filter and is reflected by the reflection grating to be disperse into continuous wavelength light, and projected finally onto the DLP2010NIR DMD through the focusing lens. The 900 nm wavelength is imaged at one end of the DMD, and the 1700 nm wavelength is imaged at the other end. It sequentially disperses all other wavelengths of light in the middle and continuously obtains single-wavelength light by adjusting the DMD device, which reaches the single-point InGaA detector through the condenser lens. The optical signal that reaches the single-point InGaA detector is outputted into digital signal through the signal amplifier and analog-to-

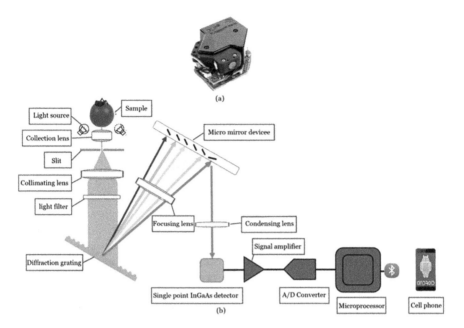

Light source
Collection lens
Slit
Collimating lens
light filter
Diffraction grating
Sample
Micro mirror devicee
Focusing lens
Condensing lens
Signal amplifier
Single point InGaAs detector
A/D Converter
Microprocessor
Cell phone
(a)
(b)

Fig. 9.4 (**a**) DLP NIR scan nano physical image and (**b**) internal optical path and signal transmission diagram

digital converter, then transmitted to the microprocessor board. This spectrometer is small in size, low in price, low in power consumption, superior in performance, and convenient in secondary development, and suitable for the development into a portable instrument for detection.

The software of the portable detection device is developed using various forms of software such as desktop computers, tablet PC, and mobile phones. The following introduces an Android application APP, which uses the development kit of the DLP NIR scan Nano to perform secondary development on the Android Studio development platform. The APP is modularized according to the functions required by the detection device.

1. **Android Permission Configuration**

When the mobile phone is installed with the Android application APP, the mobile phone system would automatically remind the user of the various permissions required to install the APP. Permission is a security protection mechanism and is mainly used to inform users that the operation of the software requires the use of certain specific functions and component access between applications. The Android system has more types of permissions compared to other operating systems. If users need to access a certain permission when developing an APP, they must declare the corresponding permission, otherwise an error may occur when debugging and running the application. The APP used by this device needs to use the relevant permissions for the Bluetooth function, access network, access location, and read and write external storage data.

2. **User Registration and Login Module**

The user needs to access the system with a legal identity required for this software to operate the device. Therefore, it is necessary to design and develop the user registration and login function. The user can enter login information in this interface, and click the remember password check box, so that when entering next time, the user does not need to re-enter the account password and can log out of the unnecessary account. The users can also modify the user information such as the password of the registered user. Unregistered users should register after clicking the register button and enter the required information in the registration interface. In the user registration interface, the users can fill in the user name and user password information, then submit the completed registration information. It verifies the information filled in the database. If the user information is already registered in the database, it would prompt that the user has been registered. If no user information is submitted in the database, it would show that the registration is successful. In the user login interface, it compares the entered login account and password with the information in the database to confirm whether it is correct. If it is correct, the login is successful, otherwise the login fails. If the login or registration is successful, toast in Android is used to show reminders. Toast has no focus, and it can be used in the program to notify users of some short messages. These messages would disappear automatically and do not occupy any operation interface space.

3. **Bluetooth Communication Module**

Under the system platform of Android version 4.3 or above, the BLE core functions could be easily realized by the provided API. Android connection to Bluetooth devices generally requires four steps: register Bluetooth-related broadcasts, search for Bluetooth devices, pair Bluetooth, and establish a connection. In this device, the Android mobile phone is used as the central device, and the miniature NIRS is used as the peripheral device. After the BLE pairing connection between the Android mobile phone, the NIRS successfully reads the built-in spectral correction data, and interactive communication can be carried out. The mobile APP could control the miniature NIRS, wherein the spectrum data obtained by the spectrometer can be transmitted to the mobile phone in real time.

4. **Positioning Module**

According to actual needs, the developed device needs to have the function of locating and detecting the location, so the positioning module needs to be configured in the software development. Although the Android system provides developers with a positioning function library that specifically supports the development of applications based on positioning services, it is compatible with third-party positioning services such as AutoNavi Maps, Google Maps, and Baidu Maps in terms of positioning accuracy and mode selection. After comparison, this software selects AutoNavi Maps Android positioning SDK service interface for positioning development. AutoNavi Maps Android positioning SDK supports Android version 2.2 or above systems, and it provides GPS positioning and network positioning (Wi-Fi positioning and base station positioning) methods. The GPS positioning and network positioning methods are encapsulated and opened in three modes. If the developer does not make specific settings, the high-precision positioning mode is automatically selected. How to use: First step is to import the jar package of AutoNavi map, then apply for the key value on AutoNavi official website. After obtaining the key value, positioning-related permissions in the AndroidManifest.xml file is added along with the key value in the <application> tag Information.

5. **Spectral Data Processing Module**

   After the miniature NIRS inside the device is successfully paired with the Bluetooth of the Android mobile phone to achieve communication, the acquired spectral data is transmitted to the Android mobile phone through the Bluetooth low energy (BLE) module. The spectral data processing module in the mobile APP analyzes and extracts the characteristic spectral data information. After the spectral data is properly preprocessed, the soluble solids and lycopene content of the cherry tomatoes are calculated by combining with the prediction model embedded in the software, and the results are displayed in real time on the main interface of the APP user.

6. **Detecting Data Management Module**

   When using a portable detection device to test the quality of cherry tomatoes, it is necessary to save the obtained data for later viewing and management. The software is designed to save the data to the SD card of the Android phone and set the data format to.csv format. It displays in detail the reflection, absorbance, intensity spectrum data, detection object, detection result, detection time, detection person, detection location, and other information of the sample, so that users can view historical detection data very intuitively.

9.2.2 Spectral Imaging Technology

Multispectral imaging detection technology integrates the characteristics of spectroscopy and imaging technology and collects and analyzes data in a discrete spectral range, which greatly simplifies the data [35]. The developed multispectral imager based on this technology has a fast processing speed. The reflecting light enters the imaging system after passing through a filter in a certain wavelength range, forming multispectral information of the sample.

The Blaze spectroscopic CCD camera is produced by *American PI Company*. It relies on the photosensitive chip technology and has higher quantum efficiency and lower dark noise in the near-infrared band than the traditional back-illuminated chip, as shown in Fig. 9.5a. Portable spectral imaging technology is currently dominated by multispectral imaging instrument and infrared spectral imaging. *American Altum Company* has developed a series of multispectral cameras, including multispectral, thermal imaging, RGB, and other technologies. The Altum multispectral camera is small in size and easy to carry and has powerful detection and analysis functions, as

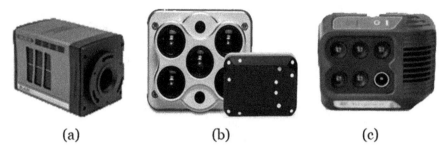

(a)   (b)   (c)

Fig. 9.5 (**a**) Blaze high-speed spectroscopic camera, (**b**) Altum multispectral remote sensing camera, (**c**) AIRPHEN channel multispectral camera

shown in Fig. 9.5b. The instrument is mainly used for irrigation and plant health analysis. The AIRPHEN is an unmanned aerial vehicle (UAV) multispectral camera system developed by *HI-PHEN Company*, as shown in Fig. 9.5c. The device is light and compact and has powerful configuration functions (wave band, field of view). It could be operated wirelessly and used in combination with an auxiliary thermal imaging IR camera and a high-resolution RGB camera. It is a UAV multispectral camera system with multiple measurement bands, fast speed, light weight, and powerful functions. It is very suitable for being used in large-area forestry, agriculture, ecological survey, and plant field phenotypic imaging research.

The traditional multispectral imaging system is mainly composed of a camera, a filter, a dispersion device, a light source, a housing, and a sample stage. Instrument developer utilizes the current high-tech technology of unmanned aerial vehicles equipped with a multispectral camera. Therefore, it can simultaneously obtain spatial information and spectral information of the sample to be tested, which is convenient without physical or chemical processing, and only takes 10–20 s to obtain information. Based on these advantages, multispectral imaging technology has received widespread attention in recent years and is designed to meet the speed requirements for industrial production. *Guangzhou Aoudun* long-distance multispectral forest fire prevention evaluation system adopts the color-to-black camera with high-performance, high-sensitivity, and high-definition, and the performance of the camera is not affected by bad weather such as fog and haze. The forest area can be monitored and warned 24 h a day in any weather, as shown in Fig. 9.6a. The MS-IR multispectral infrared thermal imaging cameras are equipped with an eight-hole high-speed rotating filter wheel, so that the field signal is divided into different spectral bands, as shown in Fig. 9.6b. The structure of the filter wheel is designed to maximize the acquisition frame rate of the camera and can be used in any fixed or rotating mode.

9.2.3   Electronic Nose

Persaud and Dodd developed the first artificial nose in 1982, which consists of various cross-responsive sensors for gas or odor. The multiple sensors are combined

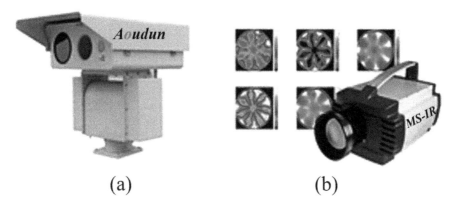

Fig. 9.6 (**a**) Aoudun long-distance multispectral forest fire prevention evaluation system, (**b**) MS-IR multispectral infrared thermal imager

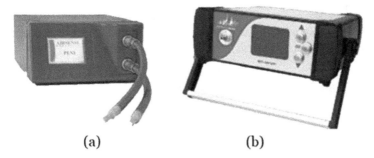

Fig. 9.7 E-nose biosensor device. (**a**) PEN3, (**b**) GC-IMS

into a sensor array, and responses of multiple sensors to one gas constitute the response spectrum of the sensor array [36]. The response signal of the sensor array is properly preprocessed, and then feature extraction is performed to obtain the feature vector of the gas. Different gases have their own different feature vectors (intuitively expressed as feature graphics or feature fingerprints). The appropriate pattern recognition analysis method would be used to process the feature vectors of responses to identify the gas qualitatively and quantitatively.

The PEN3 electronic nose shown in Fig. 9.7a is designed by the German AIRSENSE company. It is a gas and vapor detection system that can quickly identify a single compound or mixed gas after training. The design of PEN3 is small and exquisite, and the sensor response usually needs less than 1 s and the high-efficiency measurement cycle usually needs 1 min. The sensor has a protection device and has a long service life. The host has a large LCD screen, which can display the experiment progress in real time. At the same time, it can work independently in off-season without connecting to the computer and automatically store evaluation data information, which improves the instrument's mobile evaluation capabilities in the field.

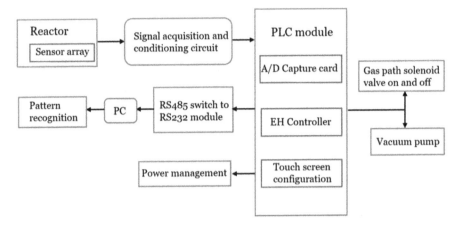

Fig. 9.8 Electronic nose system design scheme

Owing to the fast, compact, and efficient detection features of the PEN3 electronic noses, it is widely used in food rancidity analysis, molasses evaluation, type and aroma characteristics analysis, meat freshness analysis, and milk freshness evaluation [13].

The GC-IMS portable electronic nose is produced by *INNO Concept*, Germany. It has high sensitivity and good reproducibility and has the capability to detect a variety of substances such as ketones, aldehydes, alcohols, amines, and other organic substances, as shown in Fig. 9.7b. The GC-IMS system has a built-in computer unit that can be controlled independently. The self-explanatory operation menu of the GC-IMS system provides users with a very friendly interface. The operation steps and the settings of each measurement process are displayed on a 16.25-cm TFT display with a cyclic pulse decoder that could be used to execute or change the operation steps and measurement process settings. GC-IMS has two operating modes, automatic and manual, and the alarm levels of special areas of interest and predetermined compounds would be determined. Therefore, it is widely used in food safety storage condition control, food freshness evaluation, production process control, raw material evaluation, etc.

The design of the electronic nose system includes the hardware part that reacts with gas and the software part of data processing and analysis. The hardware system is composed of a gas acquisition module, a signal acquisition and conditioning circuit board, and a control system centered on programmable logic controller (PLC) electrical devices. The software part includes the data acquisition, the upper and lower computer communication, and the upper computer detection interface designed by *Delphi 7*. Figure 9.8 is a schematic diagram of the overall structure of the electronic nose system. It shows that the PLC module design constitutes the hardware core of the entire electronic nose system. Considering the influence of temperature on sensor performance; therefore, it must be controlled in the reaction chamber. The system connects the air compressor with the condensing fan, then it

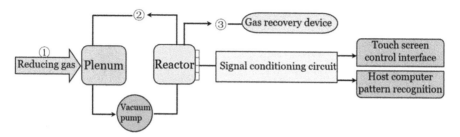

Fig. 9.9 E-nose system work flowchart

clicks the PLC control system touch screen to control the air compressor and the fan's running state, in this way, the temperature of the reaction chamber could be controlled. In terms of signal acquisition, the signal is collected through the A/D acquisition card of the PLC module and transmitted to the touch screen and the upper computer detection software. In the design of the entire electronic nose system, the employed PLC control only realizes the temperature control but also makes the operation of the test process more convenient. In addition, the use of high-precision A/D acquisition cards and anti-electromagnetic interference instrument ensures the accuracy of signal acquisition.

The design of the test loop is optimized. The response of the gas sensor to the gas essentially reflects the oxidation–reduction process of the metal oxide inside the sensor. After each test is being completed, the sensor should be restored to its original state to be repeated. Generally, it would take a long time to restore the sensor by air flow. To reduce the gas staying in the path and shorten the test time, the entire gas path was optimized. Figure 9.9 is the working flowchart of the self-made electronic nose system, ①, ②, ③are used to indicate the position of the solenoid.

Due to the diffusion characteristics of gas, some part of it is adsorbed in the gas chamber with the generation of gas residue. Generally, inert gases such as CO, H_2, and NO have no adsorption characteristics, while the degree of adsorption for SnO_2, H_2S, and NH_3 is high. The material and design of the device is designed with this background to ensure less adsorption of gases. The gas chamber is made of stainless steel, the pipeline is made of silicone products with drying effect, and a vacuum pump is installed in the gas chamber loop to accelerate the gas flow. Preliminary tests have shown that the higher gas flow rates and thinner pipes greatly reduce in the adsorption phenomenon. Finally, a vacuum pump with a flow specification of 8 L/m and a silicone tube with an inner diameter of 3.3 mm are used to form the loop. The sensor array is installed on the circular cross-section plate of the cylindrical gas chamber, and the sensitive parts of all sensors could be in sideways to the gas flow direction, to avoid excessive sensor poisoning due to airflow concentration. The structure of the reactor is shown in Fig. 9.10.

The sensor will take an amount of time to be completely restored from its initial values after the reduction reaction between the sensor array and the gas. The combination of nitrogen and oxygen in a ratio of 1:4 from the outside and the suction air flow of the internal vacuum pump is adopted to accelerate the oxidation

Fig. 9.10 Reactor structure diagram. Sectional view of gas chamber, gas sensor array

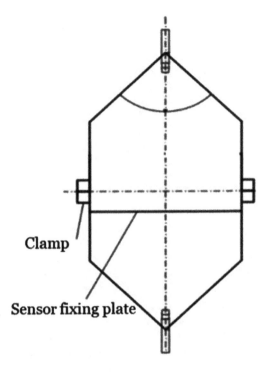

Clamp

Sensor fixing plate

reaction, and shorten the test time. The two-position two-way solenoid valve is placed at different positions of the gas circuit. It is kept on and off under different test conditions to realize the different gas isolation circuits in the reduction process and the oxidation process. In this way, it could shorten the reduction time of the sensor and reduce gas residue in the pipeline to ensure the test effect. The design sketch of the gas passage is shown in Fig. 9.11.

Temperature control system is a core part of the whole test process. Excessively high or low temperature would affect the performance of the sensor. Since most sensors are made of metal oxides and therefore require an input of heat prior working, therefore, heat emitted during the process would increase the temperature. Therefore, the entire reaction process must be thermostatically controlled. The temperature control module consists of a condensing fan, an air compressor, and a fan inside the cabinet. In the preparation stage of the test process, the initial temperature and other parameters would be set first. In the gas path control system, gas sampling is involved with the test process of volatile gas and the reduction process of the entire gas path. The realization of the entire experimental process is based on the control of the solenoid valve switch.

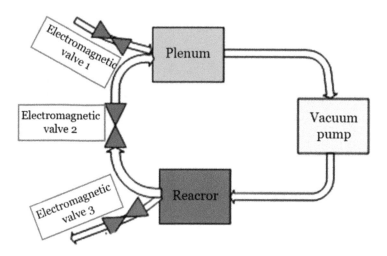

Fig. 9.11 Schematic diagram of gas path

9.2.4 Acoustic Vibration

The sound signal generates a wealth of information which cannot be perceived by sight, touch, or any other sense. The sound signal is typically correlated with the vibrational signal, and its data is collected with ease owing to the contact-free nature of the sound signals. Certificate authority (CA) technology based on general audio/environmental sound belongs to the branch of AI in the audio field. It is a very practical technology, and there are many applications in dozens of social and economic life aspects of life, in the field of medicine, manufacturing, transportation, security, storage, agriculture, forestry, animal husbandry and fisheries, water conservation, environment, and public facilities management industry.

Acoustic vibration nondestructive testing mainly includes acoustic impedance method, acoustic percussion method, and transducer acoustic resonance method. The physical (mechanical) performance parameters of the inspected object, the detection sensor, etc. are simplified into a certain mechanical system, and then analyzed and processed according to the mechanical vibration theory. Figure 9.12 shows the Japanese ACO noise tester TYPE 6236_6238 product. The design of the microphone of the front magnifying glass can eliminate its own noise, expand the lower limit of the bass, and can measure the bass below 0 dB-SPL. The "1/1 or 1/3 full audio real-time analysis card" is inserted in the tester, which has the NC histogram function with evaluation of the NC value. The 7052NR/7146NR high-precision condenser microphone is embedded, and the measurement frequency range includes all audible frequency. The ACO microphone has high seismic performance and is not easy to be damaged. It has a compact structure and is convenient to carry. The device is easy to operate and could carry out powerful measurements, and it is equipped with an RS-232 interface, connected to an external

Fig. 9.12 ACO noise
vibration tester

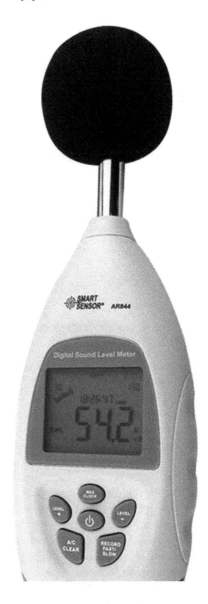

CPU for continuous operation, and the measurement results would be displayed on the LCD screen, as shown in Fig. 9.12.

The current research on the acoustic properties of agro-product refers to the relationship between a certain quality index in the agro-product and a specific resonance frequency, reflection, transmission, absorption, attenuation characteristic, propagation speed, and its own acoustic impedance. The sound wave emitted by the sound wave generator is continuously directed toward the material to be measured.

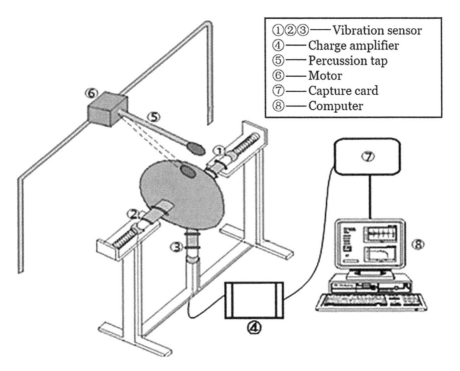

①②③——Vibration sensor
④——Charge amplifier
⑤——Percussion tap
⑥——Motor
⑦——Capture card
⑧——Computer

Fig. 9.13 Schematic diagram of the structure of the egg shell quality evaluation system

The sound wave signal transmitted, reflected, or scattered from the material is received and amplified by the sound wave sensor, then sent to the dynamic signal analyzer and computer for analysis. By this means, we would obtain related acoustic properties of agricultural products. At present, this field has begun to develop at home and abroad and has infinite and broad development prospects.

The vibration mechanics evaluation system is composed of percussion module, support module, signal acquisition, and processing module, as shown in Fig. 9.13. The percussion module provides an external pulse excitation to the egg so that the egg generates free vibration after being actuated. The primary aim of a support module is to withstand the impact without causing any damage to the egg shell. The signal acquisition and processing mainly collects the vibration response signal of the egg and performs corresponding filtering processing, and converts the analog signal into a digital signal for further processing.

The percussion module of the egg is mainly composed of an adjustable linear power supply, an electromagnet, a drive motor, and a percussion rod. The strength of percussion is driven by electromagnets, combined with linear power supply to achieve strength control. The electromagnet adopts the two-way corner permanent magnet electromagnet (SXCT38-15/18) from *Nanjing Kailing Automation Factory*. The performance of this electromagnet covers the main functions of a rotating electromagnet and a bidirectional self-holding electromagnet. Compared with the rotating electromagnet of the same size, it has the salient features in a large rotation angle (action angle up to 100°), bidirectional rotation (left and right rotation can be), fast response time, low power consumption, low temperature rise, high power, etc. [44]. Therefore, it could meet the demand for automatic and fast knocking of the inspected eggs.

When the percussion module executes the percussion command, it needs to give the electromagnet a percussion signal to execute the percussion command. At the same time, the frequency of percussion would also be adjusted according to the needs of detection. The percussion command and percussion frequency of the percussion module are controlled by the capture card. The core processor of the percussion system, USB-1208FS, is a bus-powered USB device produced by *American MCC Company*. It provides 8 single-ended inputs with 12-bit resolution, 11-bit resolution, or 4 differential inputs. The sampling rate unit provides up to 50 kilosamples/s, and there are 8 software-selectable input ranges in seconds. The device also provides two 12-bit analog outputs, a 32-bit counter, and 16-bit I/O lines.

The support module of the egg is composed of the supporting acceleration sensor, the steel plate to fix the sensor, the base and bracket of nylon material, and the movable rail. The acceleration sensor is fixed on a movable steel plate, and its shape and size are designed according to the size of the sensor, which is suitable for the placement of normal eggs. The acceleration sensor is sleeved with the steel plate and is fixed on the movable rail through a nylon material. The movable track can be adjusted appropriately according to the size of the egg, so that the longitude of the egg could be fully knocked.

The software system includes three functional modules: signal acquisition, signal threshold triggering, and signal processing. The software system sets a threshold trigger in the process of time domain and frequency domain signal conversion, that is, when the system recognizes that the collected signal is an egg strike vibration response signal instead of environmental noise, the time domain signal is transformed into the frequency domain signal by using fast Fourier transform (FFT), and the power spectrum is regarded as the statistical parameter of the frequency domain signal. The power spectrum of the response signal could also be observed in real time on the man–machine interface of the system. After obtaining the power spectrum of the response signal, the system uses a filter to filter and denoise the signal, and store the data in the computer hard disk for subsequent processing.

9.2.5 Colorimetric Sensor Array

Colorimetric sensor array (CSA) technology is a new olfactory system by the means of imitating humans and mammals. Based on the color change before and after the reaction of chemically responsive dyes with the gas, it would realize the visualization of the gas to be detected. Compared with traditional electronic olfaction

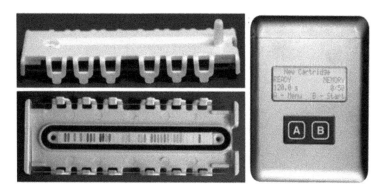

Fig. 9.14 CSA response and output array (American, Illinois)

technologies that rely on weak forces such as physical adsorption or van der Waals forces, CSA detection technology relies on the strong covalent bonds formed. The olfactory visualization sensor is mainly composed of some dyes with specific recognition capabilities [37, 38]. The color of the dye molecules changes significantly after interacting with the detected object. When these image signals of CSA is run through the computer, specific RGB data signals would be formed. Then, regression analysis would be performed in pattern recognition method on RGB data with the qualitative and quantitative analyses of characteristic chemical indicators related to odor. At present, the CSA technology is still in the early scientific research stage, and commercial instruments have not been developed. Some universities and research teams have developed and continuously improved related experimental instruments. Figure 9.14 is an olfactory visualization device developed by the University of Illinois at Urbana-Champaign. The CSA array detection system is developed by Chen's group as shown in Fig. 9.15. The CSA device consists of an array, a gas enrichment module, a gas reaction module, a signal processing and output module.

The operating process of the detection system (as shown in Fig. 9.16) is as follows: Set the corresponding response time and response temperature on the touch screen, and communicate with the microcontroller through the serial port; after the microcontroller receives the signal, it supplies power to the temperature sensor; then the temperature sensor detects the reactor temperature and communicates with the single-chip microcomputer using the 1-wire protocol; after the single-chip receives the temperature signal, it displays the temperature change in real time on the human–machine interface (HMI). And the reaction time is measured by HMI timing, when the response time is reached, the HMI sends a signal to the single-chip microcomputer; after the single-chip receives the signal, the response progress display module prompts the completion of the reaction; the signal acquisition and processing module is used to obtain and process the color spectrum information.

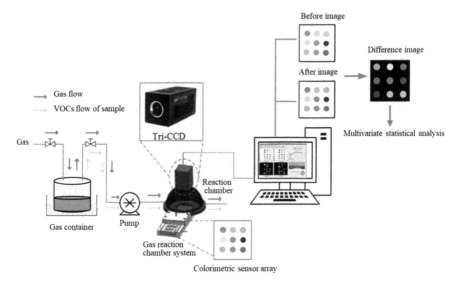

Fig. 9.15 The CSA array detection system

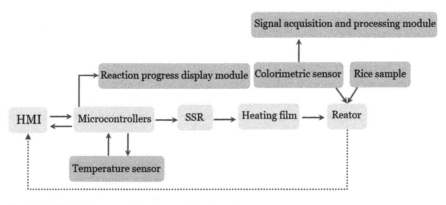

Fig. 9.16 Block diagram of rice storage time detection system

1. **Spectral Acquisition and Processing Module**

   The signal acquisition and processing module is composed of spectrometer, light source, reflection probe, detection station, and computer. In the module, the light signal generated by the light source is hit in the dye printing area of the CSA by reflection probe, then the reflected light signal is detected with a spectrometer. The detection data is saved and processed to the computer through serial communication. Spectrometer is used in optical evaluation that separates composite light into spectral lines, and the light intensities at different wavelengths are

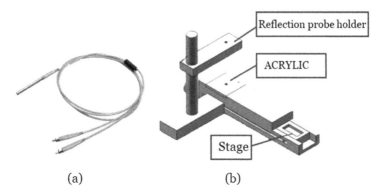

Fig. 9.17 (**a**) Reflection probe and (**b**) schematic diagram of test bench

measured using photo detectors. It is generally composed of an incident slit, a focusing element, a dispersive element, a collimating element and a signal detector. The detection system uses a USB2000+ (Vis-NIR, Ocean optics, America) for the detection of spectral information. The light source provides the optical signal for the detection system and is an important part of the system. In order to match the wavelength detection range of the miniature spectrometer, a halogen lamp (LS-HA, OTO Photonics, China) was selected as the light source. This light source could provide a stable widescreen spectral output (350–1700 nm); therefore, it is suitable for detection in the visible and NIR light range. In addition, the built-in power regulation circuit of the light source ensures the stability of the current, to allow a reliable output of the light source.

The reflection probe is used to provide the optical signal reflected by the CSA of the spectrometer, and it is constructed based on the principle of total reflection of light by the optical fiber bundle. As shown in Fig. 9.17a, the reflection probe has two SMA905 interfaces, one end is connected to the light source to provide the original optical signal, and the other end is connected to the spectrometer to obtain the reflected spectrum information. The detection platform is used to load the CSA after the reaction completing. It is composed of light aluminum alloy material and acrylic plate. The device is shown in Fig. 9.17b.

2. **Reaction Process Control Module**

The reaction process control module consists of relays, temperature sensors, heating bases, fans, and HMI. It is mainly used to control the temperature and time of the reaction chamber. The operation of each electronic component in the module is controlled by the STC89C52 single-chip microcomputer. As an important controlled parameter, accurate temperature measurement is the prerequisite for providing stable reaction temperature. Besides, considering the acquisition speed, cost performance, detection stability, environmental adaptability, and other factors, DS18B20 is selected as the temperature sensor to collect temperature information.

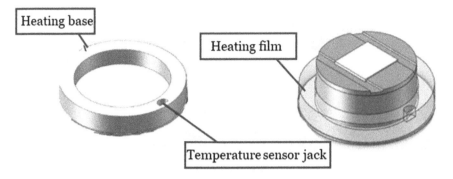

Fig. 9.18 Schematic diagram of reaction chamber heating

The CSA would have a better color rendering effect after using a free-volatile reactor, and 55 °C is the best temperature for the reaction. Since the optimal reaction temperature is higher than room temperature, heating elements are required to increase the temperature in the reaction chamber. The heating film is operated at voltage of 12 V and a heating power of 25 W and will function well and be applicable to −190 to 205 °C conditions. Figure 9.18 shows the sketch of the proposed heating reaction chamber. The heating film is packed close to the base made of aluminum alloy with geothermal conductivity to prevent inadequate heat distribution. The edge-designed socket is used for DS18B20 to read the temperature information of the heating base. In order to control the heating efficiency of the heating film so that the reaction temperature could be basically maintained at 55.0 °C, the energization duration of the heating film needs to be changed with changes in temperature. Therefore, the SRD-05VDC-SL-C relay module is used to control the on–off of the heating film circuit.

HMI is the machine–human information interface system. It is usually connected with control electronic device, mouse, touch screen, and keyboard as input units for parameter or operation command input and a display screen for data output. The reaction progress control module is mainly composed of three light-emitting diodes and a buzzer. The color of the diode has three colors of green, yellow, and red. Among them, green means the system is in an idle state; yellow means the reactor is heating up; red means the sensor is in the reaction stage. The visual interface of HMI is written by the built-in development software USART HMI, which enables the instrument operator to control the reaction process. STC89C52 is communicated with HMI through serial port to monitor the entire reaction, and the entire detection process could be displayed from the HMI interface. Figure 9.19 shows the reaction process between the CSA and the volatile gas.

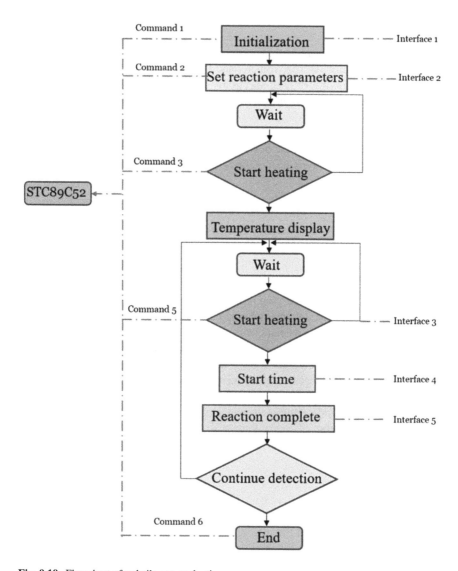

Fig. 9.19 Flowchart of volatile gas evaluation

9.3 Application in Rapid Evaluation of Agro-Product Quality

The interaction of NIR electromagnetic radiation with samples results in its absorption, transmission, reflection in the region of 650–2500 nm and measured with different models [39, 40]. The application of NIR technology has helped to evaluate the quality and protection of food and agricultural products, such as fruits, tea, meat, poultry and eggs, condiments, wine, and so on. Nordey [41] employed NIR to model and analyze the soluble solids and dry matter content of mangoes with different maturity. Malegori [42] applied PLS model and the SVM regression method to compare the miniature NIR spectrometer and Fourier NIR spectrometer. The results indicated the applicability of both spectrometers to the internal quality of cherries using SVM regression algorithm. Kim [43] used the NIR technology to establish an effective method for nondestructive testing of kiwi fruit maturity, harvest period, and storage period. Therefore, the marvelous potential of the NIR technique as a nondestructive tool has driven the development of more sophisticated NIR detection systems in quality assessment of food or agro-products applications.

9.3.1   NIR System for Quality Evaluation of Cherry Tomatoes

Cherry tomatoes are bright in color, beautiful in shape, delicious, and nutritious and are deeply loved by consumers. In the context of expanding market demand, the cherry tomato industry is developing rapidly with high consumer demands, but there are many shortcomings in the pre- and post-production fruit processing. At the same time, with the gradual improvement of people's living standards, a rise in demands for the internal quality of fruit is noticed. The research and development on the rapid nondestructive detection methods of the cherry tomatoes internal quality and design of portable devices are of great significance to improve the commercial processing levels and secure intelligent development of the cherry tomato industry. The research uses NIR technology and chemo-metric methods to establish a quantitative model of internal quality indicators of cherry tomato with high accuracy, good stability, and strong temperature adaptability and establishes a portable evaluation based on the Android system.

1. **Hardware Design**

    A miniature NIRS of the model DLP NIR scan Nano was selected. The structure of the portable detection device is shown in Fig. 9.20. Its size is $150 \times 80 \times 80$ mm, and the material used for the housing is aluminum alloy, which is lightweight, durable, and robust. The reasonable layout of the components of the device makes the whole device compact and portable. In recent years, the smartphone industry has developed vigorously and its image processing capacity, data processing speed, and data storage capacity are comparable to tablet personal computer (tablet PC) and PC. Compared with tablet PC and PC, mobile phones have the advantage of small size and low price. In order to develop a portable detection device, smartphone as a terminal is a suitable and ideal choice. The developed portable detection device is based on the Android system. It is necessary to use the development kit of the miniature NIRS to do secondary development and write APP. Considering the use of BLE for communication, the Android phone is required to support the BLE function and the Android operating version needs to be 4.3 or higher. From the current situation of Android phones on market, most of the Android phones could meet these requirements.

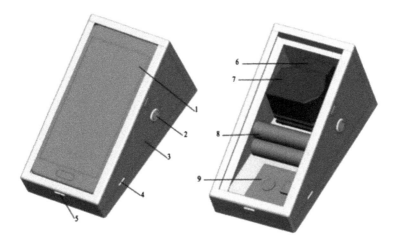

Fig. 9.20 Structure diagram of portable detection device. 1. Android phone, 2. Switch, 3. Shell, 4. Power charging port, 5. Mobile phone charging port, 6. Integrated light source, 7. Miniature near-infrared spectrometer, 8. Power supply, 9. Voltage regulator board

2. **Software Design**

The software of the portable detection device mainly include user registration and login, Bluetooth communication, positioning, spectral data processing, detection data management, and user main interface. The APP enables wireless communication between the Android mobile phone and the miniature NIRS through the BLE module. The NIR spectrum data of the cherry tomato are transmitted to the Android mobile phone to evaluate the indicators of sample test results in real time. The user login interface is shown in Fig. 9.21a. Users can also change information such as the registered user's password; the interface for changing passwords is shown in Fig. 9.21b; the input registration interface is shown in Fig. 9.21c.

Data detection files are saved according to the name of the custom file and shown in the data detection list, as shown in Fig. 9.22a. The files are arranged in the order of detection time. Users can easily locate the required files using the search history tool, click to open the detection data file, and can jump to the detection data viewing interface, as shown in Fig. 9.21b.

3. **Performance Verification**

After completing the hardware design and software development of the portable detection device, the device is finally processed and assembled. It is important to verify the performance of the device, including detection accuracy and stability, in order to determine the actual usage of the device. The soluble solid content and lycopene content of the samples at different temperatures were predicted by using a portable detection device. A paired t-test was performed on the predicted value and standard value of 15 samples, and the results are shown in Table 9.1, it can be seem that the $-t$ values are all less than $t_{0.05(14)}$, indicating no

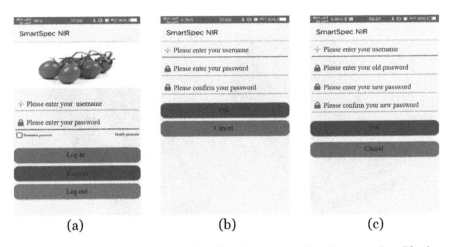

Fig. 9.21 User interface. (**a**) Login interface, (**b**) registration interface, (**c**) password modification interface

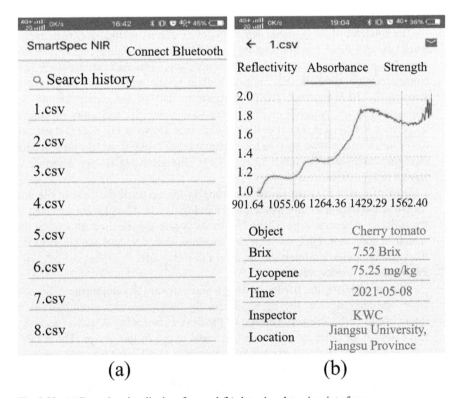

Fig. 9.22 (**a**) Detecting data list interface and (**b**) detecting data view interface

Table 9.1 *t*-test results of standard and prediction values

Quality index	Sample numbers	Standard value	Prediction value	T value	P value	Correlation
Soluble solids	15	$6.83 \pm 0.79°$Brix	$6.97 \pm 0.74°$Brix	1.841	0.087	Not significant
Lycopene	15	79.49 ± 15.28 mg/kg	77.15 ± 15.27 mg/kg	1.238	0.236	

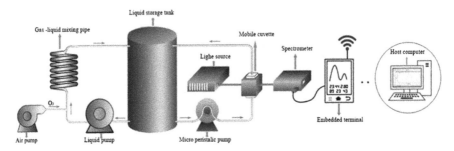

Fig. 9.23 Scheme of evaluation system for oxidation process of summer and fall tea

significant difference between the predicted value and the standard value. The above results reveal good detection accuracy and stability of the portable detection device for samples under different temperature conditions.

9.3.2 NIR System for Summer and Autumn Tea Fermentation Analysis

During the fermentation of summer and autumn tea, tea polyphenol content, total sugar content, and color of the three physical and chemical indicators would have a significant change. Compared with other nondestructive detection (NDD) methods based on mechanical, electrical, or acoustic characteristics, spectral detection system has a wider application range and stronger robustness. Researchers have developed a Vis-NIR detection system. The obtained Vis-NIR spectrum contains the color and composition information of the measured object to realize the real-time nondestructive evaluation of the tea polyphenols, total sugar content, and color in the summer and autumn tea fermentation broth.

1. **Hardware System Design**

   The basic principle of the hardware system is shown in Fig. 9.23. It mainly includes a fluid conveying-mixing device and a Vis-NIR evaluation device. The fluid conveying-mixing device mainly includes liquid storage tank, air pump, liquid pump, micro peristaltic pump, gas–liquid mixing pipe, conveying pipe, etc., and the device is responsible for the oxidation, sampling, and recovery of summer and autumn tea extracts; Vis-NIR evaluation devices are composed of light source, mobile cuvette, spectrometer, and terminal, which is used for collecting and processing spectrum data of the oxidation process during summer and autumn tea extract.

2. **Software System Design**

   The software system runs on the Windows system and is employed to collect, view, and save data, predict result, and calculate fermentation parameter as well as set spectrometer parameters (such as integration time, collection times, smoothing window size), and so on. The communication module of the software system is required to meet the same protocol as the lower computer PLC, and the data format is compatible with each other. The human–computer interaction interface of the software is required to be simple, clear, and cover all operable contents. The development of the software system in this study is based on the Windows. Using the object-oriented computer C++ language, Visual Studio 2015 programming, and the Microsoft Foundation Classes Library (MFC) as the interface development template, the developer combined the ocean optical Spectrometer Development Kit (SDK) and the Modbus communication protocol library to design software system. The user interface of the terminal has the function of output display and simple control. The terminal interface is written in the Qt framework and mainly contains some display parts and buttons, as shown in Fig. 9.24.

3. **Performance Verification**

Performance verification is mainly the pre-precision verification of the three indicators including tea polyphenols, polysaccharides, and color of the fermentation broth in the oxidative fermentation of summer and autumn tea. Specifically, the online evaluation system would test the accuracy and stability of predicting tea polyphenols and total sugar, and the accuracy of predicting the color value of fermentation broth. The prediction performance method of tea polyphenols and total sugar is based on choosing the fermentation broth with different nodes in the fermentation process, record the three parallel test results of each fermentation nodes, and test the coefficient of variation. Similarly, the color value accuracy test method would also select the fermentation broth with different nodes, record the color value ($L^*a^*b^*$) detection results and detect the color value of the current fermentation broth by using standard instrument. The detection data of the color value ($L^*a^*b^*$) is shown in Table 9.2. The relative and absolute errors between the L^*, a^*, and b^* values of the two devices are further calculated. The absolute errors of L^*, a^*, and b^* values are all less than 0.5, and the relative errors are all less than 2%. In summary, the test results show that in this analysis, the summer and autumn tea fermentation evaluation system could accurately produce the contents of tea polyphenols, total sugars, and color values in the fermentation broth.

9.3.3 Vibration Mechanics System

Different agro-products have different surface structures or internal qualities. When agro-products vibrate freely under external excitation, the generated sound response signals would have different characteristics [45–47]. The surface vibration method has been widely used in the chemical, electric power, transportation, and metallurgical industries. Its main principle is as follows: under the action of external excitation, the generated surface vibration wave is affected by the surface structure or internal quality of the object [36]. By choosing suitable sensing methods, the

Fig. 9.24 User interface of embedded terminal

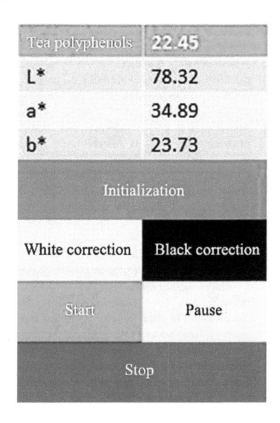

Table 9.2 Comparison of $L^*a^*b^*$ data collected by standard instrument and this research instrument

Number	Standard instrument			Spectral detection system		
	L^*	a^*	b^*	L^*	a^*	b^*
1	28.68	14.04	46.39	28.94	13.68	46.20
2	29.28	15.75	30.16	28.78	16.13	30.58
3	23.30	14.85	22.80	23.75	15.13	23.15
4	26.96	15.19	22.34	27.27	15.19	22.74
5	31.72	16.32	31.45	31.85	15.86	31.61
6	25.38	15.33	36.95	25.57	15.68	36.62
7	27.29	16.99	22.96	27.50	17.23	23.19
8	48.62	11.36	122.21	48.96	11.44	122.22
9	27.62	17.28	27.82	27.99	17.57	28.06
10	28.18	18.83	23.17	28.07	18.33	22.91
11	28.62	15.82	26.12	28.41	16.08	25.78
12	29.06	15.06	35.15	28.65	14.95	34.71
13	30.48	14.42	32.18	30.43	14.66	32.14

vibration signals are translated into electrical signals, and the characteristic parameters of vibration are obtained through signal analysis and processing. Different quality of agro-products could be evaluated by the reflection, scattering, transmission, absorption, attenuation coefficients, and propagation speeds as well as their own acoustic impedance and natural frequency, which is the distinctive features of the percussion vibration response signal under the action of sound waves [47–49]. Goodrum et al. [50] used machine vision technology to detect egg cracks; Cho et al. [51] used a CCD camera to obtain an image of the surface of a poultry egg shell illuminated by a white woven lamp; Jenshin et al. [52] and Lawrence et al. [53] have researched and developed a full-automatic eggshell detection system, which collect images when the eggshell is under certain pressure.

The egg could be regarded as a rigid body with a thin-shell ellipsoidal structure. The surface vibration wave gradually spreads from the striking area to other areas and weakens until it disappears. The surface structure of the egg will be broken when the surface of the egg cracks, impacting the distribution of vibration waves on its surface This difference in vibration response signal characteristics of percussion is utilized to distinguish cracked eggs. By collecting the vibration response signal generated by the mechanical excitation of the egg, analyzing its distribution, diffusion, and attenuation on the egg shell surface, a set of basic research platform for egg shell quality is designed. On this platform, the relevant hardware and software systems are optimized to provide a theoretical basis for identifying cracked eggs with one tap.

1. **Vibration Mechanics Structure**

   The hardware system consists of three parts: percussion module, support module, signal acquisition and processing module. When the egg percussion vibration response signal is transmitted to the host through the capture card, the software system reads the signal. Then, a two-dimensional diagram for time-amplitude of the vibration response strength is obtained, and the real-time change of the time domain signal is displayed on the software human–machine interface. The software system based on *Labview* language is shown in Fig. 9.25.

2. **Collection and Processing Flow**

   The percussion vibration response signal collection and processing steps for each egg shell are discussed hereafter: A response signal generated by the external mechanical impact of the egg. The analog signal is collected by the acceleration sensor in flexible contact with the egg shell, is processed by the amplifier. The amplified signal enters the acquisition card for A/D conversion. The software system program obtains the analog time domain signal, and the frequency domain signal of the egg response signal through the threshold trigger. Afterwards, through the software filter denoising, the required data is obtained and stored in the computer for further analysis. The test material was 100 brown eggs of 1–2 days after production in Zhenjiang Farm, Jiangsu. With the help of a magnifying glass, these eggs were carefully inspected and were in good condition [45]. The shape of the test eggs is fairly regular, and its size is also relatively

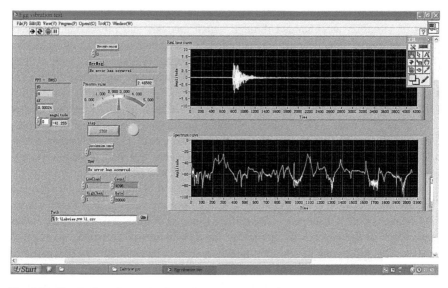

Fig. 9.25 Egg shell quality evaluation software system interface

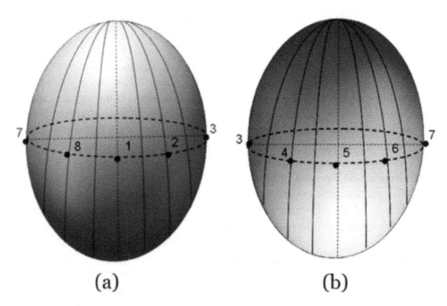

Fig. 9.26 Schematic diagram of egg striking point. (**a**) Percussion point on the front of the egg, (**b**) percussion point on the back of the egg

uniform. In order to study the size of the crack area detectable by the acceleration sensor, the egg is divided into eight points along the largest horizontal diameter, which are numbered 1–8 in turn, as shown in Fig. 9.26.

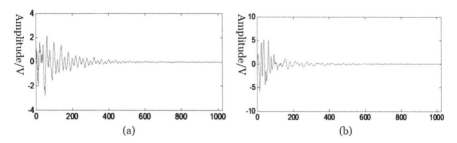

Fig. 9.27 Time domain signal in response to egg percussion. (**a**) Intact egg percussion response signal diagram, (**b**) cracked egg percussion response signal diagram

Table 9.3 LDA discriminant result of egg signal

Position	Discriminant result	
	Training set (100%)	Prediction set (100%)
1	**0.8657**	**0.8485**
2	0.7910	0.7121
3	**0.7910**	**0.8333**
4	0.7836	0.7121
5	**0.8284**	**0.8182**
6	0.6940	0.7727
7	0.7239	0.6061
8	**0.7761**	**0.8333**

Bold indicates that the model established with the spectral variables extracted by UVE-Si-PLS with best performance

According to the difference between the intact and cracked egg knock response time domain signals obtained in the experiment is shown in Fig. 9.27, the independence of each characteristic parameter is completely taken into account. Five characteristic parameters including the maximum value of time domain signal, standard deviation of time domain signal, the maximum value of first derivative of time domain signal, the maximum value of second derivative of time domain signal, and the standard deviation of first derivative of time domain signal are selected.

3. **Performance Verification**

A linear discriminant analysis (LDA) model of intact and cracked eggs is established with the above five characteristic parameters as characteristic variables. Table 9.3 shows the identification results of egg cracks at different positions. It can be seen from Table 9.3 that the designed egg shell crack detection system can realize the collection and analysis of intact and cracked egg signals. When the distance between the striking position and the crack is 1 cm, most of the intact and cracked eggs could be distinguished; when the distance is increased to 2 cm, the egg recognition rate is reduced by more than 10%; when the distance is increased to 4 cm, less than 60% of egg samples could be distinguished. There-fore, in the real-time online detection of egg cracks, in order to ensure that the crack features of the egg could be completely extracted, the position excited by the percussion rod and the egg crack would be kept within 1 cm as far as possible.

9.3.4 CSA Array System

A CSA array (CSA) technology was proposed by Suslick and Neal [54] which can be used for "odors and volatile organic compounds (VOCs) visualization." Odor imaging detection is a new branch in the research field of artificial olfaction technology [78]. The CSAs used in this technology employs volatile gases to be captured by the chemo-responsive dyes, and eventually, the obtained response is converted into an image. The rapid, economical, nondestructive identification of food quality and safety can therefore be realized [55, 56]. Imaging detection technology has been proved to have great application prospects in environmental monitoring, food and beverage quality monitoring, disease diagnosis, and other fields. In recent years, imaging detection technology has gradually been applied to the odor detection of some highly volatile foods such as vinegar, white wine, and mildew of grains [9, 11, 57–71].

1. **Hardware and Control Design**

   The miniaturized spectrometer is used for the extraction of spectral informa-tion of the CSA. Its small size, light weight, and low price provide the possibility to build a portable detection system. HMI has the functions of recording real-time data information, displaying real-time data trends, and generating and recording alarms. With the continuous development of HMI, the expression of information has evolved from simple flat expressions such as single-color graphics and tables to diversified colors, 3D graphics, and multimedia animation playback. Using the characteristics of HMI, it communicates with STC89C52 through the serial port to control the reaction parameters (temperature and time) and reaction process; display the real-time temperature status of the reactor and control the on-off status of various electronic devices. In order to realize the construction of spectrum detection system of the rice storage time, the STC89C52 single-chip microcom-puter is used as the controller to control the time and temperature of the sample reaction, and the human–machine interface (HMI) is used to display the reaction process and communicate with the single-chip microcomputer. On this basis, a Windows desktop application software is developed for the analysis and processing of spectral data. The built system is employed to classify different degree of rice aging. The schematic diagram of the rice storage time detection system is shown in Fig. 9.28. The system uses STC89C52 as the controller, which mainly includes a spectrum acquisition and processing module, a reaction flow control module, a reaction progress display module, and a power supply module. Figure 9.29 shows the HMI interface of CSA system detection of rice storage time.

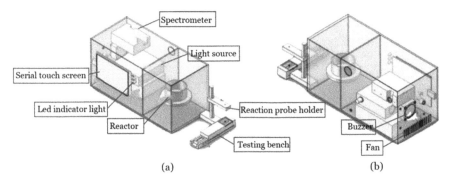

Fig. 9.28 Schematic diagram of rice storage time detection system. (**a**) Front view of device, (**b**) back view of device

2. Assessment Results for Rice Storage Time

Vacuum-packed *Su Ruanxiang* rice was selected as the research object and purchased from a local supermarket. In a constant temperature and humidity box, the fresh rice is kept and stored under conditions of a fixed temperature of 40 °C and a relative humidity of 80%. The rice samples are divided into five groups (storing for 0, 1, 2, 4, 6 months), 30 samples are taken for each storage period, and 8 g is weighted for each sample for each sample, for a total of 150 samples. Each group of samples is randomly allocated into a training set and a prediction set at a ratio of 2:1. The training set samples (100) are used to build the storage time prediction model, and the prediction set (50) is used to evaluate the performance of the discriminant model. The built-up visible/near infrared (Vis-NIR) spectrum-CSA detection system is used to predict the storage time of rice. Table 9.4 shows the results of LDA classification using this device combined with three different variable screening algorithms. After establishing the prediction model with the spectral variables extracted by UVE-Si-PLS, when the number of principal components is 9, the correct recognition rate of the training set is 98.00%, and the correct recognition rate of the prediction set is 96.00%. In the pre-judgment set, only two fresh rice samples were misjudged as a storage period of 1 month.

9.3.5   Electronic Nose System

Electronic nose (E-nose) is a method based on simulating the human nose. E-nose technology uses a conductive sensor to convert useful chemical signals into easily measurable electrical signals that can be identified and detected using characteristic chromatograms or spectra to obtain characteristic VOCs [72, 73]. Therefore, E-nose technology has proven potential for the identification and quantification of chemical constituents of food or agro-products. This method is widely used to quickly examine the variety, quality, pests, freshness, mildew degree of food, etc. [74–76]. Paolesse et al. [77] used the E-nose detection technology to analyze the degree of mold contamination in wheat. The moldy volatile organic gas detected by GC-MS

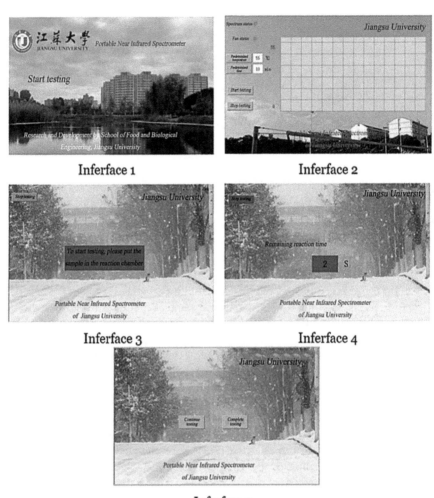

Fig. 9.29 HMI interface diagram

Table 9.4 LDA classification results of three different variable selection algorithms

Variable extraction method	Number of spectral variables	Optimal number of ingredients	Recognition rate, %	
			Training set	Prediction set
UVE-Si-PLS	**68**	**9**	**98**	**96**
GA-Si-PLS	69	7	92	92
ACO-Si-PLS	20	10	98	90

The use of bold is to distinguish the contents in the upper and lower rows

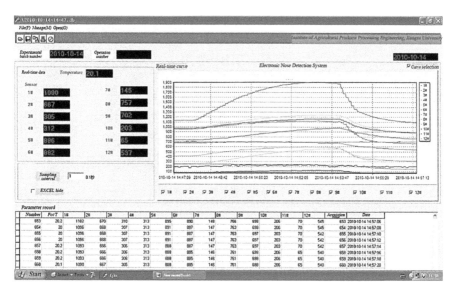

Fig. 9.30 E-nose signal diagram

was compared with the data of the E-nose to verify the E-nose's detection of VOC. Sung et al. [78] demonstrated the combined use of the E-nose with mass spectrometry, and the scheme was able to identify rice that had been processed for various storage periods with an accuracy higher than conventional sensory methods.

The test teas were from the first, second, and third *Biluochun* tea produced in *Dongting* (Mountain), Suzhou. There were 3 grades with 20 samples each, and a total of 60 tea samples. Since the flavors of agro-product are generally more diverse and sensor could only detect a specific or several gases with similar properties, a fabricated sensor array of 12 sensors was used to detect tea aroma. Furthermore, because the tea soup and the foliage fundus contain a lot of water vapor, since there is a lot of water vapor in, some desiccants such as silica gel would be placed in the air collection chamber to reduce the impact of water vapor. Figure 9.30 is the acquired Enose signal diagram.

The distribution area of different grades of tea samples in the three-dimensional principal component analysis diagram is obviously different, which indicates that the sensor array has obvious differences in response to different grades of tea soup and foliage fundus. All three-dimensional principal component analysis diagrams display the same distribution, obtained by combining grade of foliage fundus tea soup with the characteristic variables. However, there is a certain overlap between the principal component distribution diagram of different tea types, therefore, to distinguish them further, a pattern recognition methods is employed. KNN model is established to identify the fusion of the characteristic variables of foliage fundus and tea soup and has the best effect. When PCs = 5 and $K = 6$, the recognition rate of the KNN model is the best, and the recognition rates of both the training set and the prediction set reach 83.33%. Table 9.5 shows the cross-validation recognition rate of

Table 9.5 KNN model training and prediction results

Type	Principal components (PCs)	K value	Discrimination		Recognition rate	
			Sample			
			Training set	Prediction set	Training set (%)	Prediction set (%)
Foliage fundus	1	9	40	20	83.33	72.22
Tea soup	6	3	40	20	80.95	77.78
Foliage fundus and tea soup	5	6	40	20	83.33	83.33

the KNN model for information fusion samples under different principal component factors and K levels.

9.4 Conclusion and Future Trends

To ensure rapid testing goods, there are still several problems like the absence of quality assessment standard that will not be solved soon. For portable nondestructive testing instrument, as a result of the miniaturization, simplification, and maintenance of test results, the development of these instruments in the market for rapid tests is extremely promising. The application of new technologies affects the development of rapid testing products. The recent developments in nanotechnology, molecular biology, etc., might help to increase the long-term availability of rapid detection methods. In general, the future trends will be focused on the design of rapid the rapid, sensitive, specific, accurate, portable and traceable data tools for food safety.

It is important to store all kinds of agro-products which involves integrating detection and control into portable equipment, such as the NIRS and E-nose sensors. The general aim of food safety traceability is the supervision of food safety across the world. In recent years, cloud data have been widely used in food safety supervision. A challenging issue to address is how to integrate the rapid detection data into the traceability framework. Therefore, the data, network, and query results of rapid detection products are therefore the key guidelines for their development. We intend to build a workstation with on-site data collection and backend cloud computing (Fig. 9.31). The detection data of agro-product or food collected by portable nondestructive testing instrument (such as NIRS, E-nose, electronic tongue, CSA, etc.), the site is transmitted easily in 5G form to the backstage, and the database will be named to find matching models and rapidly generate qualitative or quantitative discrimination results. The output results give an instruction that the detected the agro-product or food is to be consumed or discarded.

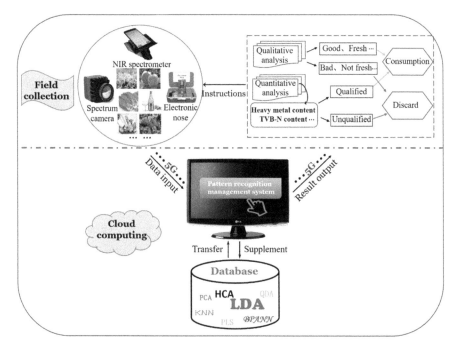

Fig. 9.31 A workstation based on on-site data collection and backend cloud computing

References

1. Moreno-Martinez E, Jiménez S, Vázquez ME (2000) Effect of *Sitophilus zeamais* and *Aspergillus chevalieri* on the oxygen level in maize stored hermetically. J Stored Prod Res 36 (1):25–36
2. Nishi SM, Viero LM, Soares RM et al (2000) Clinical microbiology reviews. Rev Bras Parasitol Vet 14(2):296–326
3. Zhang D, Li P, Zhang Q et al (2009) Production of ultrasensitive generic monoclonal antibodies against major aflatoxins using a modified two-step screening procedure. Anal Chim Acta 636 (1):63–69
4. Jayas DS (2012) Storing grains for food security and sustainability. Agric Res 1(1):21–24
5. Shahin MA, Symons SJ (2011) Detection of Fusarium damaged kernels in Canada Western Red Spring Wheat using visible/near-infrared hyperspectral imaging and pricinal component analysis. Comput Electron Agric 75(1):107–112
6. Zhang H, Paliwal J, Jayas DS et al (2007) Classification of fungal infected wheat kernels using near-infrared reflectance hyperspectral imaging and support vetor machine. Trans ASABE 50 (5):1779–1785
7. Cheng J, Dai Q, Sun D et al (2013) Applications of non-destructive spectroscopic techniques for fish quality and safety evaluation and inspection. Trends Food Sci Technol 34:18–31
8. Du C, Sun D (2014) Recent developments in the applications of image processing techniques for food quality evaluation. Trends Food Sci Technol 15:230–249

9. Elmasry G, Kamruzzaman M, Sun D et al (2012) Principles and applications of hyperspectral imaging in quality evaluation of agro-food products: a review. Crit Rev Food Sci Nutr 52:999–1023
10. Gowen AA, O'Donnell CP, Cullen PJ et al (2008) Hyperspectral imaging—an emerging process analytical tool for food quality and safety control. Trends Food Sci Technol 18:590–598
11. Chen Q, Liu A, Zhao J (2013) Classification of tea category using a portable electronic nose based on an odor imaging sensor array. J Pharm Biomed Anal 84:77–83
12. Chen Q, Liu A, Zhao J et al (2013) Monitoring vinegar acetic fermentation using a CSA array. Sens Actuators B Chem 183:608–616
13. Kondo N (2009) Robotization in fruit grading system. Sens Instrumen Food Qual 3:81–87
14. Liu C, Liu W, Lu X et al (2004) Application of multispectral imaging to determine quality attributes and ripeness stage in strawberry fruit. PLoS One 9(2):e87818
15. Qin J, Chao K, Kim MS et al (2013) Hyperspectral and multispectral imaging for evaluating food safety and quality. J Food Eng 118:157–171
16. Lin J, Puri VM, Anantheswaran RC (1995) Measurement of eggshell thermal-mechanical properties. Trans ASAE 38(6):1769–1776
17. Gulich A, Fitzgarald J (1964) USDA egg grading manual, no. 75. Agricultural Marketing Service, Washington, DC
18. Patel V, McClendon R, Goodrum J (1998) Color computer vision and artificial neural networks for the detection of defects in poultry eggs. Artif Intell Rev 12(1):163–176
19. Dehrouyeh M, Omid M, Ahmadi H et al (2010) Grading and quality inspection of defected eggs using machine vision. Int J Adv Sci Technol 17:23–31
20. Garcia-Alegre MC, Ribeiro A, Guinea D et al (2000) Eggshell defects detection based on color processing. Mach Vision Appl Ind Inspection VIII 3966:280–287
21. Li H, Kutsanedzie F, Zhao J et al (2016) Quantifying total viable count in pork meat using combined hyperspectral imaging and artificial olfaction techniques. Food Anal Methods 9:3015–3024
22. Chen Y, Liu B, Chang Y (2010) Bioactivities and sensory evaluation of Pu-erh teas made from three tea leaves in an improved pile fermentation process. J Biosci Bioeng 109(6):557–563
23. Zhang W, Yang R, Fang W et al (2016) Characterization of thermophilic fungal community associated with pile fermentation of Pu-erh tea. Int J Food Microbiol 227:29–33
24. Zhang L, Li N, Ma Z et al (2011) Comparison of the chemical constituents of aged Pu-erh tea, ripened Pu-erh tea, and other teas using HPLC-DAD-ESI-MSn. J Agric Food Chem 59 (16):8754–8760
25. Gao L, Bian M, Mi R et al (2016) Quality identification and evaluation of Pu-erh tea of different grade levels and various ages through sensory evaluation and instrumental analysis. Int J Food Sci Technol 51:1338–1348
26. Ye J, Wang W, Ho C et al (2016) Differentiation of two types of Pu-erh teas by using an electronic nose and ultrasound assisted extraction dispersive liquid-liquid microextraction gas chromatography mass spectrometry. Anal Methods 8:593–604
27. Zhu Y, Luo Y, Wang P et al (2016) Simultaneous determination of free amino acids in Pu-erh tea and their changes during fermentation. Food Chem 194:643–649
28. Chen C, Chan H, Chang Y et al (2009) Effects of bacterial strains on sensory quality of Pu-erh tea in an improved pile-fermentation process. J Sens Stud 24:534–553
29. Zhou B, Ma C, Ren X et al (2000) LC-MS/MS-based metabolomic analysis of caffeine-degrading fungus Aspergillus sydowii during tea fermentation. J Food Sci 85(2):477–485
30. Wang C, Zhang C, Shao C et al (2016) Chemical fingerprint analysis for the quality evaluation of deepure instant Pu-erh tea by HPLC combined with chemometrics. Food Anal Methods 9:3298–3309
31. Yi T, Zhu L, Peng W et al (2015) Comparison of ten major constituents in seven types of processed tea using HPLC-DAD-MS followed by principal component and hierarchical cluster analysis. LWT-Food Sci Technol 62:194–201

32. Cen H, Bao Y, He Y (2006) Pattern recognition of visible and near-infrared spectroscopy from bayberry juice by use of partial least squares and a backpropagation neural network. Opt Appl 45:7679–7683

33. Burns DA, Ciurczak EW (2007) Handbook of near-infrared analysis. CRC Press, Boca Raton

34. Sheng, R., Cheng, W., Li, H., et al. Model development for soluble solids and lycopene contents of cherry tomato at different temperatures using near-infrared spectroscopy. Postharvest Biology and Technology, 156, (2019):110952.

35. Zhao J, Lin H, Chen Q et al (2010) Identification of egg's freshness using NIR and support vector data description. J Food Eng 98(4):408–414

36. Jon RA, Zheng L, LaGasse MK et al (2016) An optoelectronic nose for identification of explosives. Chem Sci 7:199–206

37. Rakow NA, Suslick KS (2000) A CSA array for odor visualization. Nature 17:710–713

38. Askim JR, Suslick KS (2015) Hand-held reader for CSA arrays. Anal Chem 87:7810–7816

39. Zareef M, Chen Q, Hassan MM et al (2020) An overview on the applications of typical non-linear algorithms coupled with NIR spectroscopy in food analysis. Food Eng Rev 12:173–190

40. Lin H, Zhao J, Sun L et al (2011) Freshness measurement of eggs using near infrared (NIR) spectroscopy and multivariate data analysis. Innovative Food Sci Emerg Technol 12 (2):182–186

41. Nordey T, Joas J, Davrieux F et al (2017) Robust NIRS models for non-destructive prediction of mango internal quality. Sci Hortic 216:51–57

42. Malegori C, Nascimento M, Emanuel J et al (2017) Comparing the analytical performances of micro-NIR and FT-NIR spectrometers in the evaluation of acerola fruit quality, using PLS and SVM regression algorithms. Talanta 165:112–116

43. Kim JG, Park Y, Shin MH et al (2018) Application of NIR-spectroscopy to predict the harvesting maturity, fruit ripening and storage ability of ca-chitosan treated baby kiwifruit. J Stored Prod Postharvest Res 9(4):44–53

44. Lin H, Zhao J, Chen Q (2010) Eggshell crack detection based on acoustic impulse response combined with kernel independent component analysis and back propagation neural network. Intelligent Autom Soft Comput 16(6):1043–1050

45. Sun L, Bi X, Lin H et al (2013) On-line detection of eggshell crack based on acoustic resonance analysis. J Food Eng 116(1):240–245

46. Sun L, Cai J, Lin H et al (2013) On-line estimation of eggshell strength based on acoustic impulse response analysis. Innovative Food Sci Emerg Technol 18:220–225

47. Lin H, Zhao J, Chen Q et al (2009) Eggshell crack detection based on acoustic response and support vector data description algorithm. Eur Food Res Technol 230(1):95–100

48. Lin H, Zhao J, Chen Q et al (2009) Eggshell crack detection based on acoustic impulse response and supervised pattern recognition. Czech J Food Sci 27(6):393–402

49. Lin H, Zhao J, Sun L et al (2011) Stiffness measurement of eggshell by acoustic resonance and PLS models. J Food Eng 103(4):351–356

50. Goodrum J, Elster R (1992) Machine vision for crack detection in rotating eggs. Trans ASAE 35:1323–1328

51. Cho H, Kwon Y (1996) Crack detection in eggs by machine vision. Trans ASAE 39(3):777–784

52. Jenshin L, Lin Y, Hsieh M (2001) An automatic system for eggshell quality monitoring. Trans ASAE 44(3):1028–1034

53. Lawrence KC, Yoon SC, Jones DR et al (2009) Modified pressure system for imaging egg cracks. Trans ASABE 52:983–990

54. Suslick KS, Neal A, Rakow AS (2004) CSA arrays for molecular recognition. Tetrahedron 60 (49):11133–11138

55. Radi CS, Litananda WS, Rivai M et al (2016) Electronic nose based on partition column integrated with gas sensor for fruit identification and classification. Comput Electron Agric 121:429–435

56. Deshmukh S, Bandyopadhyay R, Bhattacharyya N et al (2015) Application of electronic nose for industrial odors and gaseous emissions measurement and monitoring—an overview. Talanta 144:329–340

57. Jiang H, Xu W, Chen Q (2019) Evaluating aroma quality of black tea by an olfactory visualization system: selection of feature sensor using particle swarm optimization. Food Res Int 126:108605

58. Chen Q, Sun C, Ouyang Q et al (2014) Classification of vinegar with different marked ages using olfactory sensors and gustatory sensors. Anal Methods 6(24):9783–9790

59. Chen Q, Li H, Ouyang Q et al (2014) Identification of spoilage bacteria using a simple CSA array. Sens Actuators B Chem 205:1–8

60. Gu H, Huang X, Chen Q et al (2019) Prediction of the property of CSA Array based on density functional theory. Sens Mater 31(10):3067–3073

61. Huang X, Xin J, Zhao J (2011) A novel technique for rapid evaluation of fish freshness using CSA array. J Food Eng 105(4):632–637

62. Huang X, Zou X, Shi J et al (2014) Determination of pork spoilage by colorimetric gas sensor array based on natural pigments. Food Chem 145:549–554

63. Lin H, Man Z, Guan B et al (2017) In situ quantification of volatile ethanol in complex components based on CSA array. Anal Methods 40:5873–5879

64. Lin H, Man Z et al (2018) A novel CSA array based on boron-dipyrromethene dyes for monitoring the storage time of rice. Food Chem 268:300–306

65. Lin H, Kang W, Kutsanedzie FYH et al (2019) A novel nanoscaled chemo dye–based sensor for the identification of volatile organic compounds during the mildewing process of stored wheat. Food Anal Methods 12(12):2895–2907

66. Lin H, Duan Y, Yan S (2019) Quantitative analysis of volatile organic compound using novel chemoselective response dye based on Vis-NIRS coupled Si-PLS. Microchem J 145:1119–1128

67. Lin H, Yan S, Song B (2019) Discrimination of aged rice using CSA array combined with volatile organic compounds. J Food Process Eng 42:13037

68. Lin H, Wang Z (2000) Identification of rice storage time based on CSA array combined hyperspectral imaging technology. J Stored Res Prod 85:101523

69. Lin H, Kang W, Jin H et al (2019) Discrimination of Chinese Baijiu grades based on CSA arrays. Food Sci Biotechnol 39:1039–1043

70. Ouyang Q, Zhao J, Chen Q et al (2013) Classification of rice wine according to different marked ages using a novel artificial olfactory technique based on CSA array. Food Chem 138 (2–3):1320–1324

71. Salinas Y, Ros-Lis JV, Vivancos JL et al (2014) A novel CSA array for monitoring fresh pork sausages spoilage. Food Control 35(1):166–176

72. Grossi M, Di LG, Arru M et al (2015) An opto-electronic system for in-situ determination of peroxide value and total phenol content in olive oil. J Food Eng 146:1–7

73. Ouyang Q, Zhao J, Chen Q (2013) Classification of rice wine according to different marked ages using a portable multi-electrode electronic tongue coupled with multivariate analysis. Food Res Int 51(2):633–640

74. Xu L, Yu X, Liu L et al (2016) A novel method for qualitative analysis of edible oil oxidation using an electronic nose. Food Chem 202:229–235

75. Zhang L, Tian F, Pei G (2014) A novel sensor selection using pattern recognition in electronic nose. Measurement 5:31–39

76. Banerjee R, Tudu B, Shaw L et al (2012) Instrumental testing of tea by combining the responses of electronic nose and tongue. J Food Eng 110(3):356–363

77. Paolesse R, Alimelli A, Martinelli E et al (2006) Detection of fungal contamination of cereal grain samples by an electronic nose. Sens Actuators B Chem 119(2):425–430

78. Sung J, Kim BK, Kim BS et al (2014) Mass spectrometry-based electric nose system for assessing rice quality during storage at different temperatures. J Stored Prod Res 59:204–208

Chapter 10
Nondestructive Detection Technologies for Real-Time Monitoring Food Quality During Processing

Abbreviation	Full name
Agro-product	Agricultural product
Bp-ANN	Back-propagation artificial neural network
CSA	Colorimetric sensor array
DC resistance	Direct current resistance
DSP	Digital signal process
ELM	Extreme learning machine
E-nose	Electronic nose
GC-MS	Gas chromatography-mass spectrometry
HSI	Hue-saturation-intensity
HPLC	High-performance liquid chromatography
NIRS	Near-infrared spectrometer
SG	Savitzky–Golay
SSC	Soluble solid content
TAC	Total acid content
TFs	Theaflavins
TRs	Thearubigins
PLC	Programmable logic controller
PLS	Partial least squares
Vis-NIRS	Visible near-infrared spectrometer

10.1 Introduction

When agricultural products (agro-products) and food have been purchased and processed, they are offered for sale after commercialization or further processing in the area of primary processing. Information technology, automation technology, and smart development have gradually taken hold in the agro-products circulation,

Q. Chen et al., *Advanced Nondestructive Detection Technologies in Food*,
https://doi.org/10.1007/978-981-16-3360-7_10

primary processing testing equipment and classification of agro-products and have attained advancement and perfection. These advanced technologies have been utilized in cleaning, grading, packaging, drying, livestock, and poultry slaughtering process as processing equipment to enhance commercialization, market value, and the market competitiveness. In addition, during the food processing, some important physical and chemical indicators are constantly changing, which needs to be constantly monitored via regulating the changes in indicators. However, due to the difference of raw materials, processing season, environmental conditions, and other factors, variations exist between different batches of food. The storage period and conditions (container, temperature difference, humidity, and ventilation) induce changes in the quality, and alters volatile substance present. Therefore, in order to ensure consistency in quality inspection of these batches in food processing, some important indicators needs to be monitored in real time. Hence, based on the current problems faced in the area of quality control, this chapter specifically elaborates the application of online monitoring technology in the primary processing of agro-products, solid-state fermentation food processing, dairy products, and meat products processing. Commercially processed agro-products are mainly concentrated in eggs, fruits and vegetables, meat products, dairy products, etc. Nevertheless, the quality and nutritional values are still the main concern of agro-products because of their necessity and extensive sources. Therefore, besides the need of processing techniques such as drying [1, 2] cooling [3–6], and freezing [7] to maintain their quality, there are also needs for smart and onsite methods to monitor these agro-products processing.

10.1.1 Quality Authentication of Eggs

The quality monitoring of eggs could be divided into external and internal quality. The external quality mainly includes the shape, dimensions, color, purity, and weight of the eggshell. The internal quality includes the freshness of eggs, blood spot, and the rate of meat spot [8]. European Union, American, and Japanese standards are the three major standards in the quality detection and grading of eggs. According to the outline size, weight, surface defects of eggshell (crack, stain area), egg core color, blood spot and meat spots, and other indicators are graded. For example, in the American poultry egg quality grading standard, eggs are graded according to indicators such as eggshell, albumen, egg yolk, air chamber, and embryo. Qualified eggs are classified into three levels: AA, A, and B. Unqualified eggs are classified as Class C [9, 10]. The sanitary standards of fresh eggs in China mainly set requirements for the sensory indicators, physical and chemical indicators of fresh eggs, and stipulate eggshells, albumen, egg yolks, foreign bodies, Huff units, weight, etc. In the process of detection and grading of eggs, the appearance quality of eggs is mainly detected and graded by machine vision technology, and structural stiffness such as cracks and eggshell hardness of eggs are detected and graded by vibration mechanics analysis methods. Near-infrared spectrometer (NIRS) rapidly inspects quality authentication of the freshness of eggs and other internal and external quality indicators, establish light, sound, and electrical transmission characteristics and to achieve rapid detection and sorting of eggs.

10.1.2   Quality Classification of Fruits and Vegetables

After the fruits and vegetables are picked and purchased, they undergo a series of preliminary commercialization treatments and classifications. Commercial processing includes rapid weighing of fruits, size grading, color grading, and maturity discrimination, nondestructive testing of soluble solids (such as sugar content), hardness testing, and male and female grading, etc. Light, sound, and other signals are usually used in the projection of the appearance of the fruit or the transmission in the tissue. Based on that, machine vision technology, near-infrared spectroscopy technology, pressure sensor signal analysis technology quickly detects the internal and external quality of the fruit. Therefore, it establishes the correlation model of the light and sound transmission characteristics and its internal and external quality indicators, and realizes the rapid detection and classification of the fruit's initial processing [11–13].

10.1.3   Quality Assessment for Grain

The higher water content of harvested wheat, rice, maize, and other agro-products is not conducive to maintaining their quality. Therefore, the newly harvested grain and oil agro-products need to be dried immediately. High-temperature hot-air drying is the most widely used drying method. Different control conditions during the drying process influence the quality of grain, oil, and agro-products with influence on moisture content, waist burst rate, whole rice rate, fatty acid value, germination rate, etc. Using high heating speed, and elongated drying time, for too long can lead to increased burst and fatty acid value, decreased germination rate, moisture content, and increased drying cost. Therefore, during the drying process, it is necessary to monitor the temperature and the moisture content of the agro-products to ensure that the drying process is scientifically driven, according to the design, and to maintain the original quality of the grain, oil, and agro-products to the greatest extent [14]. NIRS microwave absorption, direct current resistance (DC resistance), and capacitance methods could be used for rapid detection of grain, oil, and agro-products. Among them, NIRS is more suitable for online monitoring in the drying process. NIRS technology uses the principle that water molecules could absorb infrared light at a certain frequency for quality assessment and has been popularized and applied in rapid grain detection in China, the United States, and Canada.

10.1.4   Quality Evaluation of Fermented Food

Flavored food (such as wine, vinegar) undergoes a series of component changes during solid-state fermentation. It is vital for the quality to routinely monitor the components during food processing to ensure its safety. Alcoholic food is generally brewed with rice, millet as the main raw material, through cooking, koji, saccharification, fermentation, pressing, filtering, decoction (sterilization), storage, blending, and other processes. Take rice wine as an example, the brewing of rice wine is a process of saccharification and fermentation. A small amount of organic acid is formed as a result of the yeast fermentation and conversion of sugar into alcohol. At the same time along with the lactobacillus fermented sugar conversion into organic acids, the decomposition of proteins and lipids results via a complex process of biochemical changes [15, 16]. The changes in total sugar, alcohol content, acidity, and other content in the fermented mash, as well as environmental factors such as temperature and oxygen, reflect the dynamics of the entire fermentation process of rice wine. Some essential physical and chemical indicators need to be constantly monitored throughout the rice wine fermentation process. Through machine vision, NIRS, and biomimetic sensing technology, the indicators of rice wine fermentation process can be monitored to optimize the quality of brewed rice wine. The various methods of fermentation of acetic acid could divide the vinegar fermentation process into fermentation in the solid state and fermentation in the liquid state. The solid-state fermentation process is a process in which multiple strains participate in the fermentation. It is also the conventional vinegar brewing method in China. Solid-state fermentation of vinegar generally uses glutinous rice, sorghum as the main raw materials. After alcohol fermentation, the raw materials of vinegar are steamed to gelatinize the starch; then under the action of glucoamylase, the gelatinized starch is decomposed into glucose. Then, glucose produces alcohol under the action of yeast fermentation to provide precursor material for acetic acid fermentation, followed by fermentation. The fermentation of acetic acid mainly relies on the action of acetic acid bacteria to oxidize ethanol to acetic acid, and then seal the mash and vinegar to obtain finished vinegar. In the process of vinegar brewing, both alcohol and acetic acid are volatile, unless not properly managed, the wine vinegar will be lost, in case when alcohol is exhausted, the bacteria will further oxidize the acetic acid into carbon dioxide and water, resulting in a decline in vinegar production. Through NIRS and colorimetric imaging technology, the composition changes and volatile substances in the vinegar fermentation process can be smartly evaluated to ensure the quality of brewed vinegar.

10.1.5   Quality Analysis of Tea Products

The tea processing goes through multiple procedures in real-time, which is useful for ensuring product quality and stability, reducing labor costs and production energy consumption, and improving the automation and intelligence level of the tea processing. Take black tea as an example. Black tea is a fermented tea made with suitable fresh tea leaves as a base material, through conventional processes such as withering, kneading (twisting), fermentation, and baking, and has special qualities such as color, fragrance, and taste. It is famous for its red soup color and leaves after brewing, and it forms a large amount of active ingredients during processing. Fermentation is a key step in the production of black tea [17, 18]. In fermentation, various biochemical reactions occur on the surface or inside the leaves that damage the leaves by the action of large number of active enzymes, producing new functional substances, and leading significant changes in the outer color of the leaves, gradually changing from green to yellow, red-yellow, red and dark red. The essence is the decomposition of chlorophyll components in vacuoles, and the oxidization of tea polyphenols and other components into theaflavins (TFs) and thearubigins (TRs) and other colored substances. Conventional black tea processing companies mainly rely on experienced tea masters to judge the degree of fermentation and make rough perceptual judgments by observing the changes in leaf color. Although this method is simple and effective, the judgment result is easily disturbed by environmental factors, and the random error is large, which makes it difficult to guarantee the quality of the finished black tea, and is not suitable for industrial online mass production. Therefore, it is important to control the quality and safety of black tea using the machine vision and spectrophotometry to monitor changes in the color components during the fermentation process of tea leaves [19, 20].

10.1.6   Quality Assessment of Meat Products

Livestock and poultry meat are usually processed as the main raw material into various meat products, such as sausages and cured meats. The fermentation of conventional sausages mainly includes three stages, namely preparation stage, fermentation stage, and maturation stage. The key steps of the preparation stage are to control the pH of the lean meat used to make sausages to be low (5.4–5.8) and ensure that the fat portion is as large as possible to choose strong fat meat, such as pig back fat, and control the ambient temperature of minced meat and enema at a low temperature. The fermentation stage mainly allows lactic acid bacteria to rapidly multiply under suitable temperature and humidity, decompose carbonated hydrate to produce lactic acid, decrease the pH. In the mature stage, the protein gel is enzymatically hydrolyzed, and organic aldehydes, esters, ketones, amino acids, small peptides, and other substances increase to form the special flavor of sausage. In summary, the production process of sausages is accompanied by changes in pH, protein, fatty acids, amino acids, and other substances [21]. Therefore, according to the absorption characteristics of these substances in the NIR-visible light spectrum, the processing could be controlled by the use of NIR-visible light or hyperspectral technologies. Real-time assessment of the types and contents of organic substances is used in the chemical industry.

10.1.7　Quality Detection for Dairy Products

Dairy products are rich in protein and essential amino acids and are popular among consumers, including liquid milk, milk powder, condensed milk, cheese, and other dairy products. Taking yogurt as an example, the raw milk is cleaned, decontaminated, and standardized. The solid content in yogurt is 12.0–12.5%, and the fat and protein content is 3.0–3.5%; it is produced via mixing and removing gas homogenization, sterilization, cooling, adding fermentation bacteria, fermentation, demulsification, and cooling. During the fermentation process, protein, fat, and lactose are hydrolyzed and decomposed, respectively. Protein is gradually decomposed into polypeptides and amino acids, and fat is hydrolyzed into glycerin and fatty acids. Lactose is metabolized by bacteria into fructose, pyruvate, and lactic acid followed by aseptic canning, product cooling, storage, and transportation. During the production process, the quality and physical properties are the important safety monitoring indicators of yogurt, such as amino acid content, fatty acid content, lactic acid content, acidity, fragrance, viscosity. These indicators in processes could be detected rapidly through spectroscopy technology, electronic nose technology, and machine vision technology.

10.2　Instrumentation

10.2.1　On-Site Assessment Devices Based on NIR

At present, NIR is among the most widely used technology with advanced instrumentation. Infrared rays with a wavelength of 0.8–2.5 m are commonly referred to as near-infrared. The NIR spectroscopy method uses statistical methods and computer technology to analyze the composition and physical and chemical properties of the sample in the near infrared region. The principle is based on the interaction of these near-infrared radiation with the sample, where part of it is reflected and part absorbed. It is possible to determine the composition's content associated and dependent on particular absorption wavelength. By detecting the specific absorption wavelength associated with the composition, the composition content can be calculated.

For example, in the late 1980s, Mitsui Metal Corporation of Japan developed an online acidity detection device for citrus fruit, which is mainly composed of light source, optical sensor, and data processing. With this device, the sugar acidity can be obtained without any damage to the citrus fruit. The schematic diagram of the device is shown in Fig. 10.1.

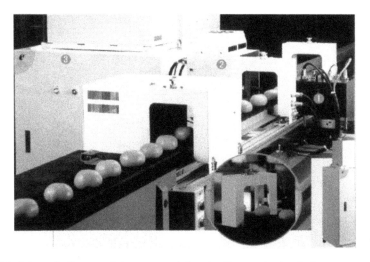

Fig. 10.1 Schematic diagram of citrus at near-infrared online monitoring device

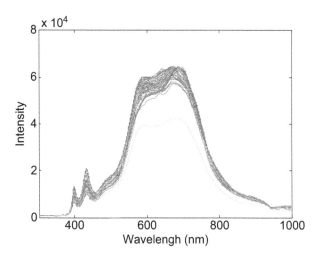

Fig. 10.2 Portable spectrum detection system

NIR online devices for monitoring the rice wine is mainly composed of near infrared detection box body, near infrared spectrometers, light sources, peristaltic pump, optical fiber, hoses, Siemens programmable logic controller (PLC), metal switch, miniature solenoid valve, tee joint, flushing cistern, cistern, liquid pool, power board, and computer [15, 16]. Figure 10.2 shows the spectrum of the mash sample in the rice wine fermentation process by portable visible near-infrared spectrometer (visible-NIRS) detection system. The test control area includes peristaltic pump, PLC, solenoid valve, relay, tee pipe, hose, and system switch. The software composition of the system is shown in Fig. 10.3. In order to meet market

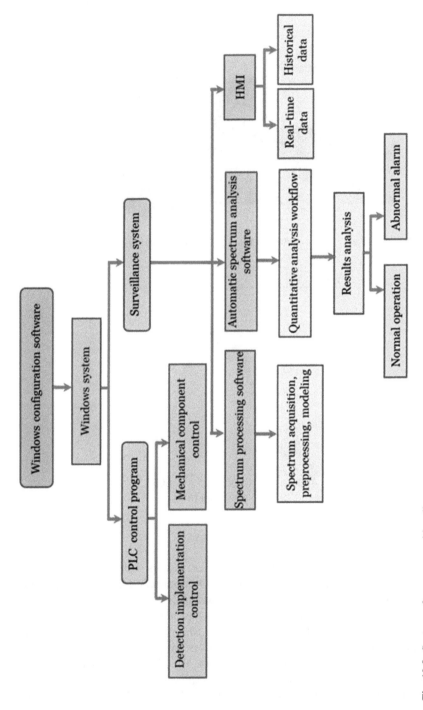

Fig. 10.3 System software composition diagram

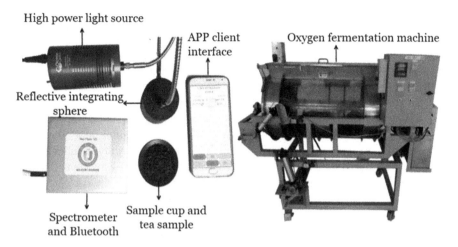

High power light source

APP client interface

Oxygen fermentation machine

Reflective integrating sphere

Spectrometer and Bluetooth

Sample cup and tea sample

Fig. 10.4 Physical image of online monitoring of black tea fermented leaf color

demand, some commercial portable NIR analyzers were developed. The current mainstream products include the Phazir Rx series from Thermo Fisher Scientific Company of America, the SupNIR-1000 series from Concentrator Technology (Hangzhou) Co., Ltd., and the FT20 and FQA-NIR Gun from FANTEC Company of Japan.

The system includes the following five parts: (1) light source (halogen lamp & LED, Taiwan OTO Optoelectronics); (2) optical fiber adjustable attenuator (FVA-UV, Shanghai Fuxiang Instrument Co., Ltd.); (3) transmission module (adjustable optical path, Hangzhou Jingfei Technology Co., Ltd.); (4) micro spectrometer (SD1200, Taiwan OTO Optoelectronics Co., Ltd.); (5) computer. The optical fiber tunable attenuator and the transmission module, as well as the transmission module and the spectrometer are connected by an optical fiber (Hangzhou Jingfei Technology Co., Ltd.). A quartz cuvette with a light path of 5 mm was selected as the sample pool for collecting the spectrum of samples during the fermentation process of rice wine. During spectrum acquisition, the integration time is set to 50 ms, the number of acquisitions is 5, and 11-point Savitzky–Golay (SG) smoothing preprocessing is used to express the spectrum value of the sample by light intensity.

Color has been applied in the food industry to determine the quality of different products, as color has come to be a significant marker of product quality. Using the principle of spectroscopic color measurement, combined with NIRS, the reflectance spectra of fermented leaves is collected by a monitoring system as shown in Fig. 10.4. The spectral data of the blade surface were collected, and the reflectance was calculated. This method uses optical fiber to realize long-distance monitoring of the color change during the fermentation of black tea. The indirect contact between the optical sensor and the high temperature and high humidity environment ensures the accuracy and stability of the monitoring results.

10.2.2   On-Site Assessment System Based on Electronic Nose

At present, the commercial devices that are widely used in the field of quality evaluation in the processing of agro-products mainly include PEN3 from AIRSENSE Company in Germany, SuperNose from ISENSO Company of America, and Bosin CNose Company from Shanghai. In response to market needs, optimizing the sensor and hardware design and developing handheld products are actively explored topics in the electronic nose (E-nose) research. The online odor monitoring device of E-nose for heating and drying fruits and vegetables mainly includes E-nose of quartz crystal oscillator, condenser, control circuit, microwave oven, Teflon container, gas storage tank, and other devices. The flowchart of the specific device is shown in Fig. 10.5.

10.2.3   On-Site Assessment System Based on Machine Vision

In view of the fact that machine vision technology has been shown to be an efficient method of eliminating human error and increasing quality detection and classification accuracy, as well as increasing production efficiency in egg quality examination. Therefore, a set of detection devices based on machine vision technology were developed aiming at the indicators of egg shape index, weight and surface cleanliness of eggshell [22, 23]. The device mainly includes a camera, a computer, a light box, a poultry and egg conveying line, an infrared trigger device, and a USB to I/O circuit. Its 3D rendering is shown in Fig. 10.6. The egg conveying line drives the egg

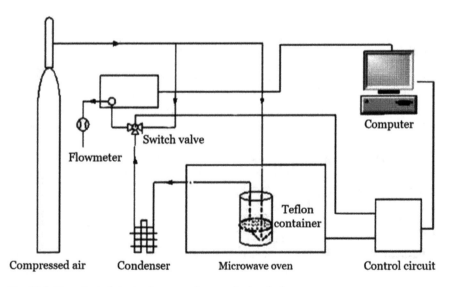

Fig. 10.5 Flowchart of electronic nose online monitoring device

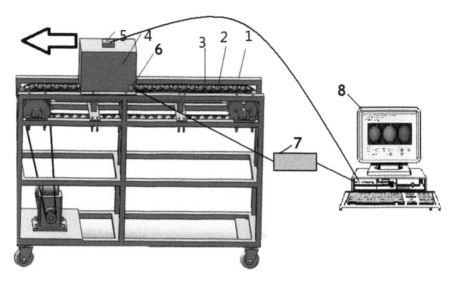

Fig. 10.6 Machine vision system for on-line inspection of external quality of poultry eggs. 1. Frame; 2. Eggs; 3. Nylon roller; 4. Light box; 5. The camera; 6. Infrared optoelectric switch; 7.USB to I/O circuit; 8. Computer

rolling forward. When the egg passes through the detection station, the infrared trigger device sends out a high level signal to the computer. The visual system software triggers the camera to shoot the egg image and extract the required characteristic parameters. The main frequency of the computer is 2.20 GHz. The camera model is Lu075C from Lumenera, USA. The lens is produced from Computar, Japan. The focal length is 8 mm, the exposure time is set to 7 ms, and the image size taken is 640 × 480. The infrared optoelectronic switch, model OA-D3224PA, adopts the opposite-type optoelectronic switch produced by *Germany Julong Group*. The commercial version of the USB interface is 8-way digital IO input and output module, and its main chip model is CH341.

The smart detection device for poultry eggs mainly includes a camera, a computer, and a light box, a production line for dynamic detection of poultry eggs, an infrared trigger device, and a USB to I/O circuit. The dynamic egg production line drives the egg to roll forward. When the egg passes the infrared triggering device station, it sends a high-level signal to the computer and triggers the camera to take the image of the egg through the vision system software. Real-time digital image processing is used to extract the necessary characteristic parameters from the egg's image.

The software system mainly completes image collection, processing, feature extraction, and feature data storage. The software system interface is shown in Fig. 10.7, which mainly includes user login module, equipment test module, image acquisition and display module, detection parameter display module, and database module. In the software, the infrared signal is detected by the timer interrupt function. When the signal is detected as a rising edge, that is, the egg has just entered

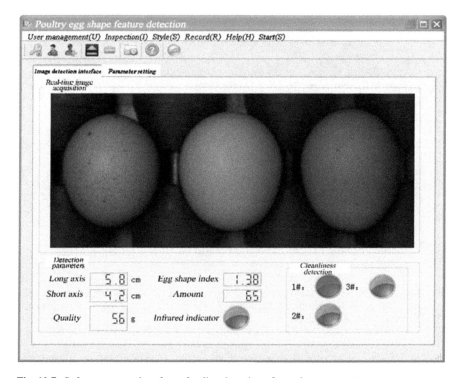

Fig. 10.7 Software system interface of online detection of egg shape parameters

the range of the infrared photoelectric switch, the camera is turned on through the software trigger method for image acquisition, and the outside of the egg is extracted through the digital image processing method features, and use the database to store the feature parameters for easy reference and analysis.

10.2.4   On-Site Assessment System Based on X-Ray Imaging

X-ray has a strong penetrating ability, and its images can directly reflect the internal defects, structural changes, and other quality conditions of food and agro-products. X-rays can be used to detect defects such as damage, decay, and water heart in apples, holes in potatoes and watermelons, wrinkled skin in oranges, and pests and diseases in agro-products. Therefore, someone has built an online X-ray monitoring device based on X-ray, which consists of X-ray generator, X-ray detector, mechanical transmission device, X-ray protection device, image acquisition card, master computer, image acquisition, processing software, etc. The experimental device for X-ray imaging detection is shown in Fig. 10.8.

Among them, (1) the X-ray generator is mainly used to generate and control the energy required by the test sample. It consists of a high-pressure generator, an X-ray

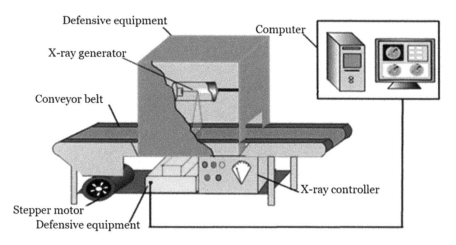

Fig. 10.8 Schematic diagram of X-ray imaging detection experimental device

tube, and an X-ray controller. Among them, the X-ray tube is the core of the X-ray source. Its basic structure is a high vacuum diode, which consists of a cathode, an anode, and a glass shell to maintain a high vacuum. (2) The X-ray detector is mainly used to receive the X-ray of test sample, the screen will receive X-ray into visible light, the photosensitive diode detector unit under visible light irradiation can produce a voltage signal, the signal through the processing of integrated circuit into a digital signal and transmitted to the computer. (3) The PCI slot on the motherboard of the master computer of the image acquisition card communicates with the master computer through PCI bus. (4) The part of the ray protection device has been calculated theoretically and must fully comply with the latest national ray protection standards. Except for the import and export part of the samples to be tested, the rest are made of 3-mm composite lead plate as the lining protection material, and the support part is made of 2-mm steel plate. (5) The transmission device consists of three parts: transmission belt, variable frequency three-phase asynchronous motor, and frequency converter. The online monitoring of general agro-products adopts two transmission modes: roller type and belt type. For the requirement of X-ray stable imaging, the band transmission method is adopted. The motor drives the belt to move, and the transmission speed is controlled by the frequency converter. The conveyor belt adopts food-grade packaging plastic as the sample contact surface. (6) Flip-flop. In this case, e3X-DA11-N digital fiber amplifier manufactured by Omron of Japan is used.

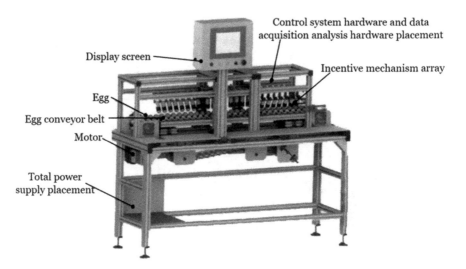

Fig. 10.9 Schematic diagram of mechanical structure of online crack detection system for poultry eggs

10.2.5 On-Site Assessment System Based on Acoustic Technology

Schematic diagram of the mechanical structure is shown in Fig. 10.9, which mainly includes motor, egg conveying mechanism, power module, excitation mechanism array, control and signal acquisition hardware system and related sensors. The transmission chain is driven by a motor with a power of 180 W, whose model is OTG-61K180RGN-CF. The maximum speed of the conveyor chain can be up to about $6\,s^{-1}$. The supporting roller, which constitutes the conveying mechanism, rolls while advancing, and roll the eggs placed between the two rollers. The excitation mechanism array is composed of 24 identical excitation mechanisms, which run along the direction of the conveyor chain. The spacing of the excitation mechanism is the same as the wheelbase of adjacent rollers. According to the different percussion direction, the 24 incentive mechanisms were divided into 3 groups with 8 in each group [22, 23].

The hardware system for online detection of poultry egg cracks mainly includes a control system and a signal acquisition and analysis system. The hardware system framework is shown in Fig. 10.10. The control system includes a total control circuit, a drive circuit of 24 excitation mechanisms and various sensors, including photoelectric switch, proximity switch, encoder, and displacement sensor. Its main function is to detect the location of eggs by sensors, the presence of eggs, eggs size, and other important parameters, through the precise control of the incentive mechanism so that the device can be based on the size of each egg to make appropriate percussion action.

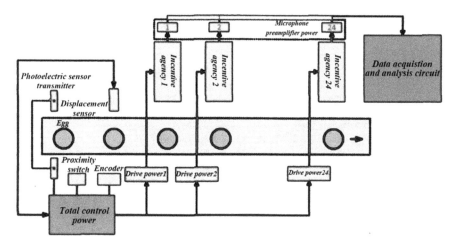

Fig. 10.10 Schematic diagram of the hardware system framework for online inspection of egg cracks

In addition, another scholar developed a set of eggshell quality online detection system based on the analysis of the response signal of the knock vibration [24]. The system can complete the automatic knocking, signal collection, and processing of the tested eggs. Herein, several factors influencing the online quality detection of eggs and eggshells are also analyzed. The quality detection model is established to realize the online smart detection of eggs and eggshell quality (cracks and strength of eggshells).

The schematic diagram of the online inspection system for the quality of poultry eggs and eggshells is shown in Fig. 10.11. The system consists of supporting roller, chain, engineering plastic platform, eggs, percussion sticks, infrared receiver, infrared transmitter, pickup, digital signal process (DSP), computer, etc.

The main operation process of the system is shown as follows: first, the egg rolls forward along with the production line. When the station is hit, the infrared photoelectricity switch will issue a trigger command to the DSP. The electromagnet is driven by the motor drive circuit to execute the knock action. The knock response audio signal generated by the knock is converted into an electrical signal through the pickup. The signal is collected by DSP after the signal conditioning circuit, then the signal is analyzed and processed by DSP software system, and the corresponding discrimination result is calculated. Through serial communication, the results will be uploaded to the upper computer for display, realizing human–computer interaction.

10.2.6 On-Site Assessment System Based on CSA

CSA is a kind of bionic sensor. At work, the smell sensor acts like an organism's olfactory cell, which will attach to the measured gas. Each sensor in the sensor array

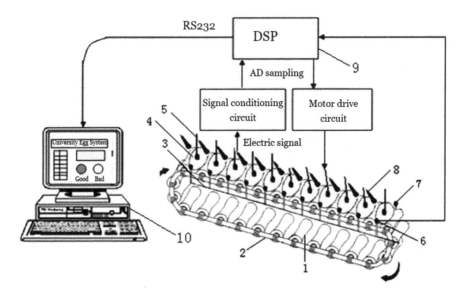

Fig. 10.11 Schematic diagram of online inspection of poultry eggshell quality (1. Supporting roller; 2. The chain; 3. Engineering plastic platform; 4. The eggs; Knock stick; 5. Infrared receiver; 6. Infrared receiver; 7. Infrared transmitter; 8. Microphone; 9. Digital signal process; 10. Computer)

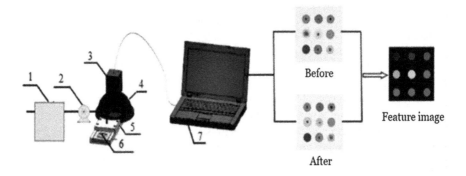

Fig. 10.12 Schematic diagram of portable odor sensor system (1. Gas collection chamber; 2. Vacuum pump; 3. Camera; 4. Light source; 5. Reaction chamber; 6. Color-sensitive sensor array; 7. Computer)

has different sensitivity to different gases, thus producing different response signals, and finally gas analysis is realized by means of pattern recognition. Portable color sensor system is mainly composed of gas collection chamber, vacuum pump, camera, light source reaction chamber, and computer. Figure 10.12 is a schematic diagram of portable odor sensor system [17].

The method of image acquisition is 3CCD camera. The prism in 3CCD camera can divide the light source into red, green, and blue primary colors, and there are three independent image sensors in 3CCD camera, which can process the red, green, and blue primary colors separately to improve the accuracy of color acquisition and image quality. In addition, the three CCD image sensors of 3CCD camera have large light acquisition areas, which makes the 3CCD camera to have a high signal-to-noise ratio, good sensitivity, and wide dynamic range compared with the CCD camera. The light source is opt-RID150 integrated spherical diffuse reflection LED light source, which acquires good image uniformity. Planar contact reaction gas chamber is selected to build the final system platform. The flow field of the reaction chamber is uniform and the flow rate is moderate, which is suitable for experimental operation. The main operation procedure is to place the sample in the gas collection chamber and collect the images of the CSA, before the reaction, by using the image acquisition device, when the gas collection is completed. After that, the vacuum pump is turned on to send the volatile organic gas into the reaction chamber, where the chemo-responsive material fully interacts with it. Finally, the CSA is obtained, and the pattern recognition is carried out by computer, so as to analyze the tested samples.

10.3 Online Sorting in Primary Processing of Agro-Products

Simple primary processing is typically required after the purchase of food products, including washing, grading, packing, drying, livestock and poultry slaughter, refrigeration, and preservation of food products. Common agro-products and agricultural by-products need to be classified during processing by various sensors. For example, Japanese used light sensors to measure the sugar content of fruits, and ranked them to ensure the taste and sweetness of the fruit. In addition, the sensors used in the food processing include humidity sensors, temperature sensors, and moisture sensors. The processing of grains, medicinal materials, and tea are inseparable from the various sensors mentioned above. Nondestructive testing technology can detect and classify the size, color, defect, and other quality of food products online to realize the commercialization and intensive processing of agro-products in the place of production, increase the value, and enhance the market competitiveness of food products.

10.3.1 Smart On-Site Assessment for Poultry Eggs Sorting

The long axis, short axis, egg shape index, weight, and surface defects of the egg are important characteristics of the egg quality, and it is also a main criterion for the grading of the egg. Since machine vision technology offers many advantages, such as high detection speed, an enormous amount of information, powerful functions, and easy application, it has been widely adopted. It has been commonly used in the

Table 10.1 Long- and short-axis models under different methods

Detection method	Index	Correlation coefficient R	RMSECV	F value	Regression model
Fitting ellipse	Long axis	0.9582	0.7103	1165.2886	$Y = 0.1995X + 10.8237$
	Short axis	0.9149	0.5078	533.9422	$Y = 0.1731X + 12.5835$
Rectangle method	Long axis	0.9584	0.7078	1174.3426	$Y = 0.1972X + 11.2863$
	Short axis	0.9074	0.5285	484.9213	$Y = 0.1689X + 13.2926$

identification of agricultural product shape parameters, as well as other visual effects through electronic perception and image comprehension. The application of machine vision technology to the quality inspection and grading of poultry eggs can effectively eliminate human subjective interference, quantitatively describe the detection indicators, reduce detection and grading errors, and improve grading accuracy and production efficiency. Therefore, the use of machine vision technology for nondestructive testing and sorting of poultry egg quality has high practical worth. V. Patel [24] employed machine vision technology to detect eggs with blood and stain spots, where the average accuracy of the two neural network models reached to 85.6% and 80.0%; Garcia-Alegre et al. [25] processed the RGB image of the egg through digital image processing and detected the shape and surface defects, which has a better detection accuracy and sorting efficiency. A nondestructive assessment system with machine vision recognition method was designed by Wen Youxian et al. [26] to detect duck egg size and center color. The established model yielded in a grading error equivalent to ± 3 g with a color grading accuracy over 90%. Dehrouyeh et al. [27] developed hue-saturation-intensity (HSI) of color space algorithm based on image processing to detect stains on the surface of eggs through machine vision. The detection rate of clean eggs in stain detection is 86%.

1. **Egg Size**

A one-variable linear regression model was established for the reference size of poultry eggs and the pixel size of image processing, and the model was verified by the prediction set samples. The form of the regression model and its regression coefficients are shown in Formula (10.1). The regression model, correlation, and significance index values established by MATLAB are shown in Table 10.1. By fitting ellipse method and rectangle method, the two methods have a good correlation between the long- and short-axis pixel values and the reference value. Moreover, the regression coefficients of the regression model established for the long axis or the short axis have very small deviations, indicating that the egg image sizes extracted by the two methods are relatively close.

$$Y = \alpha + \beta \cdot X \tag{10.1}$$

where

X—the pixel size value of the horizontal axis and the vertical axis of the image extraction.

Y—the reference value of the horizontal or vertical axis of the egg, in mm.

α, β—egg horizontal or vertical axis linear regression model coefficients.

2. **Egg Shape Index**

The egg shape index is a parameter used to describe the shape of an egg, which is defined as the ratio of the long axis of the egg to its short axis. Although the egg shape has no effect on the practical value of the egg, it is related to the seed value, hatching rate, and broken rate. The standard egg shape index of poultry eggs is between 1.3 and 1.35, the ones greater than 1.35 are elongated, and those less than 1.30 are approximately spherical, so the egg shape index is often used as an important standard for egg grading. Both the long axis and the short axis of the egg can be calculated by digital image processing, and the detection value of the egg shape index can also be quickly calculated.

3. **Egg Weight**

The volume of the egg could be estimated using the long and short axis of the egg. The volume of the egg is highly correlated with LB2, and the density of the fresh egg is also proved to be basically the same. Therefore, the direct regression model and regression coefficient design of the egg weight value and the size value are determined using Formula (10.2).

$$w = \alpha + \beta L \cdot B^2 = \alpha + \beta(\alpha_1 + \beta_1 L_P) \cdot (\alpha_2 + \beta_2 B_P)^2 \tag{10.2}$$

where W represents egg weight, in g; and α, β, α_1, β_1, α_2, β_2 are the regression model coefficients, and L is the egg long-axis size, in mm; B is the egg short-axis size, in mm; L_P is the pixel size of egg long-axis image, in pixel; B_P is the pixel size of egg long-axis image, in pixel.

The coefficients of the regression model are sorted out, and the weight detection model using digital image processing methods would be transformed into a multiple linear regression model. The model expression is shown in Formula (10.3.

$$W = a + bB_P + cL_P + dB_P^2 + eL_P B_P + fL_P B_P^2 \tag{10.3}$$

where a, b, c, d, e—regression model coefficients.

$a = \alpha + \beta\alpha_1\alpha_2^2$, $\quad b = 2\alpha_1\alpha_2\beta\beta_2$, $\quad c = \beta\beta_1\alpha_2^2$, $\quad d = \beta\alpha_1\beta_2^2$, $\quad e = 2\alpha_2\beta\beta_1\beta_2$, $f = \beta\beta_1\beta_2^2$

To explore the correlation between the size of the long axis and the short axis of the egg and its weight, the Matlab multiple linear regression algorithm was used to establish a multiple linear regression model between the reference value of the egg weight and B_P, L_P, B_P^2, $L_P B_P$, $L_P B_P^2$. In the calculation process, in order

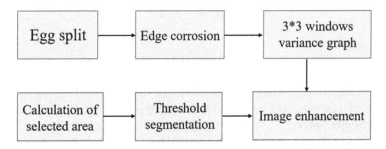

Fig. 10.13 Stain recognition process

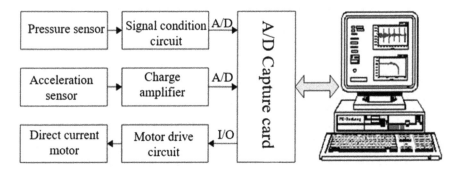

Fig. 10.14 Schematic diagram of eggshell quality measurement system

to ensure the consistency of the order of magnitude between the variable and the egg quality reference value, each variable is divided by the corresponding multiple.

4. **Egg Surface Stain**

The stains or spots on the surface of the eggshell in the egg image are often darker than the surface of the egg, and the brightness is partially discontinuous, so the hue-saturation-value (HSV) color space is generally employed. The H parameter represents the color information, that is, the position of the color in the spectrum; the S parameter represents the saturation, which is the ratio between the purity of the selected color and the maximum purity of the color; the V parameter represents the brightness of the color. The stain recognition process is shown in Fig. 10.13.

5. **Eggshell Crack**

The egg could be regarded as a rigid body with a thin-shell ellipsoidal structure. The surface vibration wave gradually spreads and weakens from the striking area to other areas until it disappears. Once crack appears on the surface of the egg, its structure would be shattered affecting the propagation of waves on the surface of the egg. Taking advantage of the difference in surface vibration wave signals generated by intact eggs and cracked eggs under external impact, the system block diagram of the egg crack detection system based on surface vibration waves is shown in Fig. 10.14, the pressure sensor is used to return the squeezing force between the egg and the acceleration sensor; the acceleration sensor is used to detect the vibration of the egg under the excitation of the impact force; the function of the DC motor is to drive the striker to achieve the percussion of the egg surface; the main function of the A/D acquisition card is to collect sensor signals and realize the control of the motor; the computer is the main processor of the system, and the acquisition card is controlled through LabVIEW software to realize the acquisition, storage, and processing of the signal.

10.3.2   Smart On-Site Assessment for Fruit Classification

Size, shape, color, and surface defects are the main indicators of apple that determine the appearance of the fruit. Visible light imaging technology is widely used at home and abroad to quickly and nondestructively detect the appearance of apples. Even though apples cannot be entirely identified with a camera, information omission occurs in the process of information acquisition. Therefore, this example is based on apple appearance inspection method with three-camera system. *Sofu* et al. [28] have designed an automatic sorting and quality inspection system for apples. The system could collect and analyze four images of apples rolling on the roller conveyor. With dual channels, 15 apples per second would be sorted on average. *Aguilera* et al. [29] collected olive endocarp images from three different perspectives and obtained feature vectors based on image histograms. Discriminant analysis and neural network were used to perform classification tests. The three-camera fruit appearance inspection system is composed of three parts: a conveyor line, a lighting room, and a three-camera image acquisition and processing system. The system includes three mutually independent cameras. The distance between each camera and apple is 58 cm, distributed at 60°. Three computers form a small local area network that can exchange data in real time. Since the field of view of the camera is set to three-wheel distances, the camera can capture three apples at a time, and the same apple can be captured three times in three consecutive positions. That is, one camera can get three images of each apple, and three cameras can get nine images of the apple, as shown in Fig. 10.15.

The image taken from the production line can be simply divided into two parts, namely the apple and the background (as shown in Fig. 10.16). Through the histogram, it can be analyzed that there is a relatively distinct difference between the apple and the background in the captured image, and the global threshold method can be used to segment the apple image from the background. It can be seen that the background has been completely removed, but it is found that the fruit stalk is almost the same color as the background, due to the presence of rust in the center of the first apple. The upper and lower clamping algorithm is applied to solve this problem. The image background is removed by the up-and-down clamping method, and the apple area is completely preserved, and the stem, calyx, or defect will not be removed as the background. At the same time, it also effectively eliminates or reduces the noise after removing the background. The process saves time and speeds up detection.

Figure 10.17 is the real shot images of different grades of apples. It can be seen from the above analysis that apples of different color grades have different chromaticity histogram peaks. The higher the grade, the smaller the chromaticity value

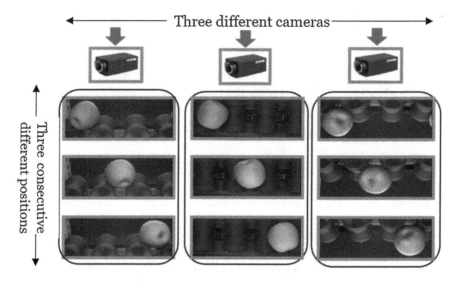

Fig. 10.15 Nine images of an apple obtained online by three cameras

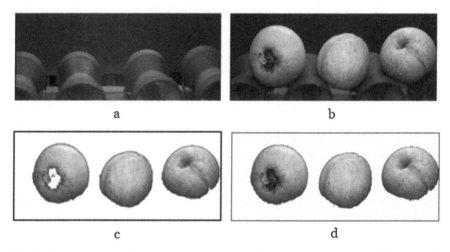

Fig. 10.16 Acquisition of apple image. (**a**) Background scroll wheel image. (**b**) Online detection of fruit images. (**c**) Apple image after threshold segmentation. (**d**) Apple image after removing the background by pressing up and down

corresponding to the peak. Therefore, using the chromaticity value corresponding to the peak value as the threshold value, the color characteristics of apples can be distinguished more accurately.

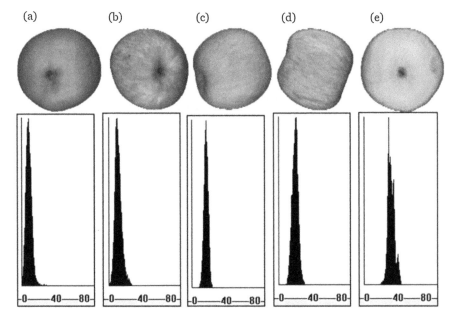

Fig. 10.17 Real shot images and histograms of apples of different grades. (**a**) First-class. (**b**) Second-class. (**c**) Third-class. (**d**) Fourth-class. (**e**) Fifth-class

10.4 Application for Real-Time Monitoring Food Quality During Processing

In the various procedures of food processing, some important physical and chemical indicators are constantly changing and need to be continuously monitored. During the processing, there are many processing steps and parameters that need to be adjusted appropriately according to the real-time changes in indicators. However, due to the influence of many factors such as the difference in raw materials, the processing season, and environmental conditions, always differences between different batches of food exist in the processing steps. During storage, it is affected by storage conditions (container, temperature difference, humidity, ventilation) and storage time, which also causes some changes in its quality, such as changes in volatile substances. Therefore, in order to ensure the consistency of production batches in the food processing steps, some important indicators need to be monitored in real time. Nondestructive testing technology can smartly monitor the changes of color, flavor, size, chemistry, and other indicators during food processing in real time to control the quality and safety of food and agro-products.

10.4.1 On-Site Evaluation for Alcoholic Drink Fermentation Process

Fermented alcoholic drink mainly consist of wine, Chinese rice wine, beer, etc. It is essential to evaluate on-site fermented alcoholic drink quality since it would influence taste and popularity. *Liu* et al. [30] and *Di Egidio* et al. [31] analyzed wine by NIRS based on the main quality of alcohols, pH, soluble solids content (SSC), trace metals, and some other quality attributes. In addition, NIR spectroscopy has been used for the authentication of age, distilled alcoholic beverage, origin, and alcohol concentration. Here, Chinese rice wine is used as an illustration object to show the online quality monitoring research of the fermentation process of alcoholic drink. *Xinliang M* et al. [15] used solvent extraction and assisted evaporation combined with gas chromatography olfactometry method to determine the aroma and flavor substances of rice wine. A total of 39 flavor substances in rice wine, such as ethyl caproate, 1-octen-3-ol, and phenylacetaldehyde, were detected. *Yu H* et al. [16] used electronic tongue to analyze the taste attributes of Chinese Rice wine and used high-performance liquid chromatography (HPLC) to measure the taste active substances in rice wine. The establishment of a partial least squares (PLS) model showed that ten kinds of taste active substances including proline, lactic acid, glutamic acid, arginine, etc. have a greater impact on the taste characteristics of rice wine.

The brewing process of rice wine adopts the conventional brewing method. The main fermentation period is carried out in a large tank, and the post-fermentation period is carried out in a large tank. The main fermentation period and post-fermentation period of rice wine fermentation process (as shown in Fig. 10.18). Thirty-five (35) days before and after the fermentation process indicated an initial rapid change, then slowly and eventually stabilized. The fermented mash of rice wine is very dense and turbid. In order to ensure the consistency of spectrum collection and chemical test conditions in the online monitoring, the test carried out filtering and centrifugal pretreatment of the mash sample to obtain better test results.

Figure 10.19 shows the trend in the total sugar and alcohol content of the mash sample during the fermentation process of rice wine (the average value of each parameter sampled on the day). It can be seen from Fig. 10.19 that the total sugar content of the fermented mash is gradually reduced, while the alcohol content is gradually increased, indicating the process of sugar conversion into alcohol in the fermentation process of rice wine. In the process of model establishment, the synergy-interval partial least squares algorithm optimization model was used to screen the characteristic wavelength variables of each quality parameter. The

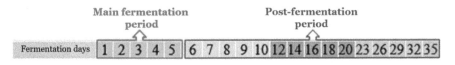

Fig. 10.18 Number of days for sample collection

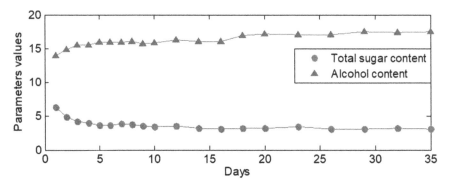

Fig. 10.19 The change trend of total sugar content and alcohol content of mash during the fermentation process of rice wine

Vis-NIR system is verified to be feasible to assess the changes in total sugar and alcohol content during the fermentation of rice wine.

10.4.2 On-Site Classification for Vinegar Brewing Process

During the vinegar fermentation process, the material (vinegar mash) is in a semi-solid form, and microorganisms such as acetic acid bacteria oxidize the alcohol into acetic acid and other flavor substances. The volatilized odor during the acetic acid fermentation stage of vinegar is the coexistence of a variety of fragrant substances. In the early stage of fermentation, alcohol and ester substances are the predominant ones. As the fermentation progresses, the middle and late stages gradually transition to acid substances. In the fermentation process, it is necessary to consider changes in the content of the main volatile gas substances such as alcohol, ester, acid to grasp the main condition of the vinegar ferment, and also consider the detection and characterization of other volatile gases to obtain the odor of vinegar. *Su* et al. [32], *Marín* et al. [33], *Callegòn* et al. [34], and *Durán* et al. [35] respectively analyzed the volatile gas components of red wine vinegar, sherry vinegar, Shanxi aged vinegar, and Zhenjiang balsamic vinegar by solid phase microextraction combined with gas chromatography-mass spectrometry (GC-MS). Extreme learning machine (ELM) and back-propagation artificial neural network (BP-ANN) were attempted to measure total acid content (TAC) based on NIR by *Chen* et al. [36] et al., and RMSEP of ELM was 0.2486 g/100 mL, which showed possibility to monitor the quality of vinegar fermentation process in real time. Application of CSA technology by *Guan* et al. [37] revealed in Fig. 10.20 that the two-dimensional scatter chart constructed with the first and second principal components successfully distinguished vinegar mash at different stages of fermentation. The cumulative variance contribution rate of the first two principal components reaches 55.44% (PC1 = 34.52%, PC2 = 20.92%).

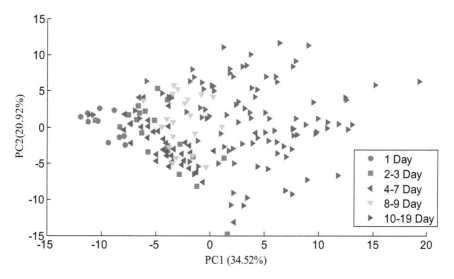

Fig. 10.20 Two-dimensional distribution of main components of vinegar in five stages

In the solid-state layered fermentation of aromatic vinegar, ethanol can be used as an indicator of the acetic acid fermentation stage, and its accurate determination in the vinegar is of great significance for the analysis and control of the acetic acid fermentation process. CSA technology was used to characterize the volatile gases of vinegar mash at different fermentation stages as a whole by *Guan* et al. [18], and then GC-MS technology was used to quantitatively analyze the ethanol content of vinegar mash with different fermentation days. It is afterward combined with some pattern recognition methods to quantitatively detect the vinegar mash of different acetic acid fermentation days, which provides a theoretical basis for real-time online monitoring of vinegar production. Figure 10.21 shows the change of ethanol content during the acetic acid fermentation. In the production process, we can control the acetic acid fermentation process by measuring the alcohol content during the acetic acid fermentation process. If the alcohol is too low, workers would replenish the mash in time to avoid insufficient ethanol, and the produced acetic acid continue to be oxidized by acetic acid bacteria to produce carbon dioxide and water. If the alcohol is too high, workers would add bran raw materials appropriately to increase the output of vinegar and increase the factory benefit. In summary, this technology is not only easy to operate but also has a wide range of applications with its reliable results. It can be used as a method for detecting the alcohol content of solid vinegar during acetic acid fermentation.

10.4.3   On-Site Analysis for Tea Fermentation Process

The real-time online monitoring of quality changes in the processing of tea plays an important role in ensuring product quality and stability, reducing labor costs, energy

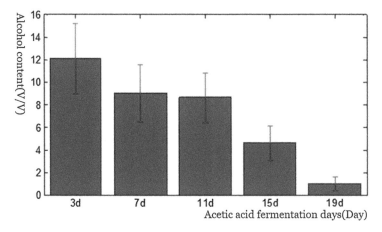

Fig. 10.21 Changes in ethanol content during acetic acid fermentation

consumption, and improving the automation and smart level of tea processing steps. Taking black tea as an example, it is a fermented tea made by withering, kneading and cutting (twisting), fermentation, baking, and other conventional processes, with special color, aroma, taste, and other quality characteristics. It's known for the color of its soup and the red leaves that appear after brewing, and it contains a lot of active compounds. *Zhong* et al. [38] evaluates that the unique aroma substances of black tea are mainly composed of phenethyl alcohol, benzyl acetate, benzaldehyde, and others with strong floral and fruity aroma. In the field of tea quality inspection, changes in color and content of these substances have been introduced into optical sensor technology for digital representation. *Pan* et al. [39] and *Bhattacharyya* et al. [40] relied on electronic nose technology to describe the changes in the aroma of black tea. The whole process of aerobic fermentation of black tea is accompanied by the constant changes of the outer color and internal effective components of the leaves. *Qian* et al. [41] use the changes in leaf color and tea polyphenol content as the main indicators for the degree of aerobic fermentation of black tea.

Chen's group [14] researched three tea samples with different fermentation levels to evaluate the online monitoring of the aerobic fermentation process of black tea based on the color value. From the beginning to the end of fermentation, the process was divided into five fermentation nodes of 0, 1, 2, 3, and 4 h. The actual color change in the leaves is shown in Fig. 10.22a. The overall color change trend of all samples in the $L*a*b$ three-dimensional space is shown in Fig. 10.22b. L refers to light intensity, from black to white; *a value represents color information, from green to red; *b value also represents color information, from blue to yellow. L value, $*a$ value and $*b$ value all can reflect the changing law of leaf surface color to varying degrees. It is necessary to analyze the change trend of $L*a*b$ value separately.

Fig. 10.22 The color change diagram of leaves under different fermentation time (**a**), the overall color change trend diagram (**b**)

Table 10.2 Evaluation results of the monitoring system

Species	Number	Number of identification accuracy	Number of identification errors
Moderate fermentation	30	25	5
Insufficient fermentation	15	13	2
Over-fermented	15	13	2

When the black tea is fermented to a moderate degree, the $L*a*b$ value of its leaf color should be in a certain area in the color space, and this area is the criterion to be established. On this basis, *Zhang* et al. [42] tested the different fermentation levels during the aerobic fermentation of black tea, and used the color monitoring system to read the $L*a*b$ value. The fermentation level of the sample was determined by the distance from the center of the sphere (the average value of L, $*a$, *and* $*b$). After calculation, judgment, and referring to the conclusion of manual review, the evaluation result is shown in Table 10.2. There are nine errors in 60 samples, and the correct recognition rate is 85%.

10.4.4 On-Site Assessment in Meat Product Processing

Livestock and poultry meat are usually processed into a variety of meat products as the main raw material, such as sausages, cured bacon, bacon, sauced meat, barbecued meat, etc. The fermentation steps in the production of traditional sausages mainly include preparation phase, fermentation phase, and maturation phase. This process is accompanied by changes in pH, protein, fatty acids, amino acids, and other substances. Therefore, the absorption characteristics of these substances in the near-infrared-visible light spectrum would be determined.

Guan et al. [18] used the E-nose to conduct qualitative research on the flavor of meat products, and this technology has limited online monitoring indicators. *Chen Xiaodong* et al. [43] used hyperspectral rapid nondestructive testing to predict the content of nitrite in the incense field based on the principal component analysis algorithm. *Sun Xia* et al. [44] have studied that microorganisms such as biogenic amines produced in fermented sausages have a bad effect on the quality of the product and cause serious harm to human health. *Zhang Hutao* et al. [42] used an electronic nose to detect the correlation between odor fingerprint parameters and volatile base nitrogen in the fermentation process of Chinese sausages and constructed a PLS prediction model.

10.4.5 Smart Evaluation in Dairy Product Processing

Rapid detection of the indicators in the table of representative studies are shown in Table 10.3. Among the following online monitoring technologies, NIR is a technology that has been studied extensively. The principle of this technology is the qualitative and quantitative detection and analysis of a specific chemical component in the tested sample for a certain segment or distinct near-infrared spectrum absorption characteristics based on its unique absorption characteristics. NIR shows the integrated waveband and harmonic band, which are the absorption frequency harmonics generated by C–H, O–H, N–H, and S–H molecular groups, and the frequency overlap, its multiplication and combined nature in case their groups can also be collected.

Table 10.3 Research on online monitoring technology of milk quality

Technology	Detection indicator	References
NIR	Protein, fat, lactose	[45]
Hyperspectral	Bacillus cereus or other bacteria	[46]
Electronic tongue	Apparent milk viscosity	[47]
Adenosine triphosphate bioluminescence	Detergents, sanitizers, or other chemicals	[48]
Polymerase chain reactions	*S. aureus* strains	[49]
Electrochemical methods	Bacterial numbers and activity	[50]
Fourier transform infrared spectroscopy	Polymers, protein, lipid, and carbohydrate	[51]
Colorimetric detection	Microorganisms	[52]
Electronic nose	Freshness and level bacterial spoilage	[47]

10.5   Conclusion and Future Trends

In the food safety guarantee system that implements management from farmland to table, assessment work should be continuously improved following the revision of standards. As an important means of internal self-monitoring and external supervision in the links of food raw materials, product processing, transportation, and marketing, assessment work directly affects the quality and safety of food. With the gradual reduction of the limit values of food safety and health indicators, higher requirements are put forward for testing technology. Inspection and assessment would move toward integration of assessment and control, rapid testing, portability, multi-technology integration, and information sharing. Figure 10.23 shows the realization of the integration of control and detection from raw materials to "dining table."

Common online monitoring equipment involves optics, electronic sensors, biosensors, radiation, and other technologies. The advantages of these online monitoring equipment is their fast nature, no requirement of the pretreatment steps, and

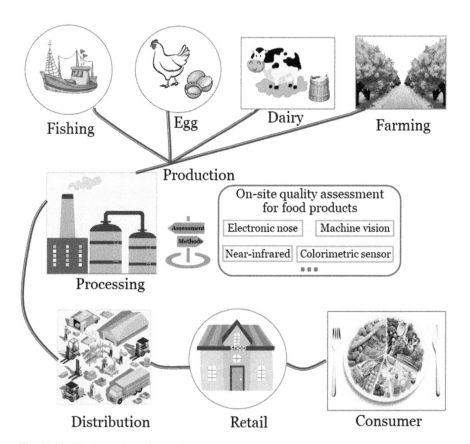

Fig. 10.23 The integration of control and detection from raw materials to "dining table"

environmental protection. Food quality monitoring involves external attributes such as color, form, and texture, as well as internal attributes such as fat and protein. Compared with conventional and complex static and manual detection technology, online monitoring devices can realize fast, nondestructive real-time monitoring of the quality changes of food or agro-products during processing. While these online monitoring or testing equipment is not yet completely industrialized in industry and agriculture due to the absence of uniform and standardized standards or requirements in the domestic food or agro-product testing field.

References

1. Cui Z, Sun L, Chen W et al (2008) Preparation of dry honey by microwave-vacuum drying. J Food Eng 84(4):582–590
2. Sun D, Woods JL (1994) Low-temperature moisture transfer characteristics of wheat in thin-layers. Trans ASAE 27(6):1919–1926
3. McDonald K, Sun D (2001) Effect of evacuation rate on the vacuum cooling process of a cooked beef product. J Food Eng 48(3):195–202
4. Sun D (1997) Solar powered combined ejector vapour compression cycle for air conditioning and refrigeration. Energy Convers Manag 38(5):479–491
5. Wang L, Sun D (2004) Effect of operating conditions of a vacuum cooler on cooling performance for large cooked meat joints. Food Eng 61(2):231–240
6. Zheng L, Sun D (2004) Vacuum cooling for the food industry—a review of recent research advances. Trends Food Sci Technol 15(12):555–568
7. Kiani H, Zhang Z, Delgado A et al (2001) Ultrasound assisted nucleation of some liquid and solid model foods during freezing. Food Res Int 44(9):2915–2921
8. Erdogdu F, Sarghini F, Marra F (2017) Mathematical modeling for virtualization in food processing. Food Eng Rev 9(4):295–313
9. Kuchida K, Fukaya M, Miyoshi S et al (1999) Nondestructive prediction method for yolk: albumen ratio in chicken eggs by computer image analysis. Poult Sci 78(6):909–913
10. Gulich A, Fitzgarald J (1964) USDA egg grading manual, no. 75, vol 78. Agricultural Marketing Service, Washington, DC, pp 909–913
11. Ouyang Q, Zhao J, Chen Q (2014) Instrumental intelligent test of food sensory quality as mimic of human panel test combining multiple cross-perception sensors and data fusion. Anal Chim Acta 841:68–76
12. Blakey RJ (2016) Evaluation of avocado fruit maturity with a portable near-infrared spectrometer. Postharvest Biol Technol 2016(121):101–105
13. Iqbal SM, Gopal A, Sankaranarayanan PE et al (2016) Classification of selected citrus fruits based on color using machine vision system. Int J Food Prop 19(2):272–288
14. Chen Q, Liu A, Zhao J (2013) Classification of tea category using a portable electronic nose based on an odor imaging sensor array. J Pharm Biomed Anal 84:77–83
15. Mo X, Xu F, Fan W (2010) Characterization of aroma compounds in Chinese rice wine Qu by solvent-assisted flavor evaporation and headspace solid-phase microextraction. J Agric Food Chem 58(4):2462–2469
16. Yu H, Zhao J, Li F et al (2015) Characterization of Chinese rice wine taste attributes using liquid chromatographic analysis, sensory evaluation, and an electronic tongue. Chromatogr B Anal Technol Biomed Life Sci 997(1):29–35
17. Guan B, Zhao J, Cai M et al (2014) Analysis of volatile organic compounds from Chinese vinegar substrate during solid-state fermentation using CSA array. Anal Methods 6(23):9383–9391

18. Guan B, Zhao J, Lin H et al (2014) Characterization of volatile organic compounds of vinegars with novel electronic nose system combined with multivariate analysis. Food Anal Methods 7 (5):1073–1082
19. Ouyang Q, Zhao J, Chen Q (2013) Classification of rice wine according to different marked ages using a portable multi-electrode electronic tongue coupled with multivariate analysis. Food Res Int 51(2):633–640
20. Ouyang Q, Zhao J, Chen Q et al (2013) Classification of rice wine according to different marked ages using a novel artificial olfactory technique based on CSA array. Food Chem 138 (2–3):1320–1324
21. Li H, Kutsanedzie F, Zhao J et al (2016) Quantifying total viable count in pork meat using combined hyperspectral imaging and artificial olfaction techniques. Food Anal Methods 9:3015–3024
22. Sun L, Bi X, Lin H et al (2013) On-line detection of eggshell crack based on acoustic resonance analysis. J Food Eng 116(1):240–245
23. Sun L, Cai J, Lin H et al (2013) On-line estimation of eggshell strength based on acoustic impulse response analysis. Innovative Food Sci Emerg Technol 18:220–225
24. Patel V, McClendon R, Goodrum J (1996) Detection of blood spots and dirt stains in eggs using computer vision and neural networks. Appl Eng Agric 12(2):253–258
25. Garcia-Alegre MC, Ribeiro A, Guinea D et al (2000) Eggshell defects detection based on color processing. Mach Vis Appl Ind Inspect VIII 3966:280–287
26. Wen Y, Wang Q, Chen X et al (2001) Development of automatic grading system software for duck egg size and egg center color. J Huazhong Agric Univ 20(4):395–399. (Chinese)
27. Dehrouyeh M, Omid M, Ahmadi H et al (2010) Grading and quality inspection of defected eggs using machine vision. Int J Adv Sci Technol 17:23–31
28. Sofu MM, Er O, Kayacan MC et al (2016) Design of an automatic apple sorting system using machine vision. Comput Electron Agric 127:395–405
29. Aguilera PD, Martinez-Gila DM, Gamez GJ et al (2015) Sorting olive batches for the milling process using image processing. Sensors 15(7):15738–15754
30. Liu F, Jiang Y, He Y (2009) Variable selection in visible/near infrared spectra for linear and nonlinear calibrations: a case study to determine soluble solids content of beer. Anal Chim Acta 635(1):45–52
31. Di-Egidio V, Sinelli N, Giovanelli G et al (2010) NIR and MIR spectroscopy as rapid methods to monitor red wine fermentation. Eur Food Res Technol 230(6):947–955
32. Su M, Chien P (2010) Aroma impact components of rabbiteye blueberry (Vaccinium ashei) vinegars. Food Chem 119(3):923–928
33. Marın RN, Mejıas RC, Moreno MVG et al (2002) Headspace solid-phase microextraction analysis of aroma compounds in vinegar: validation study. J Chromatogr A 967(2):261–267
34. Callejón RM, Tesfaye W, Torija MJ et al (2009) Volatile compounds in red wine vinegars obtained by submerged and surface acetification in different woods. Food Chem 113 (4):1252–1259
35. Durán E, Palma M, Natera R et al (2010) New FT-IR method to control the evolution of the volatile constituents of vinegar during the acetic fermentation process. Food Chem 121 (2):575–579
36. Chen Q, Ding J, Cai J et al (2012) Rapid measurement of total acid content (TAC) in vinegar using near infrared spectroscopy based on efficient variables selection algorithm and nonlinear regression tools. Food Chem 135:590–595
37. Guan B, Zhao J, Cai M et al (2014) Analysis of volatile organic compounds from Chinese vinegar substrate during solid-state fermentation using a CSA array. Anal Methods 6 (23):9383–9391
38. Zhong Q, Lin Z, Chen C et al (2016) Changes on biochemical composition of Chungui fragrant black tea during fermentation process. Acta Tea Sin 57(2):85–91

39. Pan K, Feng L, Chen J et al (2015) Analysis of aroma compounds in black tea ventilated with oxygen for different durations during the fermentation process by head space-solid phase micro-extraction coupled with gas chromatography-mass spectrometry. Food Sci 36(8):181–186
40. Bhattacharyya N, Bandyopadhyay R, Bhuyan M et al (2008) Electronic nose for black tea classification and correlation of measurements with "tea taster" marks. IEEE Trans Instrum Meas 57(7):1313–1321
41. Qian Y, Yang Y, Zhou X et al (2012) Current research situation in black tea fermentation technology. Sci Technol Food Ind 33(23):388–392
42. Zhang H, He J, Wang Z et al (2014) Physicochemical quality index detection method of Chinese fermented sausage based on electronic nose. Food Ind 40(7):205–211
43. Chen X, Guo P (2016) Extracting hyperspectral image features based on principal component analysis to detect nitrite content in sausages. Meat Res 30(12):22–27
44. Sun X, Yang Y, Song Y et al (2015) Research on the microorganisms that produce amines in fermented sausages and their detection methods. Food Ind Sci Technol 36(12):379–384
45. Ghosh S, Bhattacharya K, Nayak S et al (2015) Identification of different species of bacillus isolated from nisargruna biogas plant by FTIR, UV-Vis and NIR spectroscopy. Spectrochim Acta Part A Mol Biomol Spectrosc 148:420–426
46. Valous NA, Mendoza F, Sun D (2010) Emerging non-contact imaging, spectroscopic and colorimetric technologies for quality evaluation and control of hams: a review. Trends Food Sci Technol 21(1):26–43
47. Ciosek P, Brudzewski K, Wroblewski W (2006) Milk classification by means of an electronic tongue and support vector machine neural network. Meas Sci Technol 17(6):1379–1384
48. Lomakina GY, Modestova YA, Ugarova N (2015) Bioluminescence assay for cell viability. Biochemistry 80(6):701–713
49. Montville TJ, Matthews KR (2005) Food microbiology: an introduction, 3rd edn. ASM Press, Washington, DC
50. McMeekin TA (ed) (2003) Detecting pathogens in food Cambridge. Woodhead Publishing Limited, Sawston, pp 147–162
51. Lu M, Shiau Y, Wong J et al (2013) Milk spoilage: methods and practices of detecting milk quality. Food Nutr Sci 4(07):113
52. Cavallo JA, Strumia MC, Gomez CG (2014) Preparation of a milk spoilage indicator adsorbed to a modified polypropylene film as an attempt to build a smart packaging. J Food Eng 136:48–55

Lightning Source UK Ltd.
Milton Keynes UK
UKHW020723310822
408113UK00001B/1